STO

ACPL ITEM
DISCARDED

 Y0-ACG-002

1-21-74

TALKING BACK:
CITIZEN FEEDBACK AND
CABLE TECHNOLOGY

The MIT Press
Cambridge, Massachusetts,
and London, England

TALKING BACK:
CITIZEN FEEDBACK AND
CABLE TECHNOLOGY

edited by
Ithiel de Sola Pool

Copyright © 1973 by
The Massachusetts Institute of Technology

All rights reserved. No part of this book may be reproduced in any form or
by any means, electronic or mechanical, including photocopying, recording,
or by any information storage and retrieval system, without permission in
writing from the publisher.

This book was typed with an IBM Executive

by Inge Calci,

printed on Finch Textbook Offset

by Halliday Lithograph Corp.,

and bound in Columbia Milbank Vellum

by Halliday Lithograph Corp.

in the United States of America.

Library of Congress Cataloging in Publication Data

Pool, Ithiel de Sola, 1917-   comp.
    Talking back.

    Includes bibliographical references.
    CONTENTS: Mayer, M. Cable and the arts. —Mendelsohn, H. The neg-
lected majority: mass communications and the working person. —Tate, C.
Community control of cable television systems. —[etc.]
    1. Community antenna television—United States—Addresses, essays,
lectures.   I. Title.
HE8700.7.C6P65   384.55'47   72-13886
ISBN 0-262-16056-0

CONTENTS

1785822

vi

PUBLISHER'S NOTE

The aim of this format is to close the time gap between the preparation of
certain works and their publication in book form. A large number of significant though specialized manuscripts make the transition to formal publication
either after a considerable delay or not at all. The time and expense of detailed text editing and composition in print may act to prevent publication or
so to delay it that currency of content is affected.

The text of this book has been photographed directly from the author's
typescript. It is edited to a satisfactory level of completeness and compre-
hensibility though not necessarily to the standard of consistency of minor
editorial detail present in typeset books issued under our imprint.

The MIT Press

About one home in ten now receives television by cable. With each passing
year the number slowly rises. If public policy so dictates, cable may become
the standard American mode of television transmission.

If that should happen, the American communication system will be improved
in important ways. CATV can offer services not easily provided over the air.
These include copious numbers of channels, pay TV, and two-way communication. With two-way cable the viewer can signal his preferences and responses
to the head end. He can, that is, "talk back" to obtain, at his demand, the particular information or program material that he wants. Such interactive on-
demand telecommunication is the subject of this book.

This book is addressed to those who are concerned with CATV as part of a
larger professional interest. It is not a handbook for communication practi-
tioners. It is directed to political scientists, sociologists, educators, journa-
lists, policy makers, and civic-minded members of the general public who
want to know what the technical possibilities of two-way cable are, how close
they are upon us, and what their social impact may be.

There is much current enthusiasm about how CATV may encourage citizen
participation, raise the level of culture, and provide education and services
to the public. Popular magazine articles and lay spokesmen sometimes give
the impression that a new era of citizen access is just around the corner.
Debunkers can easily demolish such unrealistic hopes, but the debunking can
go too far. Broadband two-way communication does promise to change society
in profound and perhaps desirable ways. Public policy will determine whether
we use the options wisely.

The papers in this book present a variety of points of view. Some have appeared before, while others are published here for the first time. Several of
the papers were originally background papers for the Sloan Commission on
Cable Communications.

NOTES ON CONTRIBUTORS

Herbert E. Alexander is Director of the Citizens' Research Foundation, Princeton, New Jersey, and an expert on the financing of politics.

Gilbert Cranberg is an editorial writer on the Des Moines Register and Tribune.

Kenneth Dobb, from Canada, is a graduate student in Political Science at the Massachusetts Institute of Technology.

Susan C. Greene, senior staff associate at Urban Communications Group, was formerly with Children's Television Workshop, producers of "Sesame Street."

Ephraim Kahn is a free-lance writer and journalist based in Washington, D.C.

William T. Knox, a chemical engineer, is Director of the National Technical Information Service, Department of Commerce. During his two and a half years in the White House Office of Science and Technology, he was responsible for national information systems; he has also served as Vice-President for Planning and Information Systems of McGraw-Hill.

Theodore S. Ledbetter, Jr., is founder and president of Urban Communications Group, a communications consulting firm, publisher and editor of Black Communicator, a monthly journal of minorities and the media, and adjunct professor of communications at Howard University.

Noam Lemelshtrich, from Israel, has a Master's degree in Industrial Engineering from Stanford University and is working for a joint Doctorate in Mechanical Engineering and Political Science at the Massachusetts Institute of Technology.

Martin Mayer, author of About Television (1972), is a free-lance writer based in New York City.

Harold Mendelsohn is Professor and Chairman of the Department of Mass Communications, University of Denver, and President of the American Association for Public Opinion Research.

Charles Murray has done field research in Thailand and is a graduate student in Political Science at the Massachusetts Institute of Technology.

Ithiel de Sola Pool is a Professor of Political Science at the Massachusetts Institute of Technology and specializes in the study of communications.

Thomas B. Sheridan is a Professor of Mechanical Engineering at the Massachusetts Institute of Technology.

Charles Tate is a member of the staff of the National Cable Information Service in the Urban Institute in Washington, D.C.

John E. Ward is Lecturer in Electrical Engineering and Deputy Director of the Electronic Systems Laboratory at the Massachusetts Institute of Technology.

ACKNOWLEDGMENTS

Six of the papers in this book were originally written for the Sloan Commission on Cable Communications. In their original form they were background papers for the commission, whose report, <u>On the Cable</u>, was published last spring (McGraw-Hill, New York, 1972). These chapters are those by Gilbert Cranberg, Ephraim Kahn, Martin Mayer, Harold Mendelsohn, Ithiel Pool, and Herbert Alexander, and John Ward's "Present and Probable CATV/Broadband-Communication Technology." We appreciate the Alfred P. Sloan Foundation's permission to reproduce them here.

Two other chapters were prepared, in earlier versions, at the MIT Center for Space Research under contract with NASA (NASW-2197). Those chapters are by Ithiel Pool, Charles Murray, Kenneth Dobb, and Noam Lemelshtrich. The chapter by Thomas Sheridan is drawn from the AFIPS Fall Joint Computer Conference <u>Proceedings</u> of November 1971 (39:327-336). The chapter by Charles Tate is drawn from his volume published by the Urban Institute, <u>Cable Television in the Cities</u> (Washington, D.C., 1971, Chap. 2). The chapter by Theodore S. Ledbetter, Jr., and Susan Greene was originally published in the <u>Yale Review of Law and Social Action</u> (vol. 2, no. 3, Spring 1972) in a special issue called <u>The Cable Fable</u>. The chapter by William T. Knox is an abridged version of a report (of April 1971) by the Council on Communication to the Smithsonian Institution.

We are flooded by a torrent of communication in contemporary society, and yet we hear everywhere that there is a breakdown in communication. The average citizen spends more than four hours a day with mass media, while increasingly he doubts that his government listens to him or that what it tells him is credible.

The trend of growing alienation is well documented. For several years the Harris poll has been asking people whether they agree or disagree with each of the following statements. The trend of increasing agreement is frightening.

Statement	Percentage Agreeing			
	1966	1968	1971	1972
What you think doesn't count very much.	39	42	44	53
The people running the country don't really care what happens to people like yourself.	28	36	41	50
You feel left out of things around you.	9	12	20	25

Alienation has many roots. Failure of communication is but one, and perhaps more often a symptom than a cause. But one situation that clearly reduces the citizen's sense of potency is that the flood of communication is one way. The citizen watches TV, reads newspapers, and listens to radio, but he has no way of talking back. He hears, but he is not heard. At least that is the way that he feels.

In the momentary crisis of confidence, one hears a recurrent demand that the country's communication system be made more responsive. Witness the increase in organized pressure on the media. The broadcasters have never felt such a barrage of criticism. In many cities public-interest law firms and community groups are challenging broadcast renewals. Irate mothers have organized Action for Children's Television, and the FCC has held hearings on their petition. The Surgeon General had a commission review the effects of TV programs on children. Ecology groups have gone to court to try to oblige TV to carry counter-advertising debunking claims in the commercials. And community groups have been demanding access to the airwaves.

The print media have also been challenged. Journalism reviews are being published in various cities giving newsmen a chance to hit back at their employers. A national news council seems to be on the verge of creation by the Twentieth Century Fund. Newspapers are providing op-ed pages for the presentation of views different from the paper's own, and some papers are employing ombudsmen or press critics on their staffs.

Perhaps most important of all is the emergence in some circles of a new interpretation of the First Amendment that includes the right to "access." Best stated by Barron,[1] the argument is that where there is monopoly or oligopoly of the media, freedom of the press should include the right to get views into those dominant media. According to this argument, refusal by the networks or community newspapers to carry a view constitutes censorship, just

as much as does a government ban. This view, quite reasonably, alarms the
media. What is the limit? Must the <u>Pilot</u> carry atheist propaganda or SDS
papers carry the government's views? Clearly not, but how shall the line be
drawn?

The media fear the access argument, for if pressed to the limit it would
turn the media, so they say, into ill-edited bulletin boards where anyone could
tack up his views, without regard to clarity or presentation. The electronic
media have lived with a fairness doctrine for a long time, but professional
producers at the networks have decided how to package the various views in
such a way that the public can understand them and the ratings can remain
high. The broadcasters dread the idea of blocks of time out of their control
that would become dreary impositions of partisan tirades upon the public,
confusing rather than informing it, and satisfying only the speaker and not the
millions in the would-be audience. The networks sell 30-second spots out of
their control and accept it as a necessary evil, but 30-minute blocks are an
entirely different matter.

The printed press sees a similar danger. The argument is now being made
that the fairness doctrine is just as relevant to the print media as to the electronic ones. Originally the argument for regulation of the electronic media
was the scarcity of the spectrum; there could be only a few TV stations. But
most American cities now have more TV stations than newspapers. The other
argument for regulation of the electronic media was that they enjoy a grant of
a public resource; but print media, it is now argued, also enjoy the second-
class postage subsidy and some exemptions from antitrust regulations.[2] So the
print media, too, feel threatened with being treated as bulletin boards.

If the only means for increasing the public's access to the media of communication were by such legal redefinitions, the prospects would be discouraging. The public policy debate would become a search for a balance between
unpalatable alternatives. The proposed reforms, at least many of them, present problems as severe as the abuses they are supposed to cure. But fortunately for private citizens, the choice society faces is not limited to alternative ways of handling communications monopolies. The technology of modern
communication is moving in a quite different direction. There is an important
revolution in the offing. For a hundred years or more the technology of communication has favored ever larger, ever more concentrated media of communication. But that trend seems to be about to change.

Today's communication devices are largely mass media. They send messages one way from a few sources to audiences of millions. The growth of
such mass media, since the penny newspaper a century and a half ago, is accounted for by the economies resulting from mass production. The printing
press and broadcasting make it possible to deliver uniform messages to millions of people at pennies per head.

Now new communications technologies have appeared on the horizon that
may reverse the trend toward uniformity in communication and thereby help
close the gap of alienation. The communication technologies based on the computer, the videorecorder, and cable TV promise to permit individualized communication to become economically competitive with mass communication.

In another two decades the American telecommunication system will probably consist of two basic services reaching into homes and work places. One

will be narrowband (5kc) and fully switched, an offspring of the telephone system. The other will be broadband but only partially switched and only partially two way, the offspring of CATV. The latter system will be most suitable for delivering those messages to the public that are sent by the few to the many.

It is not at all clear that this dual system is an optimum solution that one would pick in the 1970s if one were designing a national communication system from scratch. It is, however, one solution to a key resource allocation problem. That problem is the costliness of trunk bandwidth and switching. In popular magazine articles one sometimes reads a science-fiction fantasy of a near-future communications system in which each individual can transmit a full-scale TV picture in telephone fashion to any other individual anywhere in the country. In that fantasy, citizens spend much of their lives using such individualized video communication for work, pleasure, shopping, and education.

The trouble with that fantasy is that each linked pair of persons communicating ties up 6MHz of bandwidth (the equivalent of 1000 phone calls) for all of the tens or thousands of miles between them and also ties up all the switches en route. If such broadband two-way communications were to be a process occupying much of men's lives, then a vast and costly resource would be continously tied up for long periods. The cameras, VTRs, and other terminal equipment for doing such things are expensive, too, but the big expense is the bandwidth of signal for long distances and the associated switches.

Obviously what the public will be offered must be something more economical than that. There are various corners that can be cut, reductions of service that can be made, to make a national communication system that is economically feasible in this century. Each such trim hurts some services and not others. It follows that the particular simplification that one is willing to make depends upon one's image of the kind of service that is important to society. Any system that is less than the luxurious system of magazine fantasy will be differentially limiting on different uses of the system, such as person-to-person conversation, entertainment, citizen feedback, interactive computing, data and text transmission, marketing and business uses, and education. What kind of limited system one favors is apt to reflect implicitly the kind of use to which one gives priority.

The telephone company offered one solution to the bandwidth problem that reflected the assumption that what people wanted most was to see each other's faces while making person-to-person phone calls. The Picturephone provided a reduced size, low-resolution picture that would be transmitted at 1 MHz over wires instead of 6 MHz over coaxial cable. For people who are interested either in education in the home or in widening the variety of enjoyable entertainment available on TV, the Picturephone is a virtually useless device. Its small screen is inadequate for interesting visual displays, and using a switched system with twisted wire pairs rather than a tree system with frequency division multiplexing of cables makes the cost for protracted connection by any one viewer exorbitant.

Most entrepreneurs in the CATV industry advocate a different solution predicated on the assumption that the important service that people want is better mass entertainment and news in real time without delays. By multiplying the number of channels available to the viewer and improving picture defini-

tion they can provide those services. In such an entertainment and news system, feedback can be limited to a few bits of information from each home.
The system requires just enough upstream bandwidth to make accurate polls
of who is turned on and to make pay TV possible. Three or four upstream
video channels at most might be provided to facilitate pickup of program material from theaters, sports arenas, news sites, schools, and neighborhood
studios. For people who are interested in citizen feedback, individualized
small-group communication, powerful information services, or education in
the home, however, that type of CATV system is a grossly inadequate facility.
For such purposes one needs more individualized two-way feedback capacity.

Other groups, such as the airlines (in designing their reservation systems)
or the Plato Project at the University of Illinois, have designed special-purpose dedicated communication systems that provide such interaction. The
Plato system would provide in-school computer-aided instruction over a substantial region. Clearly, such dedicated systems do not do, and were never
intended to each do, all the other various things that other people want to do.

The Mitre system solves the bandwidth scarcity problem by having people
share a single TV channel by means of a device called a frame grabber. It
displays a still picture, one of the 30 frames a second shown by a TV set. If
the still picture is to be changeable only once in 10 seconds, then 300 persons
could share a channel. The goal of individualized instruction in the home is
well served by such a system. But it is not clear whether still pictures would
have much appeal for either entertainment or social interaction, and the extent of consumer demand for them is dubious.

Sometime in the twenty-first century bandwidth scarcity may give way to
such abundance that all of the different communication needs may be met at
once. For the present, however, despite a growing abundance that places
marvelous new communications possibilities on our doorsteps, we still must
make choices. The immediate issue for society is which communication services it will choose to have.

The papers that follow deal with these problems and options. We have divided the book into three parts. In the first part we talk about the social context into which new technologies for individualized communication will fall.
We look at what the public wants that the cable may help bring, partly by providing many more channels and thus space for small group access and partly
by providing some degree of audience feedback to the transmitter. Martin
Mayer explores the prospects for a variety of programming in the arts. Technology is only one of many factors limiting the offering of a fuller and better
menu in the arts. Mendelsohn asks what it is that the ordinary working person
wants on his set. Tate, Pool, and Alexander consider citizen feedback through
politics. Tate raises the issues of black community access. Pool and Alexander ask more generally how multichanneled cable systems could be used in
politics and what political uses there are to electronic feedback. Knox deliber-
ates about what communications facilities contemporary cities need to improve
the quality of life in them. Cranberg looks at a leading public sector application, public safety.

The second part of the book deals with the technology of cable, or more accurately, of sophisticated cable systems. With minimal two-way facilities pay
TV becomes possible; Ledbetter and Greene outline that possibility. But pay

TV is only the first step in interaction between source and audience. How
much interaction is possible is a question of engineering and economics. In
his first paper, John Ward deals with the present technology of cable television, particularly the question of how many channels different technologies
can offer at what costs. CATV today can provide enough abundance of channels so that time can be made available at low cost to audiences numbering
mere hundreds instead of only to mass audiences of millions. It thus permits
use of TV in the way that periodicals and pamphlets are now used by tiny
special-interest groups. Ward's second paper raises the problem of limitations on the cable and discusses the feasibility and appropriateness of various uses. Noam Lemelshtrich outlines several techniques of screen feedback
using the TV set as a home terminal.

As of yet (1972) no existing CATV system has provided facilities for user
response. There is much development and planning activity these days in designing technical means for making two-way individualized communication
possible. The Sloan Commission, from whose efforts a number of the papers
in this volume stemmed, recommended a requirement of a minimum of 40
channels on CATV with some devoted to upstream use. The FCC on March 31,
1972, adopted rules requiring all new cable systems in metropolitan areas to
have a minimum of 20 channels and, in addition, always to provide more channels than are required by current demand. The systems must also provide
two-way capability within five years. Experiments by RCA, Hughes, and the
Mitre Corporation in which two-way CATV will be installed in test cities have
also been launched. The Mitre experiment, as we have noted, emphasizes use
of a frame grabber to permit the sharing of a single TV channel by still pictures. The other systems allow two-way interaction in digital mode, as occurs with an interactive time-shared computer terminal, offering only one-
way video but in the abundance that cable allows.

The technical possibilities and the social applications of two-way interaction are considered in Part III. Thomas B. Sheridan describes a citizen feedback technology that enables people at a meeting or lecture to silently signal
the speaker about their reaction, questions, and understanding of what is going
on. Ephraim Kahn explores the uses for cable in the private sector. Pool's
closing paper considers the economic and other problems involved in the introduction of these or any other kinds of telecommunication that are sufficiently interactive to allow people in the audience individually to control what programs or messages they receive. The chapter analyzes these possibilities
and attempts to estimate how long it will be before there will exist the kind of
information utility that allows the citizen at home to dial up whatever entertainment, course of study, news, or other information he wants whenever he
wants it.

The social effects of interactive two-way cable technology are our central
interest in this book. Providing citizens with increased participation in the
running of their own communities is a priority goal. The thesis of this book
is that the communication technologies that can most deeply affect the character of community interaction and community structure in the decades ahead
are those that permit communication among medium-sized groups of persons,
with two-way interaction among them. The past decades have been an era of
both mass media and a one-to-one telephone system. The coming communica-

tions revolution may complete the spectrum by providing two-way communication at a distance for special-interest communities who number only in the hundreds or thousands. That is the level at which most citizens feel that they can significantly influence the course of social events.

One should not assume, however, that the net effect of the technological revolution in communications that increases the facilities for citizen feedback will be to knit society more closely together. Its effects could be the opposite. The effect of the mass media that grew up over the past century, so social critics have argued, has been to make for a homogenized society. Economies of scale were achieved by standardizing the information output. The number of different newspapers published in the United States declined from 2,042 in 1920 to about 1,750 today, and they get most of their nonlocal news from two wire services. A dozen magazines with circulations of 4.5 million to 16 million dominate the popular field. Any evening, in prime time when as many as 60 percent of the families have their TV sets turned on, the only choice most of them exercise is to view the entertainment offered by one of three networks, all of which offer fairly similar programs.

There can be no doubt that this standardized fare is a powerful force toward conformity. Wherever one goes in the United States, the same fads, the same styles, the same scandal of the week, the same ball scores, the same entertainments are on people's lips.

However, we are predicting that the effects that technology has had over the past hundred years are quite different from what is to come. Increasingly, communications devices will be adapted to individualized use by the consumer where and when he wants, on his own, often without the cooperation of others; he will use machines as an extension of his own capabilities and personality, talking and listening worldwide, picking up whatever information he seeks.

One simple example of this sort of thing is the evolution of text reproduction from a technology of mass production toward devices that produce exactly the text the user wants. The printing press made text available to the millions, while the photographic and electrostatic copiers now found everywhere enable each reader among the millions to acquire just the pages he wants when he wants them at a cost hardly greater than that of printing. In the future, television will undergo a similar transformation.

Thus, more individualized communication systems may break the mold of impersonal mass-media uniformity. But if that sounds desirable, consider how the same facts sound if we choose oppositely charged words. The mass media helped to create and integrate national societies out of their parochial fissiparous traditional forerunners. Now the breakdown of the mass media could produce a new kind of isolation in which the cohesion of society is weakened.

The present political world is created by the fact that most citizens see or read the same top news story of the day and very little else. But in the coming society with individualized information retrieval systems, the information the citizen gets will arise from his own specific concerns, whether they relate to the traffic problem in his community, legislation dealing with his business, or draft policy. Today people do read things that affect them personally when they happen to crop up in the press, but they are not in a position to ask for that particular information alone. When everyone can select his own fund of information, the political problem of gathering an effective body of support behind an issue or a candidate may become very different and much more difficult.

Perhaps one need no longer fear that 1984 will see a nation of helots cast from a single mold. But there may be other things to fear. Perhaps in the twenty-first century the sort of critic who now attacks conformity in society may be complaining of an atomized society. He may look back with nostalgia to the days when a movie star had millions of avid followers and gave unity and content to the life of the nation, whereas in his day every little clique of creative people makes its own movies incorporating its own idiosyncratic point of view. Modern technology, he'll assert, has destroyed our common cultural base and has left us each living in a little world of our own.

There is the dilemma! The mass communication system of our present society is unsatisfactory. It is impersonal, overly uniform, unresponsive to the audience, and unrewarding to individuality. New technology can attack those faults. Communication can be made individualized and responsive. But when it is made individualized and responsive, will it help society to cohere better by making group interaction more meaningful, or will it tend to tear society apart by removing the element of commonality in what is communicated? It could do either or even both at once. Which it does may depend upon many details of how the system is designed.

Communication technology is changing rapidly, and so is communication practice. We are in the process of a communications revolution. To predict what a revolution will bring is not easy. To attempt to write about the process while it is still aborning undoubtedly means that much that we write today will look foolish in five or ten years' time. One is tempted to sit back, wait, and see what happens. But if we wish to influence the outcome rather than just observe it post hoc, we must do our best to understand it better while it is still in process. That is the purpose of this book.

Notes

1. Jerome A. Barron, "Access to the Press: A New First Amendment Right," Harvard Law Review 80 (1967): 1641; also "An Emerging First Amendment Right of Access to the Media," George Washington Law Review 37 (1969): 487. A general discussion of the access problem will be found in a forthcoming Aspen Institute Publication on government and the media.

2. "Media and the First Amendment in a Free Society," Georgetown Law Journal 60 (1972): 871-1099.

Television has been less hospitable to the arts in the United States than in other parts of the world. In Britain, the BBC supports several orchestras, televises concerts from Festival Hall and Albert Hall and operas from Covent Garden, Glyndebourne, and Aldeburgh, puts on performances by the Royal Shakespeare Company, the various Old Vics, and by dance ensembles up to and including the Royal Ballet. Commercial television in Britain has in effect guaranteed the losses of off-Broadway-type theater groups in cities outside London and has carried dance presentations by both local and visiting groups. RAI in Italy maintains two "lyric orchestras" for studio opera performances in addition to symphonic ensembles and picks up occasionally from the state-supported theaters. The separate broadcasting organizations of the German <u>Länder</u> have major musical and dramatic ensembles under contract, sometimes with and sometimes without cooperation from local government; Karajan's separate Eastertime Salzburg Festival is made possible in part by deals with Südwestdeutsche Rundfunk and Oesterreichischer Rundfunk, for which he has made films of most of his major productions — films that have played one-night stands at Philharmonic Hall in this country but have been broadcast on Eurovision throughout Europe. Though French television tends to be much less supportive of the arts (for very American reasons — union contracts, plus a theoretical commitment to television art as a distinct enterprise), the week's schedule in France is likely to offer one or two hours of more serious artistic interest than a viewer is likely to find on American airwaves. In general, performing ensembles that play for real audiences are much more significant in European broadcasting than they are here; and payment for broadcast services is much more significant in the budgets of European artistic institutions than it has ever been here.

Attention to the arts has greatly declined in American broadcasting since the days of radio. Both NBC and CBS maintained radio symphony orchestras (NBC for almost two decades, with Arturo Toscanini at the helm), and both commissioned radio plays of considerable pretension. Metropolitan Opera broadcasts were (and still are) a feature of Saturday afternoon, and auditions for the opera were on the air. Apart from nationwide hookups for the New York Philharmonic and sometimes other orchestras, local stations often carried local concerts, typically with sponsorhip by a local bank or utility, which contributed to the financial stability of the orchestra. So elitist a form as chamber music got a play — the harpsichordist Sylvia Marlowe, for example, was under contract to NBC and put together groups that introduced Vivaldi (then esoteric) and commissioned pieces (still esoteric) and "swung the classics" (always shocking). Even today, the "good music" FM station has a role in many communities, though that role usually has little to do with the presentation of live performances — and almost never contributes to the income of musicians.

Hosts of reasons can be marshaled to explain the difference between European and American television. The European tradition of state patronage of the arts obviously extends to state broadcasting stations. ("The BBC," says an official document, "has long been recognized as the most powerful single influence in British musical life."[1]) Given the stature and clout of existing artistic institutions, television abroad was more or less compelled to seek ways to use them, while in the United States the limited resources available for the

promotion of serious programming went chasing the will-o'-the-wisp of a new
art form called "television" that would not use older performing ensembles.
(Thus, the Germans have brought in distinguished film directors to work with
Karajan on television projects for the Berlin Symphony, and Boulez has helped
the BBC develop techniques for superimposing a score over the performers of
absolute music, but no imagination has been applied to such efforts here.) Initial perceptions about television and the arts have proved remarkably resistant
to experience: though dance in its various forms has been the art with the most
rapidly expanding audience through the television years—and efforts as early
as those on "Omnibus" showed the television screen a surprisingly capable re-
ceptor of dance imagery—little has been done to bring dancers and image Or-
thicons together.

It should not be thought that American television has done nothing at all with
or for the arts. NBC for several years sponsored a capable opera company
(now resurrected, with the same leadership personnel, for NET); CBS covered
the opening of Lincoln Center, telecast some Sol Hurok Russians, presented a
number of children's concerts, mostly with Leonard Bernstein; for some years
the Boston and Chicago orchestras appeared regularly on many noncommercial
stations, which also offered several series of interesting "master classes" and
ad hoc television adaptations of classic theater and fiction. Bell, Firestone,
and Ed Sullivan paid fat fees to big names from the serious music world.
Moreover, commercial television's relation to American drama and film-
making has been more complicated than most critical opinion seems to assume.
The "golden age" of the mid-1950s is perhaps somewhat overadvertised today,
but the fact is that television has been for years the major market for dramatic
writing and performing in the United States. Every year since the mid-1960s,
Hollywood has made more hours of film for television than it made for theatrical use in the years of its heyday. While most of this was pretty bad by any
standards, I am by no means certain that the Broadway or off-Broadway theaters always give us better stuff than the average of NBC's "Ironsides" or
ABC's "Love, American Style" or even some of the made-for-television 90-
minute movies. Nor are the revivals of classic plays in our repertory theaters
always better than the efforts of the Hallmark Hall of Fame.

Perhaps the most discouraging aspect of broadcast television in the arts
was the failure of the noncommercial stations, while locally oriented, to utilize the talents of the repertory theaters that sprang up all around the country
in the 1950s—nurtured, as ETV was nurtured, by the Ford Foundation. It is
a big country, and there is certainly more talent out there than gets through.
It was a tragedy for American television, and perhaps for American culture,
that the arts division of Ford and the television division never made contact
with each other—that local theaters dried up for lack of audience while local
television starved for lack of program, and neither was encouraged to work
with the other. When Ford took noncommercial television big time, with Public
Broadcasting Laboratory, the emphasis was almost entirely on public affairs,
to the extent that opening night (the only program in the series that commanded
much audience) featured a play of zero aesthetic value, amateurishly performed,
selected solely because of some presumed impact on race relations.

When everything is said that can be said on behalf of the proprietors of
American broadcast television, the fact remains that they have done much less

in and for the arts than their European contemporaries. And when all expla-
nations have been considered, one overwhelms the others: European television
is programmed essentially by what the British called an Establishment (before
American usage corrupted the term), while American television is programmed
almost entirely with reference to popular taste. The most extensive study of
audiences ever made came to the conclusion that the total ticket-buying public
for the performing arts, amateur or professional, comes to about 4 percent of
the population aged eighteen or over.[2] And this is for all the performing arts
put together —concert, opera, dance, theater. Though the demographic characteristics of the audience stay much the same as one moves from art to art,
the individuals are different. For any one of the arts, the total ticket-buying
audience must be under 2 percent of the adult population.

State-financed European broadcasters can program for audiences of this
size. In Germany, François Bundy wrote in early 1971 in the New York Times,
"the main support for quality films comes from the television stations of the
various German Länder ... because of a comparatively small elite endowed
with almost dictatorial powers in running TV, which imposes its high standards."[3] Even those who would like to see American television run that way
would not be willing to say so in public.

Again, a caveat: properly sold on the right occasion, art can reach through
broadcasting to audiences much larger than those indicated by such pessimistic
statistics. Arthur Miller's Death of a Salesman racked up the largest share of
audience in its time period; the recent Hallmark Hamlet was seen in more than
ten million homes. The credo of the Service de la Recherche de l'ORTF can be
taken as at least an arguable principle by all men of good-will: "Refuser ab-
solument cette dichotomie: programme distingué pour l'élite et programme
vulgaire pour le populaire."[4] My own definition of art clearly includes the
early years of "The Honeymooners," still playing on local stations in reruns —
despite the terrible quality of the kinescopes —nearly twenty years after the
initial presentations. More commonly, though, the general law of communications applies: entertainment reaches more people when it is familiar, fashion-
able, evanescent; fewer people when it is novel, durable, artistic. Where
channels are a scarce resource, using them to seek smaller rather than larger
audiences may or may not be admirable, but it is obviously undemocratic. And
it is certainly unprofitable.

Presumably, noncommercial television could be more active in the arts,
but it clearly won't be, for a complex of reasons. As a network, PBS cannot
present less-than-professional work of local repertory companies: remote
audiences won't tolerate poor quality. With only appropriations and occasional
contributions for support (presumably Ford will phase out), it cannot often afford the very considerable costs of professional performance in the collaborative arts involving music. But beyond all that, its bias, as a "public" broadcasting service, seems to be toward larger audiences, like the bias of the
commercial networks. Minorities seem more legitimate to PBS when they can
be expressed in terms of race, creed, color, or national origin —rather than
when they are expressed in terms of taste. In fact, oddly enough, popular commercial programs cut straight across lines of race, creed, color, and national
origin, appealing about equally within such groupings. What television lacks,
in terms of diversified service, is the appeal to a variety of tastes.

If the arts have been neglected on television simply because channel capacity is a scarce resource that democracy must devote to more popular entertainment, then the multichannel cable solves the problem automatically. Obviously, the world is more complicated than that. Somebody has to produce programs, get them to audiences, and pay the bills. Let us look at the ways these missions can be accomplished.

PAY TV

Cable programming as an extension of a box office was attempted in 1960-1963 in the Toronto suburb of Etobicoke, by a subsidiary of Paramount Pictures. The cable in the system, which extended at its peak to 5,800 homes, carried only programs originated by the company: broadcast programs were still received by the householder off the air. And at the beginning there·was no fixed monthly charge for the service: payment was by coin into a box (a true box office), strictly for programs watched. Among those offered were an off-Broadway Hedda Gabler, a Broadway musical live from its theater, and a performance of Menotti's The Consul (for $1.50; Jack Gould wrote in the New York Times: "It is not too much to suggest that seeing the program, with Patricia Neway's superb tour-de-force in the heart-rending evocation of the human spirit under trial, must rank as one of the most civilized experiences in viewing that can be imagined." [5] The big audiences were drawn by movies and by the professional hockey and football games; and the whole venture went down the drain.

More ambitious in theory and in publicity was the STV pay system launched in Los Angeles and San Francisco in 1964. More than $12 million worth of hardware and telephone company cable was in place and about 12,000 homes had signed up when a state constitutional referendum prohibited pay TV in California. Sylvester L. (Pat) Weaver, former president of the National Broadcasting Company, was the head of this operation, and its consultant on cultural matters was Sol Hurok. "Pay TV," Hurok said at the time, "is the only instrument we have to use this invention for cultural enlightenment. In the long run it will be a great advantage to the artist. We use the same accounting system as records — it will be like an annuity, 7 1/2 percent royalty ..." Weaver spoke of bringing to California "the new production at the Kabuki Theater, by satellite from Tokyo."

Indeed, more than that: "When he lets himself go," I wrote that year in an article about STV for the Saturday Evening Post, [6] "Weaver envisages a custom-tailored television service, built around a video tape recorder in every home. The householder will merely call up before he goes to bed, and let his television service know what he wants to see the next day — and the television service will synchronize its transmission with the home video recorder and put a couple of hours onto tape for him in six minutes. 'You want a special stock market report,' Weaver says earnestly, 'or you want to see Maria Callas's debut as Carmen at La Scala, or take a course in nuclear physics — all you'll have to do is make a phone call.' Weaver's dream starts from the proposition that if he can add several billion dollars a year to what Americans have been spending on arts and entertainment, he will lure from their lurking places whole schools of new talent now neglected by the unimaginative businessmen who run our cultural enterprises." Plus ça change . . . .

Pay TV is a dirty word around the cable companies and in Congress. It is not, however, a dirty word around the executive offices of Madison Square Garden, where the current deal with the cable companies for the Knicks and Ranger games is regarded as nothing more than a sampling operation, drawing customers who will subsequently pay. Should the sports promoters win what will be a very bloody war, the arts impresarios and top-dop institutions will be right on their heels.

Personally, I think the attitudes that carried the California referendum in 1964 are semipermanent in the society, and that pay TV (especially on top of a monthly cable charge) is politically not viable. But the stakes are very large, and I believe a run will be made for them, especially by the Garden and the Metropolitan Opera, both of which seem likely to go broke unless a large remote box office can be generated.

CABLE SUPPORTED BY ADVERTISING

Support by advertisers would make cable a technical variant of broadcast programming. As you know, commercials are now legal under FCC rules (though Bob Bleyer, Teleprompter's director of programming, tells me they are resented in the sports events, which are the only place they appear in that service, and I imagine they would be resented even more in "serious" programming). In prime-time commercial television, advertisers now pay a total of about 3 cents per household per hour (the hour including about 7 minutes of commercial messages, counting network and station sales). They might pay more to reach the high-income levels of the arts audience (the pro football games, which also reach high-income audiences, were salable last fall at slightly more than 4 cents per household per hour); but, on the other hand, the number of messages per hour would have to be drastically reduced. The move from pay TV to sponsored programming, then, reduces potential receipts from a minimum of 50 cents a household an hour (which would be pretty cheap for an opera or concert or play for the whole family) to a maximum of 4 cents a household (less agency commissions). Audiences would have to be much larger to yield equivalent revenues.

Given unusual complaisance by the unions, minimal selling costs associated with the purchase of the advertising minutes, inexpensive carriage to head ends, and donation of the cable by its proprietors, I think we are talking about audiences of well over a million homes before advertiser-sponsored symphony concerts, operas, or ballets can enlarge rather than drain the resources of our performing companies. Serious theater is less expensive to produce and can probably be sustained on an advertising basis by a smaller audience — perhaps as small as half a million homes. All these figures assume that it will be technically possible to make satisfactory television from regular performances before audiences, and that the surcharge by those involved in the performance (actors, singers, dancers, musicians, stagehands, electricians, janitors, etc.) will be no more than, say, 150 percent over what they would be paid for an untelevised night's work. Both these assumptions could easily turn out to be false.

If live performances and television are to work symbiotically, some way will have to be found to convince television people that such programs are not a kind of slum housing. Despite the experience of the BBC, RAI, the Scandi-

navians, and even ORTF (which does two plays a month from Boulevard theaters), American television directors and producers insist that the results of
filming or taping a production designed for theatrical performance are simply
unworthy of their machinery. Yet the NET Uncle Vanya from the Birmingham
Old Vic, whatever its defects, was surely no worse technically than the single-
set Andersonville Trial made for television and much acclaimed a few months
back, and the NET Opera Abduction from the Seraglio had nothing to recommend it over the Peter Grimes the BBC did at Aldeburgh or even the Barbiere
Japanese NHK taped at the Met in New York. Fortunately, cable people are at
present less biased against taping real performances. Teleprompter in New
York has done a tape with some interesting production values (until the dis-
solver broke down) of a garden-variety piano-violin-clarinet recital at the
Washington Heights YM-YWHA, and it will presently do a pair of one-act operas the Mannes School is presenting at the 92nd Street Y. With money, cable
programmers would probably become as snooty as broadcast programmers
about the unsuitability of live performances for broadcast, but that kind of
money won't be around for a while.

Union problems are much more severe. Teleprompter was able to cablecast its tapes of the Y concert without paying the artists only because the musicians' union never heard about it. And the artists are only the beginning.
Rogers Cable in Toronto wanted to tape an amateur theatrical presentation
from the Queen Elizabeth Theatre. "The director was a professional," says
Phil Lind, the young head of programming for Rogers, "and he wanted a couple
of hundred bucks, which was okay. But the step-up fee for the stagehands —
without lighting, which would have been extra — was $1,800, and that we could
not do. We did a folk festival with the finest rock groups in Canada, at a university, and we got releases from the groups, but after we ran it one of the
groups called almost tearfully and said, 'Stop — they're going to take away our
union card.'" The St. Lawrence Arts Centre in Toronto was built in part for
television origination (especially cable), but IATSE has ruled that even debates
and speeches cannot be broadcast without a minimum stagehand and lighting
crew at a step-up fee. The centre, which is strapped, found itself in the monu-
mentally embarrassing position of having to sell an appearance by the Premier
of Canada for $350, to cover extra union costs. "Thank God," says Sandy
McKee, who handles broadcasting for the centre, "he didn't come." The experience was familiar for Miss McKee, who ran the broadcasting end of Expo
'67, from which the Canadian Broadcasting Corporation tried to carry an appearance by the Winnipeg Ballet. The stagehands' demands were so high that
CBC ultimately rebuilt all the sets in its studio and telecast a studio performance — dubbing in audience reaction from the theater to simulate a live performance (!).

By far the most ambitious plans to date have been those hatched by John
Goberman of the New York City Opera. In the fall of 1970 Goberman got within two days and a few thousand dollars of a sale to Teleprompter of a performance of Donizetti's Lucia, to be cablecast live (one time only) from the stage
of the New York State Theatre. Teleprompter offered $25,000, and Goberman
thought he had all his union deals made to come in under that price, with something left over for the City Opera. "I'd say our unions have been — well, not
flexible, but far-sighted," Goberman reports. "The musicians, for example,

would have been paid one-tenth of their commercial videotape rate for the time.
And that pushed the price higher than Teleprompter would go."

At a meeting between Goberman and representatives of Teleprompter and
Manhattan Cable, Goberman made a global offer of 44 opera and ballet evenings a year for seven years, at a fixed annual price of one million dollars.
The City Opera and City Ballet would have the right to offer these same evenings simultaneously to other cable systems, which would pay separately, and
the unions involved would participate in each payment pro rata to their participation in the original Teleprompter and Manhattan contracts.

Admittedly, much happens in the world that was undreamed of in my philosophy, but I would be truly surprised to see this deal come off. What Goberman
is asking for is about one-fifth of the total revenues Teleprompter and Manhattan will receive from subscribers in 1971. He argues that they need him:
"They should be offering their customers something broadcast television
can't. With this sort of deal, the subscriber can have cable TV as a cultural
resource in his house — like the World Book, though he never opens it. And
twenty-five thousand dollars an evening is nothing for three hours of anything."
But the public-relations and sales benefits of carrying opera and ballet from
the State Theater can be gained with the purchase of lots less than 44 evenings.
And for a one-time use of each program, the advertising revenue possibilities
just aren't there to pay even a fraction of the bills.

Actually, Goberman's deal could be a disaster in disguise, because it could
freeze into cablecasting restrictions even more onerous than those that apply
in broadcasting. Goberman's arrangements with the unions are such that it
would be illegal to make a tape of the performance: "There's a kind of para-
noia here," he says, "in their fear of being taken advantage of — if you make a
tape, somebody's going to pirate it." But live-only goes against both the logic
and the economics of cable.

The audiences estimated a few paragraphs back — at least a million homes
for opera, half a million for drama — are not out of the question for cablecasts
of serious stuff, but they almost certainly cannot be achieved on a single exposure. An advertiser can be assured an audience large enough to justify his
expenditures only if the program is made available several (perhaps many)
times, on each cable system. In theory, this makes no nevermind, because in
theory cable with its many channels to fill is a multiple-use medium for programs. (Indeed, Nathaniel E. Feldman of the Rand Corporation waxes lyrical
about the prospects: "Mere repetition [of the material on 13 broadcast chan-
nels] on other days and at other hours could consume 30 to 50 channels. TV
watching, like moviegoing, could become more discriminating ... . Note that
such extensive repetition of commercial TV would involve no additional costs
for programming preparation." [7] But in fact all TV talent contracts, commercial and noncommercial, now provide for payment of "residuals" if a program
is shown more than a stated number of times (usually twice; some noncommercial contracts are drawn to permit two uses a year for two or three years).
CBS would like to use cable systems to test pilots for new programs, but CBS
lawyers have told the research department that a single such test would probably lead to payment of residuals from the second (rather than the third) broadcast use of the material.

All the pressures on those who would take responsibility for producing arts

programs for television lead them to retain the residuals system: "We want to
do this," says William Hadley, Director of Finance at the Metropolitan Opera,
"on a basis of people being paid royalties for every use." The first significant
contracts for opera, ballet, and theater on cable will probably be signed soon,
by Goberman or others: it's valuable, even necessary public relations for an
industry that will be scandal-spattered throughout the decade because the conditions of franchising invite scandal. If these contracts are drawn on a live-
only or residuals basis, programming for cable will probably follow closely
the patterns of programming for broadcast, and there will be very little of
serious artistic value on the wire.

   All this is not to say, incidentally, that artists should sacrifice their share
of profitable programs to impresarios or cable companies. Some circulation-
based pricing structure would have to be developed out of simple fairness. But
that structure would have to be significantly different from the residuals structure if cable is to make a contribution to the professional performing arts or
the audience for them.

CABLE PROGRAMMING AS A GIFT TO SUBSCRIBERS

This approach is used by Canada, because the Canadian Radio and Television
Commission, concerned about the financial condition of broadcast TV in that
country, fears a diversion of advertising revenues to cable companies. In the
light of the previous discussion, the absence of any revenue assignable to arts
programming would seem to doom the cablecasting of professional performance, but this isn't necessarily so. In Vancouver, for example, Cablevision
originates two half-hour programs a week featuring young artists — one called
"Pianoforte," for outstanding diplomats of the local university music departments, and another called "Festival of Stars," presenting winners of the annual Kiwanis Western Canada Music Festival. Performers are paid (in the
$50-$100 range); "for some of them," says program director Vic Waters,
"it's the first dollar they've ever earned." The Vancouver Art Gallery, too,
has a half-hour a week dedicated to its use on Channel 10 and sometimes presents musical performers who are giving or have given concerts at the gallery. Teleprompter's little chamber recital from the YMHA would have been
possible on a professional basis without advertising revenue: union minimums
would have run no more than about $250 for the entire hundred minutes of the
program.

   Necessarily, such programming would be small potatoes, in terms of the
size of the performing groups or the reputation of the performers. But the
fact is that the young badly need showcases and experience before microphones
and cameras. Though one would hope that the wired nation would contribute
more to the arts than debut appearances by young professionals working for
minuscule fees, even that would be better than the expensive debut recital in
a rented hall, which is now a musician's first step toward a solo career — and
might be better than the drudge work of off-Broadway for significant numbers
of young actors and actresses. It wouldn't do much for the quality of programming on the cable — but for that purpose even ill-paid young professionals
would be more valuable than amateurs.

   The obvious source of arts programming for cable systems is the university.
At most universities, students produce dozens of plays every year, music de-

partments give scores of concerts, happenings are planned and do actually
happen. Many schools have broadcasting departments in their journalism or
speech or education divisions, training students to operate television equipment; and most today have some sort of closed-circuit capability for instructional use, so some minimal level of expertise is instantly available for cable
origination. And the cable companies, being urged or required by the FCC to
offer unique, local programs, would like nothing better than signals from the
university to put on the cable.

But the university, while glad to offer sporting events and not unwilling to
supply speakers who will tell viewers how to run their lives or the world, has
been most uncooperative about supplying entertainment or art. At Hays,
Kansas, for example, the local cable system (which is owned by the same
people who own the local broadcasting operation) has set up what is in effect
a separate head end in Fort Hays Kansas State College (5,500 students). This
facility passes through to dormitories and other college buildings whatever is
on the public cable, and can add whatever else the college would like to send
to its members, instructional or otherwise. In addition, a channel of the public system has been dedicated to the college, so that it can communicate to the
town whatever it would like to put out. There are broadcasting courses for
credit in the speech department, and a studio is built into the second floor of
the college theater: to televise anything going on in the theater, students need
merely dolly the cameras out to a separate small balcony. And in fact the
cameras have gone onto the balcony — for student productions of plays, faculty
recitals, even the staging of an opera written by a member of the music department. But nothing has ever been put on Channel 12 for the subscribing
public — or even on the "academic" cable for the students themselves. "They
never use it," says Jack Heather, who runs the broadcasting courses. "These
are thirty-five-dollar tapes, just sitting there. I've begun to toss them into
the 'to be used' bin."

Teleprompter reports that the University of Oregon in Eugene has supplied
theater pieces to the cable, but otherwise the experience of those I interviewed was almost entirely negative. "We made a community channel available to the radio-TV course at Scarborough Centennial College," says Gordon
Keeble of Keeble Cable in Toronto, "and in eight months they produced one
hour." Bill Brazeal, executive vice-president in charge of programming for
Denver's Tele-Communications, Inc. (73 systems in 22 states), reports only
a handful of hours of program from numerous university contacts: "What happens is that it's tremendously intriguing at the university when it starts, but
it's work; takes time and effort to put together a meaningful program; and
there's not much audience, and they lose enthusiasm ... ." Charitably, the
college performers are seeking to retain the limited live audiences they have;
uncharitably, the colleges are suffering a deep and perhaps justified suspicion
of the third-rateness of their efforts, which cable presentation would reveal
to a potentially unsympathetic public.

Local cable will undoubtedly receive some programming from amateur
symphonies and little theater groups, concerts and plays to be cablecast after
a delay of a day or two, to preserve the live audience and (perhaps even more
important) permit the performers to tune in on themselves. Cable is a big see-
yourself-in-the-paper medium: Cablevision in Vancouver reports success with

school and club soccer and football, taped in the afternoon and shown early in
the evening so the participants can view it.

   After the fading of the initial pleasure of seeing one's own kind in action,
amateur performances — indeed, all cable origination that relies on amateur
work — may be unable to draw audience. Canadian experience, at any rate,
argues that "community channels" are effectively dark in terms of viewing
patterns except in periods of local stress or festival. In November, 1969, A.
C. Nielsen did a special study of Middlesex County, Ontario, where cable
penetration exceeds 60 percent (80 percent in the county seat of London) and
where the local cable companies have been offering original programs for
some years. Among the results was "the fact that we were unable to find any
viewing of measurable proportions to the locally programmed cable channel."[8]
Ross McCreath, president of the Canadian TvB, which sponsored the study,
reports that the 223 diaries distributed to cable homes did not show a single
entry for the cable company's own channel in an entire week's viewing. Confirmation may be found in the Report of the Special Senate Committee on Mass
Media, which hails "the development of community programmes on cable television ... as a most welcome addition to the mass media in Canada, a new dimension that can dramatically improve the quality of life in our country,"[9] but
accepts the accuracy of the TvB study[10] and prints a separate March 1969
survey of Metro London by the Bureau of Broadcast Measurement, in which
a tabulation of some 25,000 quarter-hours of viewing by cable subscribers is
completed to 100 percent without any mention of the cable company's own
channel.[11]

   Still, it would seem likely that some audience, some fraction of one percent of the subscribers, could be found for amateur performances on a local
cable system. And the possibility of such performances might encourage the
growth of amateur activity in the arts in many communities — which brings us
to what may be the most difficult philosophical tangle presented in these pages:
the relative societal value of amateur and professional performances.

   "The musician who has been a professional critic," Bernard Shaw wrote in
1917, "knows better even than Wagner that music is kept alive on the cottage
piano of the amateur, and not in the concert rooms and opera houses of the
great capitals."[12] In any calculus of pleasure, participation in music or dance
or theater would have to come out upscale from attendance at performances,
for the society as a whole. Moreover, the support and audience for professionalism comes in large part from amateurs, who can appreciate most deeply
the accomplishments of the artist.

   But it is also true that amateurs by definition cannot develop the skills or
indeed the art that keeps forms alive and kicking with the passage of generations. From the time of Shakespeare and Molière to the present, theater has
flourished only at those moments and in those places where a critical mass of
artists could make a living at it. The maintenance of that critical mass — of a
reasonable pool of not-great executant artists — may be a requirement for the
emergence of greatness in the arts. With rare exceptions, significant execu-
tants and creators in any of the collaborative arts start serious work as
children. Because much more than just talent is ultimately required to make
a contribution in the arts, only a minor fraction of those who seem so promising in childhood and early adolescence ultimately become major artists. If

it is not possible for those who fall short of memorable performance to live
by their art, if there are to be only big winners and utter losers, the supply
of those who try will dry up. Isaac Stern uses the analogy of the prize-fighters,
which is a little unpleasant, because prize-fighting is unpleasant, but by no
means false.

There is no "correct" balance between amateur and professional in the
arts; every generation finds its own and bemoans the change from yesteryear.
But the fact is that for almost half a century the market for the lesser professional has been shrinking, and the cries of the American Federation of Musicians and Actors' Equity have been heavy on the ears. As the Rockefeller
Brothers report emphasized in 1965, the much-advertised "cultural explosion"
has been predominantly amateur.[13] The availability of talent of national caliber,
via films, records, television, crushes professionalism of merely local caliber. Meanwhile, the "economic dilemma" described by Baumol and Bowen —
the tendency of unit costs to rise rapidly in service industries when wage rates
are pegged to increasing productivity in manufacturing industries — makes the
not-quite-first-rate professional seem awfully expensive for value received.
As Baumol and Bowen put it at the end of their book, "This area lives under
the shadow of its own Gresham's law: without constant vigilance and willingness to bear the constantly rising costs of professional performance, amateur
activity will tend to drive the trained performer from the field."[14]

Ours is a time when people are very conscious of all that can go wrong with
a technological novelty and not very conscious of the human resourcefulness
that deals with things that go wrong. Thinking about a wired nation is a worri-
some activity for someone interested in the future of the arts; indeed, the better the service on the cable the worse the worrying, because the arts require
a willingness of audiences to go out at night and give their human attention and
human reaction to performers. (The cities need a willingness for people to go
out at night, too, but that's somebody else's department.) What ought to be
done and how, I don't know. What I do know is that people should be thinking
about the problem, and they're not, not really; and I hope I've got them
started.

One further topic: cable and art as seen by the counterculture. Half-inch
videotape has become the medium of choice for sections of what calls itself
The Movement. They approach the problems of this medium with the same
graceful insouciance they apply elsewhere: "The VIDEOFREEX," reads an announcement in Vol. I, No. 1 of Radical Software (summer 1970), "are involved
in television technically and artistically, intellectually and emotionally. Technical labors bring us together. We are in a web of video/audio energy flows.
We are caught in the act of electronic fucking. And we sure like to fuck. Contact us at 98 Prince Street, NYC." Nebbish.

"If you need a picture that's always clear, alternative television isn't going
to make it," says Jay K. Hoffman, an interesting impresario who has presented
attractions from baroque music to Jeanne-Marie Darré to Czech puppets and
rock groups. "If you're talking about a verité, then maybe — and maybe big.
As of 1971, the stuff is below any standards, it's inarticulate on every level,
but the seminal aspect is being ignored. Watching it gave me a feeling for the
subject — what used to be called 'heart' . . . ."

There are technical problems involved in cablecasting the half-inch tapes of

the radicals, and the Canadians say the problems can't be solved: one inch is
a minimum technical standard. Thea Sklover, who has been responsible for
securing New York State Arts Council grants for the videotape communes, believes half-inch can be made viable. Teleprompter engineers don't believe in
it but are acquiring the new Sony half-inch color equipment. "Half-inch," says
Nancy Salkin of Teleprompter, "answers an awful lot of needs; it has to work."

If the technical difficulties can be overcome, there is every reason to give
these kids their chance on the cable, and to call it "art" if they like. (They
may not like.) There is almost certainly some talent here, though probably
not much. At present, the product doesn't improve, because everybody's tape
expresses something deep within himself or herself, and the mere utterance
of a criticism is repressive, and if it doesn't bore me you aren't allowed to
be bored, not in a real democracy. Getting it out where strangers can see it
might stimulate that sense of a need for craft, which is the foundation of all
art.

I do think, though, that the legal prohibition against broadcast obscenity,
written into the Communications Act, should be maintained in cable. The ob-
jection used to defeat the First Amendment argument in broadcasting cannot
be sustained here — there is no shortage of channels to impel a supervised
franchise — but the common-carrier analogy will serve the same purpose well
enough. It's an offense to shout obscenities in somebody's ear over the telephone, even if you dialed a wrong number. Artistically, of course, the chance
of getting something worth having from these experimenters will be much improved if they are structured into situations where they must sublimate their
aggressions. In any event, it is one of the attractions of the cable that more
people will get a chance to earn the privilege (it is never a "right") of being
taken seriously.

Notes

1. The BBC and the Arts (London: British Broadcasting Corp., 1968), p. 6.

2. William J. Baumol and William G. Bowen, Performing Arts — The Economic Dilemma (New York: Twentieth Century Fund, 1966), p. 96.

3. "Munich: The Decline of Cinematic Art," New York Times, Feb. 22, 1971,
p. 23.

4. Le Service de la Recherche de l'ORTF (Paris: ORTF, 1968), p. 18.

5. "Triumph in Pay-TV," New York Times, March 19, 1961, Sec. II, p. 13.

6. Martin Mayer, "Big Play for Pay-TV's $1.50 Splendors," Saturday Evening
Post, May 2, 1964, Vol. 237, pp. 71-75. The paragraph quoted here was not
used in the article as printed. The title of the article is not mine.

7. Nathaniel E. Feldman, Cable Television and Satellites (Rand Corp. P-4171,
August 1969), pp. 5-6.

8. The Effect of CATV on Television Viewing (Toronto: Television Bureau of Canada, n.d.), p. 14.

9. Mass Media, Report of the Special Senate Committee on Mass Media, Vol. 1 (Ottawa: Information Canada, 1970), p. 216.

10. Ibid., p. 218.

11. Ibid., Vol. II, p. 391.

12. London Music 1888-1889, as Heard by Corno di Bassetto, Later Known as Bernard Shaw (New York: Dodd, Mead & Co., 1937), p. 397.

13. Rockefeller Panel Report, The Performing Arts: Problems and Prospects (New York: McGraw-Hill, 1965), p. 13.

14. Baumol and Bowen, p. 407.

AN OBSCURED MAJORITY

A 1965 study of "happiness" [5] among 2006 male residents of four Illinois communities observes that "there is a strong positive correlation between happiness and both education and income, a marked negative correlation between happiness and age, and no difference in reported happiness between men and women" (p. 10).

Examining Table 1 reproduced from this study we note in the authors' words that, "at every level of education making more money is associated with being happier, but having more education is not always related to being happier."

Persons earning the least amount of money and simultaneously being encumbered by the lowest amount of educational achievement are most likely to experience unhappiness. In general we note a positive correlation between educational achievement and happiness among persons earning less than $7,000 annually. However, this relationship is inverse among those earning more, where persons who are relatively better off both financially and educationally are the most likely to express unhappiness. Of considerable interest is the finding that persons with the lowest educational status combined with the highest income status are the least likely to be unhappy. Here, the authors of the report suggest that having achieved a status well beyond what might ordinarily have been expected accounts for these persons expressing the relatively high degree of positive feelings that they do.

The study further observes: "Among the better educated, low-income respondents, age is related to unhappiness. In this group those who are under forty are happier than those who are forty or over.... .

"Income makes little difference in reported happiness among younger respondents, but a considerable difference among respondents forty or older .... . Among the poor it might be said that 'life ends at forty' " (p. 12).

Further, the study notes that the content of personal worries varies with age and, most importantly, with socioeconomic status (S.E.S.)

As compared to their high S.E.S. correspondents, older people of lower socioeconomic status are considerably more apt to worry about growing old (21% to 12%); are more likely to worry about their health (39% to 14%) and the approach of death (14% to 7%). Similarly, they worry more often about nuclear war (8% to 4%). In short they just worry more. To be older and relatively less affluent is to worry more.

Of significance is the fact that higher S.E.S. persons aged 50 and over are far more frequently worried about work (55%) than are their lower S.E.S. counterparts (39%). For the more affluent older person continuity in work is a source of concern; for his less affluent fellow the release from work no doubt represents a pleasurable anticipation.

These findings should be kept in mind. Shortly, we shall see that two important organizing thematic principles that distinguish the life styles of working people from those of other subpopulations in America are (1) an almost pervasive anxiety about the lack of ability to control the "outside world" and (2) dissatisfaction with work and drudgery combined with a desire for dissociation from work as an integral aspect of a total life pattern. These themes are important because the fact is that "workers" and "workingmen families"

Table 1. Education, Income, and Happiness (percent "not too happy")

Education	Income			
	Less than $3,000	$3,000-$4,999	$5,000-$6,999	$7,000 or more
8th grade or less	33(359)*	13(115)	13(97)	3(32)
High school or some high school	27(142)	16(213)	10(284)	7(227)
Some college or more	21(29)	9(53)	7(107)	10(188)

*Numbers in parentheses in this table represent case base upon which percentages are based.

make up an actual majority of the American society — no matter what criteria are used to identify them.

For the purposes of this paper let us designate workers operationally as being those persons who are regularly employed in either manual occupations or in occupations that call for routine task performance and that do not entail much abstract thinking in order to be accomplished. According to the U. S. Census such persons would be employed as clerical workers, sales workers, craftsmen and foremen, operatives, nonfarm laborers, private household workers, nonhousehold service workers, and farm workers. Although the jobs held by individuals in these categories vary, there are shared attributes among these people that reflect a high degree of homogeneity. For example, these workers are reimbursed in hourly wages rather than in annual salaries. The job tasks they are required to perform do not require a college education. Their annual earnings rarely extend beyond the national median and more often than not fall below the median. Advancement in job careers is generally limited with promotions coming at very infrequent intervals, if at all, over relatively lengthly periods of employment. Finally, they tend to concentrate in and around large urban centers.

Additionally these workers share a wide range of common norms, values, attitudes, tastes, and life styles, which is clearly distinctive and which offers a considerable rationale for treating these subpopulations as one particular type of social aggregate or subculture.

In 1967, the total U. S. labor force, comprised of persons aged sixteen and over, was estimated to be 74 million in number. (Unless otherwise indicated demographic data for this chapter are taken from the U. S. Bureau of the Census's Statistical Abstract of the United States: 1968, 89th Edition, Washington, D. C., 1968.) Of this total, 8 million (11%) were nonwhite.

Fully 95% of the labor force was employed in nonagricultural jobs. More than 3.5 million (5%) of all those employed worked at more than one job.

More than three-fourths of the employed people in this country, comprising no less than 57 million persons, function as working people in one way or another.

Table 2. Socioeconomic Status, Age, and Worries* (percent worrying
"often")

| Worries | Socioeconomic Status | | | |
| | High | | Low | |
	Younger than 50	50 or Older	Younger than 50	50 or Older
Growing old	4	12	4	21
Death	4	7	5	14
Health	8	14	19	39
A-bomb or fallout	2	4	4	8
Getting ahead	39	24	46	24
Money	52	43	51	41
Personal enemies	13	9	8	6
Work	58	55	54	39
Marriage	17	8	12	4
Bringing up children	55	18	46	14
Total	612	227	450	547

*Respondents were divided into two social classes. "High" consists of people
who have at least two of the following attributes: family income of $5,000 or
more, high school graduate or more, and white collar occupation. "Low" consists of those with none or only one of the above attributes.

Consider the following figures as they pertain to the occupational distribution of the contemporary labor force, which numbers 74,372,000.

13,884,000 persons or 19% of the total labor force work as operatives (for example, truck drivers, semiskilled factory workers).

12,330,000 or 17% are employed in clerical jobs (typists, clerks, book-
keepers).

9,845,000 or 13% work as foremen or in crafts jobs (mechanics, carpenters,
electricians).

7,566,000 or 10% are employed in service work other than that performed in
private households (janitors, barbers, cooks, policemen).

4,525,000 or 6% work in sales jobs.

3,533,000 or 5% are employed as nonfarm laborers.

3,554,000 or 5% are employed in farm work.

1,769,000 or 2% are private household workers.

In all, 19,181,000 employed persons or 23% of the total labor force belonged

to a trade union in 1966. Among all persons employed in nonagricultural jobs
in 1966, 28% held memberships in trade unions. Not to be overlooked is the
substantial presence of women in the labor force. Totally, some 27 million
females are employed in the United States. Of these, two-thirds or 18,000,000
are married. In their distribution by jobs, 79% of the employed females can
be classified as working women.

Other than serving as important sociological data per se, gross as they
may be, these figures are of singular interest to the student of mass communications. Here, it is immediately apparent that there exists a majority subgroup in the population, which no doubt possesses distinctive mass communications needs. The questions remain, What are these distinctive needs, and
how are the mass media serving such needs?

Before we put the matter of pure demography aside, let us consider several additional facts.

In 1966 the estimated median income for all full-time employed persons
was $6,856.

For clerical and kindred workers median income that year was $6,542.

For sales workers in the retail trade it was $6,150.

For craftsmen, foremen, and kindred workers the median was $7,161.

For operatives and kindred workers it was $6,135.

For service workers other than household workers the median was $5,117.

For household service workers it was $5,210.

For farm laborers $2,576 represented the 1966 median income.

For laborers other than farming and mining workers it was $5,133.

No wonder that concerns about money and earning capacity play such important roles in the higher frequencies of "unhappiness" we have seen expressed by lower S.E.S. individuals.

Education is the ticket to better jobs and higher incomes in America, as
the figures in Table 3 suggest.

Put another way, employable individuals who have no more than an elementary school education are destined to a lifetime of earnings that are well
below the national median. In 1967, there were approximately 31 million
Americans aged 25 and older who had not progressed beyond grammar school.
(See Table 4.)

Persons with some high school exposure or those who have completed a
high school education pretty much reflect the national median income over a
lifetime of working. By far the largest single bloc of persons aged 25 years
or older in 1967 — approximately 50 million in number — fall into this category.

Those individuals who have been exposed to a college education or who have
received a college degree are practically guaranteed a lifetime earnings average that exceeds the national median substantially. Yet, this subgroup com-
prises the smallest number of individuals — some 20 million in all.

All things considered, we are basically a nation of high school attendees

Table 3. Lifetime Mean Income in 1966 of All Males Aged 25 Years and Older

Schooling Completed	Average Lifetime Income
Less than 8 years	$ 3,520
Eight years	4,867
Some high school	6,924
High school graduate	7,494
Some college	8,783
College graduate and beyond	11,739

who, for the most part, are cut out for a lifetime of pursuing workingmen's jobs and relatively low incomes.

Eventually, these circumstances operate to lock in the workingmen's offspring into what many observers refer to as the "stable working class." Other observers have referred to the same phenomenon as the "American working caste system."

Perhaps "caste" is too strong an ascription, but if we glance at data relating to father-son occupational status we will note why some observers are moved to use so trenchant a phrase (Table 5). Breaking into the "better," more prestigious middle and upper occupations is a meaningful alternative for only a minority of workingmen's sons. For most, they no doubt will continue to follow in the paths of their working-status fathers.

OVERWORKED, UNDERPAID, AND OVERTAXED

Exploring into the less manifest realms of working people's values, attitudes, tastes and life styles, one is struck with a single oft-repeated and overriding fact, namely, the utter feelings of boredom, anxiety, dissatisfaction, distrust, and defeatism that characterize working people's orientations to their jobs as well as to large areas of life generally. "In the working class," R. S. Weiss and David Riesman sum up, "jobs are often so unsatisfactory that there is no social pressure to say one enjoys one's work." [33, p. 582]

Limited chances for advancement, low pay, drab surroundings, and repetitious performance of routine tasks all contribute to feelings of job dissatisfaction. Perhaps more importantly than anything else, however, is the prevailing notion that in the general scheme of things holding down a worker's job stands for very little in a society that places the highest premiums on such upper- and middle-class "virtues" as getting ahead, achievement, success, and — in modern jargon — "making it big." Realizing that neither has he gotten very far in the past nor is there a strong likelihood that he will do so in the future, the workingman (along with his family) is well aware of his "locked-in" inferior social status. This awareness generates a "muffled resentment" toward all those who have been "lucky" enough to escape the workingman's fate.

This syndrome manifests itself in a variety of interesting ways. The feelings of detachment from one's work result in a compartmentalization of "the job" as simply a burdensome means toward other more satisfying life ends. These life goals relate to being able to cope "on the job" to the degree that it does not become overwhelmingly obnoxious. Getting away from the job through

Table 4. Median Years of School Completed in 1967 for All Persons Aged 25 Years and Older (99,438,000)

Less than 5 years	6.1%
5-7 years	10.2
8 years	14.8
High School:	
1-3 years	17.8
4 years	31.6
College:	
1-3 years	9.5
4 years or more	10.1
Median school years completed in toto	12.0

shorter work weeks, temporary leisure activities, and ultimate early retirement (or, for females, through marriage and child-rearing) make up meaningful pursuits for significant numbers of working people.

The buildup of "muffled resentment" to which Weiss and Riesman refer goes a long way toward establishing a tenaciously "we-they" orientation to life generally and toward the so-called upper, middle, and poor classes specifically. Resentment directed toward the upper- and middle-class power-wielders is expressed in terms of avoidance, suspicion, and distrust, privately expressed derision, and small-scale sabotage.

De Grazia's [6] observations are pertinent here: "In the factory an underground life is lived under the noses of foremen, supervisors and time-study men. They may smell it but they find it hard to see or touch. The workers live in a world apart, on its negative side slow, restrictive, inimical to supervisors, management, and other outsiders; on its positive side inventive, ingenious, and loyal to co-workers" (p. 50).

Toward the unemployed poor who threaten either to cut into or take away what little workingmen have managed to acquire and hold on to, resentment takes on much more open and direct expressiveness in terms of hostility, confrontation, and even violence.

The muffled resentment syndrome makes for a peculiar form of turning inward that results not so much in a political "class consciousness" but rather in what S. M. Miller and Frank Riessman term a "person-centered" orientation.[26] Here institutions, bureaucracies, abstract ideas issues, political platforms, and economic and social programs are considered to be irrelevant, difficult to understand, harmful, and barriers to normal social intercourse. Good guys—people—are capable of transcending such remote phenomena. You "make a deal"; you "find out for yourself from a guy who really knows what he's talking about"; you "go see a guy who can fix it"; you "fool around and have a good time"; you "size the guy up and then decide what you will do."

Contrariwise, you "take care of the s.o.b.'s" who make up the "thems" rather than the "us's." You "don't give 'em the time of day"; you "give 'em a

Table 5. Current Occupation by Father's Occupation, Noninstitutional Male Population 25 to 64 Years Old, United States (March 1962)

Father's Occupation	Current Occupation in Percentage White Collar (professional, managerial, sales, clerical)
Professional	70
Managerial	67
Sales	66
Clerical	58
Skilled workers, foremen, etc.	39
Semiskilled	32
Service (including private household)	37
Laborers (excluding mine and farm)	24
Farmers and farm managers	23
Farm laborers	14

Source: Bureau of the Census, Current Population Reports, "Lifetime Occupational Mobility of Adult Males, March, 1962" (Washington, D.C., May 12, 1964). Based on a study conducted by Peter M. Blau and Otis D. Duncan.

taste of their own medicine"; you tell them to "shove it"; you "fix their wagon"; you "take a walk"; you "knock a few heads together."

After reviewing an extensive literature on the "working-class subculture," Miller and Riessman [26] concluded: "[The worker] is traditional, 'old-fashioned,' somewhat religious, and patriarchal. The worker likes discipline, structure, order, organization, and directive, definite (strong) leadership, although he does not see such strong leadership in opposition to human, warm, informal, personal qualities. Despite the inadequacy of his education, he is able to build abstractions, but he does so in a slow physical fashion. He reads ineffectively, is poorly informed in many areas and is often quite suggestible, although interestingly enough he is frequently suspicious of 'talk' and 'new fangled' ideas.

"He is family centered; most of his relationships take place around the large extended, fairly cooperative family.... .

"... He is not class conscious although aware of class differences. While he is radical on certain economic issues, he is quite illiberal on numerous matters, particularly civil liberties and foreign policy.... .

"... He feels estranged from many institutions in our society. This alienation is expressed in a ready willingness to believe in the corruptness of leaders and a general negative feeling toward 'big shots.'

"He is stubborn in his ways, concerned with strength and ruggedness, interested in mechanics, materialistic, superstitious, holds an 'eye for an eye' psychology, and is largely uninterested in politics" (pp. 28-29).

The pressures upon the workingman's sense of security that are generated by job instability are shared equally by his spouse. In this regard Rainwater,

Coleman, and Handel [27], who studied 480 workingmen's wives in Chicago, Louisville, Trenton, and Tacoma, write as follows: "The working class wife's outlook is shaded by a fairly pervasive anxiety over possible fundamental deprivations. She is anxious about her physical safety, stability of affection, dependable income. ...She knows the threat of curtailed income, the loss of dependable funds for necessities through layoffs, strikes, reductions in the hours of work. These are things she knows <u>may</u> happen. They may come upon her with no advance warning, the result of larger forces which she has little ability to influence or control (p. 46).

"...Working class women as a group are not confident that their economic futures hold promise of improvement.... . They fear that the future 'won't be any better than the present—and it might get worse'" (p. 148).

The notion of "getting ahead" for the working person is translated into relatively small and narrow steps forward rather than gigantic leaps. The assembly line worker aspires to becoming a foreman; the bank clerical worker wishes to be a teller; the telephone operator hopes to be a supervisor some day; the clerk-typist aspires to the status of secretary. Moving into these minor advanced jobs not only spells additional income, but even more importantly, the security associated with these "better" jobs is perceived to represent something of a buffer against a precipitous lay-off.

If working people spend considerable energy in striving for security and stability it is unlikely that they will welcome change and innovation for their own sakes. In a world fraught with the threats of imminent disaster, it can be expected that whatever equilibrium is established, precarious as it may be, will be clung to most tenaciously in the face of pressures to change. Changes imposed by a world over which working people believe they have very little, if any, control are translated into additional dangers to their sense of stability, harmony, and balance. Rather than run the risk of disaster by adopting innovations readily, the working person finds it considerably less menacing simply to resist change.

In addition to high resistance to change, traditionalism among working people manifests itself in a pervasive distrust of institutions, concomitant with a heavy reliance on face-to-face primary and peer group relationships; in tightly-knit family orientations; in sex-segregated activity; and in moral codes that are arranged in rather strongly dichotomous explicit categories of good and evil.

At the base of the workingman's societal structure is the extended family or family circle that includes grandparents, siblings, uncles and aunts, and cousins in addition to the nuclear unit consisting of parents and offspring.

Observes Gans [10]: "Perhaps the most important—or at least the most visible—difference between the classes is one of family structure. The <u>working-class subculture</u> is distinguished by the dominant role of the family circle. Its way of life is based on social relationships amidst relatives. ...Work is primarily a means of obtaining income to maintain life amidst a considerable degree of poverty, and, thereafter, a means of maximizing the pleasures of life within the family circle" (pp. 244-245).

While the workingman's family is male oriented (that is, patriarchal) maintenance of the home and the upbringing of children falls predominantly within the responsibility sphere of the wife. It is from and within her family that the workingman's wife finds both her <u>raison d'être</u> and her sense of fulfillment. The family offers her emotional succorance as well as anchorage in a world that she often may find to be otherwise confusing and ever-menacing.

The data gathered by Rainwater, Coleman, and Handel [27] point out the importance of traditional family ties to workingmen's wives: "The working class wife's daily routine is centered upon the tasks of home-making, child-rearing, and husband-servicing" (p. 26).

As is true of most tradition-bound societies, activities within the working-man's family are divided along sex lines. For the most part men are expected to be the breadwinners, and women, the hometenders. Workingmen's informal and leisure relationships are predominantly male-oriented ("going out drinking with the boys"), while women in working families pretty much confine themselves to relating to other women ("having the girls over for a card game"). In those relatively rare instances when cross-sex events do occur (for example, a family Christmas dinner) a high degree of awkwardness and restraint is in evidence until the men eventually go off by themselves in the kitchen to "down a few" and the women are left to themselves in the living room to "gab and talk women's talk."

Traditionalism in morality among working people is particularly evident in areas relating to sexual propriety. Early marriage and motherhood are considered the only legitimate goals for girls. Thus, heavy emphasis is placed on traditional virtues of premarital virginity and postmarital faithfulness — at least as they apply to females. Males are exempted from strict sexual moral codes to the degree that occasional transgressions are to be tolerated provided they harm neither "nice" girls nor threaten family stability. Wives of workingmen will tolerate a certain amount of extramarital sexual activity on the part of their spouses primarily because they are overwhelmingly dependent upon their husbands in order to maintain their own identity, status, and self-fulfillment. However, when extramarital sexual transgressions become frequent, "forgetting and forgiving" no longer represent viable reaction modes, and the breakup of the marriage looms as the ultimate reality.

Miller and Riessman point to a quality (they refer to it as "intensity") of working people's life style that suggests tendencies to exhibit highly emotional reactions to a variety of events, institutions, ideas, and people. Working people's low boiling points with regard to temper; their tendencies to act first and think later; their stubbornness and tenaciousness with regard to their traditional "core" beliefs that relate to superstitions, religion, diet, sex, loyalty, education, and bureaucracy—all have been chronicled more aptly by novelists and playwrights such as Arthur Miller, James Farrell, Tennessee Williams, John Updike, and Clifford Odets than they have been by social scientists. This immediate, affective response style acts to shut off working people from intellectual flexibility (what Milton Rokeach calls "closed-mindedness"), making them relatively less open to "logical" argumentation and so-called reasonableness as a consequence. The work of Carl Hovland and his Yale University associates on persuasibility has demonstrated that immunity to persuasive communications is a function of positioning in the lower socioeconomic statuses. Getting through to them with innovative ideas encased in "logic" alone presents a nearly insurmountable challenge under these circumstances. Miller and Riessman write: "With workers, it is the end-result of action rather than the planning of action or the preoccupation with means that counts. An action that goes astray is not liked for itself; it has to achieve the goal intended to be satisfactory. It is results that pay off" (p. 3).

Within such a value context it is easy to see that intellectuality as such ranks very low in the workingman's scheme of things. To a degree, perhaps, the working person tends to denigrate intellectualism as a defense against his own educational shortcomings. As Miller and Riessman indicate further, what probably is at work here more importantly are deeply imbedded perceptions (based on a good deal of reality) that acquiring more education per se is not always visibly linked to direct benefits.

We have seen that the working person puts more faith in learning from experience and from people than he does in "learning from books." Although he may have adopted the notion that education per se may be of some utility at some future time, and that he may "respect" education for its own sake, he sees the educated subgroups in our society generally as operating in the amorphous arenas of abstractions, theories, and speculations, which do not lead to "practical" resolutions of difficult problems and, indeed, may often result in harmful consequences. What respect working people have for "brains" is reserved for ability per se. But the working person is convinced that over-all "thinkers are not doers." Eventually, the "doing" most certainly must always fall upon the shoulders of the nonintellectual—the workers. The popular longshoreman-philosopher Eric Hoffer underlines and reinforces these sentiments strongly with his basic pronouncement that power in the hands of intellectuals is corrupting, exploitative, and generally harmful to society.

Born, reared, and living in environments that are relatively tradition-bound and unstimulating, working people understandably seek stimulation from outside the person. Encumbered by the unattractiveness of both the job and home, working people pursue "excitement" via their informal interpersonal relationships, their leisure-time activities, and their consumer behavior. Here researchers have found that gossiping, sports, TV viewing, reading the "comics" and "confession" magazines, and purchasing gadgets and goods (particularly "major" commodities such as automobiles, color TV sets, and home laundry equipment) tend to alleviate the general humdrum quality of working-persons' lives. Rainwater, Coleman, and Handel [27] describe leisure activities: "The 'daily routine' of a working class wife typically includes only two activities beyond the big three of house, children, and husband. These 'other two' activities are TV watching and neighbor or relative visiting. However, 'casual visiting' as a daily activity is not mentioned by a majority of these women. Television, in contrast, ranks very high in their devotion ..." (p. 30).

For those working people (particularly the youth) who are not too concerned with the maintenance of a "respectable" community image, the pursuit of excitement or "action" often manifests itself in behaving in socially unacceptable or even criminal ways.

Generally, the working person sees himself as an "outsider"—an alien in his own land—who, when the chips are down, has to make it on his own and in his own way somehow. Middle-class guidelines set down by the "outside world" leave him cold. These guidelines for the most part tend to be abstract

and universalistic, while he is particularistic and pragmatic; they tend to
stress delay, patience, and long-term deferment, while the working person
places emphasis on the immediate; they tend to focus on the rational, whereas the working man's style calls for affective responses; they tend to stress
achievement, while the working person is concerned primarily with survival.

Question: To what degree are the mass media (particularly television) to
which working people turn with ostensible enthusiasm reflecting these sentiments in a manner that offers these patrons insight, guidance, and psychological support?

THE NEGLECTED MAJORITY

The apparent love affair that has been going on now for two decades between
commercial television and working people has been well documented by researchers such as Bogart, Steiner, Glick and Levy, Gans, Hodges, and
Rainwater, Coleman, and Handel. In a phrase, we can say that as compared
to others, working people spend more time watching (commercial) television
and appear to be more satisfied with what commercial TV has to offer.

Data from a nationwide survey of the adult population conducted by Mendelsohn [22] in 1968 corroborate the previous findings on viewing time spent
(Table 6).

Similarly, a nationwide study conducted by Lieberman Research, Inc., in
1966 [17] reaffirmed the observation that overall favorable disposition toward
commercial TV programming increases as income decreases. In other words
the more money one earns the more critical of commercial television's offerings is he apt to be. (Unfavorable criticisms were voiced by 48% in the $10,000
and over bracket as compared to 33% earning $5,000-$9,000 and 23% with incomes below $5,000).

Various reasons for the affinity that working people show vis-à-vis commercial television have been offered. Geiger and Sokol [12] suggest, for instance, that (1) there is simply less creative recreational activity going on in
working class homes (Wilensky refers to this as "low-leisure-competence");
(2) that the smaller physical space allotted to working class households forces
everyone to view when one member turns on the set; (3) that TV provides a
relatively inexpensive diversion (a 1970 survey of movie attendance conducted
by Opinion Research Corporation shows that where 73% of persons earning
$15,000 and over annually and 68% of those in the $10,000-$14,999 income
bracket paid for admissions to a movie during a given six-month period; 62%
of those earning $7,000-$9,999; 56% of those earning $5,000-$6,999 and only
33% of those with incomes below $5,000 did likewise); and (4) that middle- and
upper-class families tend to hold TV viewing as a questionable and "uncon-
structive" way to spend leisure time, thereby depressing the total time these
families spend with television.

Whatever is responsible for the bent that workingmen's families manifest
toward commercial television, it is evident that as compared to persons in
middle and upper statuses they are considerably more dependent upon the
medium for a variety of functions. In this regard the Mendelsohn study cited
previously found that where two-thirds of persons who can be described as
"workers" proclaimed that they simply "could not manage without TV," only
42% of so-called white-collar personnel and professionals responded in a similar vein.

1785822

Table 6. Frequency of Viewing Television by Education, Occupation of Household Head, and Income (in percentages; N = 3148)

	Light Viewers — Less Than 4 Hours Daily	Moderate Viewers — 4-Less Than 5 Hours Daily	Heavy Viewers — 5 or More Hours Daily
Education:			
Grade school	35	28	37
High school	35	32	33
College	52	26	22
Occupation of Household Head:			
Professional or business	51	24	24
Clerical or sales	42	37	21
Blue-collar workers	37	30	33
Farmers	36	34	29
Nonlabor force	24	31	44
Income:			
Under $5,000	35	28	37
$5,000-$9,999	35	32	33
$10,000 and over	49	28	23
Total	39	30	31

Additionally the same study indicates that:

Where 60% of the working people sampled claim they depended "a lot" upon TV for pleasure and relaxation, 46% of those in more prestigious statuses do so similarly.

Where 13% of the working subpopulation report a high dependency on TV for religious satisfaction and comfort, no more than 5% of the persons in higher statuses do so.

Where 19% among working people respondents say they depend on TV heavily for nonnews instrumental information only half that proportion (10%) among upper status subgroups indicate the same degree of dependency.

Important to note here is that we are not discussing the role of public television in the lives of working people.

In this regard, a national survey conducted in 1970 by Louis Harris and Associates for the Corporation for Public Broadcasting, Viewers of Public Television - 1970, [14] reports that within the six months preceding the survey fully 61% of the sample who had less than a high school education had never viewed any public TV (in areas where PTV was available). This is in sharp

contrast to the 44% among high school graduates and the 25% among college
graduates who reported no exposure at all to similar fare.

Note the following data relating viewing of PTV to income during the six-
month period preceding the Harris interviews:

Among persons earning less than $5,000, 59% said they never tuned in.

52% of all those earning $5,000-$9,999 reported that they never viewed PTV.

In the $10,000-$19,999 bracket the percentage reporting no PTV viewing at
all was 43.

Among the most affluent ($15,000 and over), only 30% revealed that they had
not watched any public television at all.

Additional data from the Harris report as well as from a number of previous studies reinforce the observation that public television's orientation is
fundamentally elitist and that the "serious" abstract, "cultural," "artistic,"
and "educational" tones of a good part of its offerings simply bypass the needs,
interests, and tastes of the "obscured majority" described in the first section
of this paper. As far as broadcast PTV is concerned the obscured majority
becomes the "neglected majority."

In its inception the elitist thrust of "educational" TV was a source of un-
mitigated pride, as witnessed in this statement appearing in the volume by
Wilbur Schramm, and his associates, The People Look at Educational Television: "The regular ETV audience is only a minority of all the television
audience. ...In every respect except size, the audience for educational television is the kind broadcasters dream about — the best-educated, most articulate, best informed, most upward mobile, culturally and civically most active
persons in the community" (p. 90).

Perhaps so long as "educational" television relied upon the financial support of private foundations as well as that of affluent private patrons, there
was some justification for ETV to address itself almost exclusively to the
needs (mostly needs of convenience) of the relatively small upper-white-collar
professional, college-educated minority. Even so, it is inconceivable that a
service supposedly dedicated to "education" could ignore the facts that only a
fifth of the population aged 25 and older have had some college exposure; that
a mere third have completed high school; and that slightly less than half have
had less than twelve years of schooling.

Although there are problems relating to accessibility to PTV, the essential
rub in this regard is that working-class people perceive public television fare
to be too intellectual, too highbrow, too abstract and obtuse, and too remote
from their own peer world as well as from their unique needs to warrant their
attendance. Consequently, what was considered as a matter of success in the
past now proves to be a source of considerable consternation. On this point,
the Harris study observes: "The only sharpness in public television's image
is one that we believe is unfortunate. It is (still) considered elitist ... educational, and appealing more to those with an above average education.

"Encouraging this elitist image, but also a measure of its potential strength,
is the fact that attitudes towards public television become more positive with

increasing education, while attitudes toward commercial television become
more negative."

Although there are many means by which working people can find out about
the world and through which they can experience temporary pleasure and diversion, commercial television fare rather than print materials is turned to
most frequently as research conducted by Bogart, Gans, Mendelsohn, Roper,
and Steiner has concluded.

As but one illustration, note the data that were obtained by Mendelsohn in
1967 in his study, Public and Academic Library Usage in the United States.
[21]

This study reports that fully 76% of the grade-school-educated adults surveyed nationally indicated that they had not read a book during the three-month
period prior to the interviews. In contrast 41% of the high-school educated
group in the sample, and 21% among those with some college education reported similarly. At the same time, nine out of every ten persons whose educations terminated in elementary school revealed they had not visited a public
library in the three-month period before the survey took place. This figure is
in sharp contradiction to the 69% among those with a high school education and
the 44% among the college-educated respondents who reported no public library
visits during the preceding quarter-year. The neglected majority thus is
equally overlooked by our public libraries as by our public television services.

The reasons for this neglect no doubt are manifold. Yet one cannot help but
think that much of it stems from a lack of understanding on the part of middle-
class providers of communications services of some of the fundamental thematic threads that continue to weave through the fabric of the working people's
subculture.

Ancillary "educational" services like public television and libraries are
typically repositories of middle- and upper-class values and tastes. Also,
they are typically established and manned by middle-class personnel and in
being so are regarded as yet another part of that strange, abstract, bureaucratic, demanding, unexciting, impractical "outside world" that working people
find to be generally discomforting and of little practical use.

We have already noted how working people disdain "education" as a means
for improving their personal lives. Before submitting themselves to communications that are purely of educational, cultural, or artistic merit then, the
working person must first be convinced that he will benefit in some explicitly
direct and immediate fashion from it. The notion of ars est longa, vita brevis
is totally alien to him. To acquire knowledge, to enjoy literature, painting,
and live drama for their own sakes appear to be silly, wasteful, and meaning-
less to the working person. Consequently, trying to foist uplifting "serious"
or "educational" communications fare upon the working person will not automatically stimulate voluntary exploration on his part. To the contrary, forced
feeding of serious stuff will more likely produce resentment and reinforcement of the stubborn traditional anti-intellectual streak that is already so important a manifestation of the workingmen's orientation to the "outside world."

Those of us who wish to communicate effectively with the working people of
America must make ourselves meticulously aware of the unique mass communications taste patterns that mark this particular subculture. Here the work
of Herbert Gans takes on particular importance. Gans submits the proposition

that a variety of "taste cultures" whose "values are standards of taste or aes-
thetics" operate simultaneously in our society and often within patterns of
similarities that together make up "the total array of art, entertainment, leisure, and related consumer products available in the society." Gans has identified six separate taste cultures that find their bases in various socioeconomic
strata of America. His "lower culture" subtype is of the greatest interest
here, primarily because it simultaneously accommodates and reflects the
largest single public — a public that is comprised mostly of working people.

The hallmark of this lower taste culture is its basically anti-intellectual
posture. "Like the lower-middle culture public," comments Gans, "this one
also rejects 'culture,' but it does so with more hostility. It finds culture not
only dull but also effeminate, immoral and sacrilegious; it supports vigorously
church and police efforts at censorship." (p. 591)

The "person-centered" orientations of working people color their cultural
tastes as well. Not only do workers place emphasis on the substance of cultural offerings, but they also focus upon the performers — the stars — who are
considered to be paramount above all, including the creators of these offerings. Emphasis upon the performer — rather than upon the creator of the fare
or, for that matter, the performance itself is manifested in the identifications
with "stars" that crop up through such collective behavioral manifestations
(particularly among the young) as "fans," "followings," "fan clubs," "groupies,"
and the like. Here working people audiences expect to relate to performers as
"real" people despite the variegated roles that these performers may assume —
in a process that sociologist term "parasocial interaction." Even though the
images projected by "personalities" in the popular arts and sports are mostly
contrived, working-class fans expect their favorites to "live up" to their projected reflections consistently just as though they were true. When this does
not always materialize, "fans" express deep-felt disbelief, disappointment,
and ultimate disenchantment. What appeals profoundly to working-class audiences, Gans finds, are presentations wherein traditional values of morality,
loyalty, love of family, religiosity, and individual responsibility are dramatized and are protected from and ultimately triumph over evil, sin, tempta-
tion, and unbridled impulse. The author elaborates: "The culture's dominant
values are not only expressed ... but also dramatized and sensationalized with
strong emphasis on demarcating good and evil. The drama is melodramatic,
and its world is divided clearly into heroes and villains, with the former always winning out eventually over the latter" (p. 59).

Because, as we have seen, the world of working people is rigidly sex segre-
gated, it is not surprising that the "popular culture" tastes and behaviors of
workers are similarly dichotomized. Here Herbert Gans notes that, "there
are male and female types of content, rarely shared by both sexes.... . Familial drama that deals sympathetically with the problems of both sexes at
once is rare." The author continues: "Sexual segregation and working class
values are well expressed in the Hollywood 'action' film and television program, and in the confession magazine, the stable of male and female lower
culture respectively. The action film insists on a rigid distinction between
hero and villain; the only social problems that are explicitly considered are
crime and related violations of the moral order" (p. 591).

In these "action" presentations what appeals to working-class men are ex-

plicitly-drawn hero types who possess such highly valued workingman characteristics as overt masculinity, practicality, resentment toward and distrust
of institutionalized authority and bureaucracy, and reticence in nonsexual relationships with women. Heroes who are "loners"—the private eye, the cowboy, for example—who function either on their own or with the help and support of a few close dependable friends or relatives enjoy particular accept-
ability on the part of male worker audiences. Thus, the themes of person-
centeredness, traditional morality, anti-intellectualism, excitement, and
muffled resentment directed toward the ouside world—so evident in the workingman's subculture at large—are seen to play critical roles in his preferences
for popular culture fare. Commercial TV and Hollywood-style films appear to
fill these apparent needs to some degree for the working status male.

That workingmen's popular media tastes run toward melodramatic "action"
fare has been confirmed by a recent study of television preferences that was
conducted by John Robinson of the University of Michigan's Survey Research
Center. Analyzing the TV viewing "diaries" of some 10,000 persons, Robinson found that the male audiences for most "violent" television programming
were drawn disproportionately from among lower S.E.S. populations.

For the women in the workingman's world popular culture tastes appear to
be bent toward the daytime "soap opera" and the "confession" and "romance"
magazine genres wherein conflicts between the need to be "sexually responsive
to be popular with men and remaining virginal until marriage" (Gans, p. 592)
represent the more or less standard fare for consumption.

Studies of the readership of confession and romance magazines like True
Story indicate that more than three-fourths of the women who patronize them
can be classified as members of working-class households. Further it has
been noted by Rainwater, Coleman, and Handel that readers of confession and
romance types of magazines rarely read more "serious" magazines or, for
that matter, more middle- or upper-status "women's" magazines. These
authors report that, "it has been estimated that over two-thirds of the working class housewives who read any magazines at all, read one of the 'romance'
or family behavior types." (p. 126) According to Rainwater, Coleman, and
Handel, readers of True Story are attracted to it primarily for its ostensible
sexual content. But here the interest rather than being of a prurient nature is
instead mainly instrumental in its origins. In this regard the three researchers
conclude that for True Story's working-status female readers the magazine's
contents "do not deal primarily with sex, but are concerned with a range of
topics of importance and human interest."

Readers of True Story-type magazines as well as "soap opera" fans more
often than not find a good deal of support for their own values and beliefs within the fiction that is presented. In addition these forms of working-status literature afford female audiences a substantial amount of instrumental information relating to coping with everyday life problems that they missed during
their own experiences with a limited formal education. Through exposure to
this fare, working women pick up "tips" that teach them "what they should do"
when they are faced with problems ranging from "summer romances" through
preparing "Meat Loaf à la Roquefort," to caring for a mentally retarded offspring without "falling apart."

Although reliable research data on the specific functions that the mass media serve for working audiences are sparse, it can be inferred from available studies that overall the mass media principally provide temporary psychological release and pleasure (entertainment), news, plus instrumental and supportive information of sorts to this subpopulation.

On the matter of sporadic psychological release, Bogart offers the reading of newspaper "comics" among adult workers as one example.[4] In his study, the researcher interviewed 121 working status males in New York City in order to determine the gratifications they derived from reading newspaper comic strips. Among other things Bogart found that newspaper comic strips act to reduce tension in their readers mainly by "offering variety and a recurrent focus of interest." The author continues: "Their (the 'comics') name implies that they also reduce tension through laughter. Actually, comic strip humor, simple and stereotyped as much of it is, seems to produce a grim, unsmiling kind of amusement for the most part" (p. 192).

Besides affording a minor degree of release to their working readers newspaper comics also provide them superficial fantasy opportunities, which the author found to be neither enduring nor psychologically damaging. Bogart concludes, "The comics do present their readers with fantasies of aggression, sex, and achievement, and their appeal for particular groups of readers may be understood in these terms. But there is no evidence that the reader is drawn to these strips by a lust (either conscious or unconscious) for vicarious sensation; it seems rather that he brings his normal impulses to them as he does to other life experiences, and that the fantasies which they arouse, though based on these impulses, are brief and have a low emotional charge" (p. 198).

Coming home from a workday that generates and reinforces "muffled resentment" workers understandably crave temporary surcease. Commercial TV's entertainment offerings of programs like "Bonanza," "Hogan's Heroes," "Mannix," "Mission Impossible," "Gunsmoke," and "Strange Report" afford considerable opportunity for fanciful excursions into worlds that appear to be congruent with the value systems of the workingman's own subculture. But note what is missing here. Nowhere in television entertainment fare—neither in commercial nor public TV—are working people to be seen as heroic role models. As a matter of fact the mass media's role models that are heroic in dimension are so indistinguishable by socioeconomic status attributes that when explicitly identifiable working people become focal points for dramatic or comic treatment they cause mild "sensations" in the society as a whole. In most recent times the swirlings of public commentary, debate, and controversy that engulfed the motion picture Joe and the CBS television situation comedy "All in the Family" are cases in point.

There is very little gratification to be derived by the working person attending either of these two particular entertainment offerings. For although in both instances blue-collar "heroes" (antiheroes is more apt a description) appear they are presented as stereotyped, caricatured buffoons who are ignorant, bigoted, and socially loathesome for the most part. Generally speaking, working people as such rarely appear in the mass media at all, and when they do they are cast in insignificant, peripheral, or socially unacceptable roles. Here too, then, the majority is neglected.

Given these circumstances worker audiences must make do with what media such as television and the movies decide to offer them in terms of role modeling. Gans has observed that it is practically impossible for working people to identify with middle- or upper-class hero types. Making do here means empathizing with a variety of "classless" types of mass media characters — the cowboy, the marshal, the armed services member, the policeman or detective, the athlete, the entertainer who is not yet a "star," and even the "criminal" who may have been "framed."

It would appear that Bogart's somewhat gross assertion that "the mass media represent perhaps the most powerful current by which blue-collar workers are swept into the main stream of conformity to middle class values and aspirations" [4, p. 407] is somewhat hyperbolic. To the contrary it is more likely that working people approach the media quite selectively — embracing those characters, values, and life-styles that best fit in with their fixed needs, predispositions, and own experiences, while they either ignore or reject everything else that does not fit in. If anything, the media appear to reinforce their working-status patrons rather than either to convert or to "uplift" them into middle- or upper-status value systems and life styles.

Working people turn to television as well as to the other media for news in addition to entertainment.

In 1968 Roper Research Associates found that 59% of all the adults in America claimed that they got most of their news about "what's going on in the world today" from television.[28] To keep abreast of events as they occur is a need that most Americans (52%), regardless of socioeconomic status, feel urgently in a time when their very survival depends on being informed on an up-to-the-minute basis.[23] This need is felt in particular by working people who, primarily because of lack of education, perceive themselves as being less informed about public affairs generally. On this score Mendelsohn reports that in his sample 31% of the persons with grammar-school educations considered themselves to be relatively uninformed about world events as compared to 24% with a high school education and 13% with a college education who saw themselves in a similar light. At the same time this study revealed that 62% of the sample who had but a grammar school education considered themselves as being forced to rely on others for ideas, opinions, and guidance on matters relating to a wide array of events, as compared to 54% among those with a high school education and 42% with a college education who had the same self-perception.

Given the circumstances of a high level of concern for keeping up with the news combined with feelings relating to being ill-informed plus a felt need for reliance upon others for counsel and guidance, it is not surprising that television is a primary source of news for a substantial proportion of working people in the United States.

Needless to say television alone is not the only source of news for working people. Newspapers and radio play important roles here as well. The issue at hand is not one of simple exposure to news; but rather, the question arises as to the possible consequences of such exposure. We can only guess at this point because of the unavailability of solid research evidence on this matter.

Typically the working person who tunes in the nightly half-hour TV network news program is exposed to the same amount of "news" that appears daily on the front page of the New York Times. The capsulated bits of news that he sees rarely relate events that take place within the working person's subculture—unless of course they are concerned with work stoppages. Once again, the majority is neglected. Instead, most of the TV news to which the working person is typically exposed is about the "outside world"—a world that is buffeted about by catastrophes both natural and man made; by changes that reflect the breakup of traditional values; by political and economic upheavals; by demands for "rights" among the youth and minorities; by crimes; by threats from alien ideologies. The miniscule information bits the working person viewing television news programs is exposed to are mostly <u>descriptions</u> of events, and rarely are they imbedded in <u>analyses</u> and <u>interpretations</u> of how these events are to be dealt with in terms of realistic consequences upon the viewer. Over time, repeated exposure to threatening messages that are unaccompanied by interpretation and analysis cannot help but provoke feelings of intense anxiety and helplessness relative to controlling one's environment.

Elsewhere I have written [18]: "From a socio-psychological point of view continued exposure to threatening messages that are unrelieved by interpretation or the possibility of enjoyment produces immobilization. That is to say individuals who are confronted with nothing but news of possible annihilation—which is the only 'realism' of consequence these days—over long periods of time will develop mechanisms of reaction that will render them incapable of functioning realistically. Free-floating anxiety under continued reenforcement results in attitudes of 'there's nothing I can do about it, so why bother'" (p. 515).

For working people who come to the news media with mental sets that expect the "outside world" to be menacing, who find it difficult to accept change and innovation as leading to "progress," who are oriented to the belief that the "theys" of the outside world are responsible for the difficulties working people must endure, "straight" news in the very least must serve to reinforce preexisting anxieties and to create new feelings of anxiety where they may not have existed.

Scammon and Wattenberg have postulated that the anxieties that are generated by working people's exposure to news about the disruptive changes that are taking place in the "outside world" have already had serious impacts on the political processes in this country and will continue to do so in the future, undoubtedly.[29] From analyses of a wide array of public opinion poll responses these authors have conjured up their now-famous picture of the significant American voter as being a "forty-seven-year-old housewife from the outskirts of Dayton, Ohio, whose husband is a machinist." They go on to profile her in this manner: "She very likely has a somewhat different view of life and politics from that of a twenty-four-year-old instructor of political science at Yale. Now the young man from Yale may feel that he <u>knows</u> more about politics than the machinist's wife from suburban Dayton, and of course, in one sense he does. But he does not know much about politics or psychology, unless he understands what is bothering that lady in Dayton and unless he understands that her circumstances in large measure dictate her concerns.

"To know that the lady in Dayton is afraid to walk the streets alone at night, to know that she has a mixed view about blacks and civil rights because before moving to the suburbs she lived in a neighborhood that became all black, to know that her brother-in-law is a policeman, to know that she does not have the money to move if her new neighborhood deteriorates, to know that she is deeply distressed that her son is going to a community junior college where LSD was found on the campus—to know all this is the beginning of contemporary political wisdom" (pp. 70-71).

"To know all this..." may not necessarily lead to the amelioration of working people's fears and concerns. The evidence seems to indicate that this knowledge has been used more to exploit rather than to enlighten working-status voters, if the most recent political campaigns of 1968 and 1970 are to be taken as cases in point.

If the mass media in a democracy are supposed to produce an informed electorate that votes in terms of the common good, both the intentional and unintentional uses to which the media were put in the 1968 and 1970 political campaigns indicate that this trust along with the voters of the nation suffered from neglect. Consequently, the working segments of the American electorate (comprising the largest single bloc of eligible voters by a margin of 45%) remained not only objects of neglect; but worse still, they were turned into objects of powerful propagandistic political manipulation.

"To know" the working-status voter should not become the basis for a manual on how to jockey him into one or another political camp. Ideally, knowledge of the workingman's subculture should be used by the media in order to enlighten him politically. Thus far the media, especially both commercial and public television, have failed to address themselves positively to this responsibility. This failure affords cable communications systems an opportunity that, if fulfilled, can have the most serious consequences imaginable upon the future quality of life in America.

We have noted that there is a tendency among working people to feel less-well informed about public affairs and to be more dependent upon others for non-news-related ideas, advice, and guidance. To a degree this syndrome reflects the generally lower educational achievement levels of working people. At the same time it reflects a more profound psychological state of personal insecurity in a threatening world and a diminished level of self-esteem. To the person who is unsure of himself and, simultaneously, who is wary of the abstract, obtuse mouthings of "egghead" authorities, television affords an easy-to-grasp, un-confusing and convenient "school of life." Consequently, as previously reported, we find that fully a fifth of working-status viewers report that they depend "a lot" upon television for instrumental information, which, for the most part "teaches" them about matters relating to dress, manners, courtship, romance, coping with personal crises and tragedy, health, child-rearing, etiquette, the law, and so on. As the lady in the pain-reliever commercial says, "You watch TV—you learn something."

What must be borne in mind is that most of the fare to be found on commercial television is not intentionally created for the specific purpose of presenting instrumental information per se to working-status audiences. In this regard again the requisites of the majority are woefully neglected. In constant need of being able to cope with a wide range of everyday practical problems, the working-status viewer actively attempts to sluice out for himself or herself whatever small nuggets of subjectively defined "useful" information may be buried in the fare that is being offered. Thus, "education" of an incidental nature does take place via TV, but it is not as easily to be identified as is edu-

cation without the quotation marks. Instead of being cognitive in nature, it is
usually affective. Rather than being concerned with ideas and things, it is
mostly concerned with people; rather than being abstract in its long-range
thrust, it is highly pragmatic in its immediate applicability.

Although contemporary television attempts to offer fare that may somehow
fill some of the cognitive, instrumental, and supportive information needs of
some of its audiences, it does so haphazardly, obliquely, irregularly, and,
it would appear, without either knowledge of or special regard for the particular needs of its working-status majority audience. This is not meant to be an
across-the-board "indictment" of television. Whether contemporary TV is
"good" or "bad" or neither is not at issue for the purposes of this paper. What
is at issue is the presence of a majority population whose information needs
appear to be inadequately served by television. What, if anything, can (or
should) cable communications do about this?

TOWARD A NEGLECT UNDONE

The technology of cable communications makes it quite feasible to create a
service capable of reaching America's working-status majority either as one
national subpopulation or as segments of that major subpopulation that share
commonalities of specific occupational pursuits or locales or combinations of
the two. The technological problems involved in forming more explicit linkages between TV and working people via cable will not be dealt with here. We
shall assume that the possibilities for creating a cable TV network (or networks) designed to carry relevant cognitive, instrumental, and supportive information to working-status homes are realistic. Putting aside important matters relating to technology, structure, administration, and financing of such
a system, two very important questions remain: Who is going to operate and
sustain the system, and what will the contents of its fare be? In actuality
these two considerations should not be separated, for each is a direct function of the other.

No communications system that proposes to address itself to the working
public will be enabled to do so unless it is manned and sustained by persons
of ostensible working-status backgrounds and interests. In order to build
credibility and ultimate acceptance among working audiences the proposed
system would of necessity have to be manned by visible working-status general managers, program managers, writers, newscasters, performers, directors, and producers. Unless the communications system that is envisaged
is of the working subculture throughout, it cannot purport to be for it.

In addition to establishing both legitimacy and authenticity, a number of obvious advantages flow from creating a communications system for working
people from among the working subculture itself.

First, the very emergence of a special working-status communications
system manned by persons from that subculture would offer evidence of the
genuine and serious concern that the so-called outside world has for working
people. Additionally, it would demonstrate that the outside world respects the
capabilities of working people to operate their own communications systems
without the constant support, direction, and aid of the middle- and upper-
class community.

Second, the "neglected majority" would begin to have a potentially power-

ful voice of its own—a voice for all to hear. Here the community at large
would begin to become acquainted with the needs, grievances, and problems
that are experienced by working people. Thus, a process wherein working
people start to tie in more intimately with the rest of the community and vice
versa can begin with the sheer existence of a working-status communications
system.

Third, the simple existence of such a system should in itself serve as an
alternative career opportunity model to working people who feel themselves
to be predestined to follow in the traditional footsteps of their fathers.

Currently the essential "visible" personnel manning the television industry
are not typically drawn from the working subculture; and the few who come
from such backgrounds quickly shed them. It would be imperative then to recruit and train members of the working subculture not only to man the proposed system but to supply the mass-media enterprises as a whole with well-
qualified personnel who are proficient in the various mass communications
techniques and processes. Initially cadres of working-status personnel may
have to be trained in such non-working-status milieux as colleges and universities and in commercial and noncommercial television outlets. Eventually,
though, once the working subculture cable communications systems (WSCCSs
for short) come into their own, such training functions would themselves become integral operations within the WSCCSs.

Let us suppose that we have overcome the problems inherent in building up
a WSCCS. What then?

Perhaps the most difficult task that can confront an educator or communicator is to transmit cognitive information (that is, knowledge) to individuals
who are neither prepared nor motivated to receive it. Typically we resort to
the classroom method of "teaching" when we face this problem. But we must
realize that the working-status individual's very need for cognitive information is, to a high degree, a very consequence of his having been "turned off"
by formalized classroom teaching in the past. Consequently, if a WSCCS is
to transmit cognitive information, the greatest degree of creative ingenuity
in devising new approaches is called for.

Without considerable experimentation with new modes of treatment it is
impossible to say how cognitive information of an abstract kind can best be
communicated to worker viewers. We can only guess that stereotyped chalk-
board presentations, discussion panels of experts, and learned lectures will
not do the trick. "Multimedia" devices, dramatizations, and animation types
of presentations hold some promise here. Regardless of how skilled we may
become in the presentation of purely cognitive informational materials, we
shall always have to cope with the factor of low motivation. Working people
generally are not active cognitive information seekers for many reasons that
have been discussed previously. If we wish working people to attend to cognitive information, we must first motivate them to do so. Here the possibility
of providing financial "scholarships" for attending cable-transmitted "courses"
comes to mind. Another idea worth exploring is to "pipe in" materials that
are designed to enhance job skills to places of work during regular working
hours. Workers attending these "training" sessions could be given released
time and perhaps even small "bonuses" for doing so.

The anti-intellectual bent of working people represents another barrier.

Somehow this subpopulation must be convinced of the pragmatic potentials of abstract ideas and information. Whether a WSCCS by itself can get this notion across effectively is questionable, but it should give it a good try nonetheless.

The cognitive information thrust of a WSCCS takes three directions. First, it should be oriented to the enhancement of job and career skills. Second, it should be addressed more to generating usable knowledge about the changing social world than about the physical or artistic world. Third, it should offer worker viewers news and information from their own perspectives about the special "world" with which they are most intimately familiar — that of the working subculture. Here a WSCCS news operation would serve as a supplement to the ordinary news fare that is presented by television.

We would expect that these supplemental news services would attempt to explain the reasons behind the emergence of issues and events for working people; describe the possible consequences of these issues and events upon working people as well as upon the total society; and most importantly, point out explicit alternative possibilities for reasonable reactions to issues and events. In sum, news analyses and interpretations by a WSCCS should be concerned mainly with the development of rational response modes among viewers, and in doing so they should be extra cautious about the possiblities of fostering nonrational boomerang response modes of fear, anxiety, and violent reaction.

Finally, a responsible WSCCS should avoid creating viewer dependency upon it alone for cognitive information (as well as for other types of information). Here, the WSCCS should serve as a stimulant for viewers to explore the widest possible array of information sources beyond the system itself. In this respect the WSCCS should serve as the broadest-scaled vehicle of social education regarded in its widest context.

In order to accomplish this end the proposed system cannot rest in its singular function of merely providing cognitive information per se. What to do with such information? How to relate this information to actions of consequence? How to use information to realize aspirations to redress legitimate grievances and to achieve improvements in the conditions of life generally? These questions require a considerable dedication on the part of a WSCCS to direct much energy in developing instrumental information that is tailored to the guidance needs of its special audiences.

We have noted that "getting by" is a dominant concern of working people. On the most simplistic level, then, working persons need the most commonplace types of day-to-day "coping" information that will help them hold on to their jobs, survive when unemployed, make prudent purchases, avoid getting cheated, conduct themselves properly in job interviews, diagnose simple illnesses, prepare inexpensive and appetizing meals, plan inexpensive vacations, make bank loans, and so on. Here the need is for a good deal of "how to" information that is not as "glamorous" as strictly cognitive knowledge, but nevertheless as important, if not more so.

On a more subtle plane we can envisage an instrumental information need that perhaps cannot be articulated readily by working people themselves. Here we can speculate that working people who have not linked up with the larger society may not know much about how such a linkage might be accomplished. How does one become part of the "outside world," which, although it controls one's life appears to be relatively unresponsive to one's overtures? Certainly, a WSCCS can present models for accomplishing this kind of hookup. The model that comes to mind first is concerned with voluntary organizations. No doubt, additional models will present themselves as the reader continues.

One of the most effective means for individuals in a complex, pluralistic, open society such as ours to overcome their individual senses of social, economic, and political helplessness and powerlessness is to combine forces with others in various networks of interrelationship. Creating or joining voluntary organizations is one means for accomplishing this. There are additional important functions that membership in voluntary associations serves, as Murray Hausknecht [15] reminds us: "In a modern, large-scale, democratic society, voluntary associations are means for furthering the political and economic interests of individuals. This implies that political effectiveness demands that the individual participate in the political processes as a member of an organization. His organizational membership, therefore, serves the further function of helping him transcend his routinized day-to-day activities on the job and in the family by establishing linkages with the broader community and society. By mediating between the individual and the state, associations protect the individual from the unrestrained exercise of power and, by the same token, serve to protect the "elites" who control and exercise this power" (p. 207).

Yet, the fact is, as numerous observers of the worker subculture have ascertained, that workers are far less prone than are upper-status people to join voluntary associations. Note the figures in Table 7.

By now the reader should be able to figure out for himself why workers are less likely to become involved with voluntary organizations. We have seen that the focal points of the workingman's world are mainly the traditional primary peer group types of the family circle, a few close friends, and his neighborhood. The secondary relationships that a complex business/technological society demands are given but a nodding acknowledgment on the part of the worker. To him this aspect of society represents the "outside world," which generally is viewed with anxiety, suspicion, awkwardness, discomfort, and hostility. Venturing into this alien environment — even via voluntary associations — represents a journey into unfamiliar territory — territory that harbors unknown dangers and threatens to break up routine modes of response (mostly avoidance). Additionally the "connection" between belonging to voluntary organizations and "making things happen" is not readily seen by most working people.

Hausknecht sums up: "An association brings together individuals who are strangers to one another in more or less impersonal, secondary relationships; their common bond is a specific interest. This requires of the individual a capacity to inhibit suspicion and hostility toward others, and maintaining this attitude represents a severe strain on the tolerance of blue-collar persons. If the sphere in which the association is active is highly complex, for example, influencing the local power structure, there is a tendency for the organization to become bureaucratic, and for impersonal social relationships to become the dominant mode. In any organization it is possible to establish primary relationships; indeed, this opportunity represents one of the main functions of associations. However, the more 'formal' the organization becomes, the more

Table 7. Membership in Voluntary Associations by Occupation

	Percentage Who Belong To			
			Two or		
Occupation	None	One	More	Total	
Professional	47	24	29	100	(259)
Proprietors, managers, officials	47	24	29	100	(294)
Farm owners	58	28	14	100	(265)
Clerical and sales	59	21	20	100	(240)
Skilled labor	68	19	13	100	(447)
Semiskilled labor	77	14	9	100	(492)
Service	73	18	9	100	(142)
Nonfarm labor	79	16	5	100	(155)
Farm labor	87	13	0	100	(54)
Retired, unemployed	77	11	12	100	(35)

Source: National Opinion Research Center, Survey 367, 1955.

irrelevant do these primary relationships become; the 'businesslike' atmosphere of the pragmatic and efficient organization does not nourish primary relations. Once an organization takes on this cast, it creates a difficult situation for the blue-collar individual oriented to the kind of interaction found in the family and the peer group (p. 208).

"...An association of citizens is a means of effective political participation, a means of influencing the decisions affecting one's life. The inability of the working class to use this instrument cripples its capacity for coping with the environment and seizing the opportunities it offers" (p. 214).

A WSCCS can go a long way in pointing up the necessity for working people to combine with others in pursuing common social goals by demonstrating how such concerted action benefits other subgroups in the society; by showing how to start voluntary organizations and how to run them; by pointing out existing organizations in the community that would welcome the participation of viewers; and by itself becoming a focal point for a voluntary organization that addresses itself mainly to the problems and needs of working people.

A direct offshoot of a policy of stimulating greater societal participation by workers through the vehicle of voluntary organizations would be an enlarged participation in the political process generally.

Table 8 shows that eligible voters in the lower-income brackets (that is, working people) are less likely to vote than are more affluent eligible voters by substantial margins.

Motivating working people to participate in the voting process in far greater

Table 8. Voter Participation by Income, 1968

Annual Family Income	Percentage of Eligible Voters Who Actually Voted
Under $3,000	54
$3,000-$4,999	58
$5,000-$15,000	72
$15,000 and over	84
Total percentage of votes cast by eligible voters in 1968	67.8

Source: U.S. Bureau of the Census, Voting and Registration in the Election of November, 1968 (Washington, D.C.).

numbers than they have been accustomed to in the past is another important mission in the overall linking-up thrust that a WSCCS might adopt.

The WSCCS's general efforts in stimulating greater societal participation among workers by providing them with instrumental information face an additional problem. It has been noted over and over again that working people for the most part fail to see direct connections between education and intellectualism and improvement in their own lives. Somehow a better linking model—other than holding up the outright adoption of middle- and upper-class educational-success values—must be forged. Precisely how to do this is unclear at the moment. But it is a problem that merits considerable attention and study if the WSCCS is to become an effective specialized communications medium.

For the most part, most of what the mass media present most of the time is supportive of middle- and upper-class values, beliefs, interests, and tastes. Even when the media present issues and events that may appear to be of a counter nature, the same media offer reassurance that these disruptive challenges and threats can be met and overcome through the application of reason, dialogue, fair play, logic, enhanced communication, and new forms of public policy. Generally, the middle- and upper-class patron of the mass media is reassured of the enduring stability of his institutional systems to the point that he believes firmly in his own ability to "work through the system" in order to maintain his sense of equilibrium. Because he is better educated and because he is convinced that "reasonable" change often does lead to "progress," the middle- and upper-class media patron is less tradition bound, and he shows far less intransigence when confronted with change than does his working-status counterpart. To a degree, upper- and middle-class individuals are better prepared to live in a changing world, because their stake in it is visibly more deeply embedded and more secure. They hold the power, and the media assure them that they will continue to do so by right. In other words, the mass media in America are generally supportive of the middle and upper classes, holding them up as examples of the norms by which all other segments of our society are to measure themselves and by which all others are to be judged.

But what of the working-status patron of the mass media? As far as he is concerned, with rare exceptions, he is the media's "invisible man." The fact that he is literally an absent phenomenon within the media undoubtedly convinces him that in the general scheme of things he is not important, not worthy of attention, not someone to be concerned about. In other words, while the mass media reinforce the "superior" status of the middle and upper classes, they also serve to reinforce feelings of diminished self-esteem among working people. A working subculture cable communications system can go a long way in balancing out this inversion.

As previously stated, such a system by its very existence would provide positive feelings of worth and self-esteem to working-status viewers. Additionally, explicitly designed programming that has a distinctive authentic working-status flavor about it would offer substantial psychological support to viewers. In particular, dramas that reflect working-status themes and project authentic heroes drawn from the working subculture should serve to enhance viewers' self-evaluations toward more positive feelings of worth in a general way. At the same time programming fare that is drawn from the working subculture and that projects authentic working-status heroes, values, beliefs, themes, and tastes could address itself realistically to the actual anxieties, fears, and worries that bedevil working people in an attempt to ameliorate them somewhat. Of course mass communications messages by themselves cannot do away with these symptoms of malaise. They can, if prepared conscientiously, ease them to a degree by offering remedially-oriented analogs through first presenting these conditions as not comprising unique or odd individualized experiences that are to be endured in isolation and second, through demonstrating different techniques for coping with anxieties by recognizing their sources and dynamics. At present, working people must scrounge around in the media in order to find this kind of supportive material, which more often than not turns out to be of meretricious quality. This circumstance can be remedied greatly by a WSCCS that dedicates itself to presenting expertly conceived and professionally implemented explicit supportive programming.

Specialized supportive information designed to dull the edges of suspiciousness, distrust, boredom, and fear that are exhibited by the youth, adult men and women, and older persons comprising the working subculture can be seen as charges to which a specialized WSCCS can address itself with some confidence in making a significant contribution. Here it must be remembered that the sense of helplessness and despair that is experienced by working persons is based mostly in reality. They are not to be considered as being clinically "neurotic" and amenable only to psychotherapy. What a specially designed WSCCS can accomplish in this regard minimally is to bring these feelings up into the open levels of consciousness and then to offer possible means for resolving them on the same plane.

If we can envisage the development of a specialized working-status cable communications sytem, we can readily see that its fundamental task is undoing the neglect that working people in America have endured at the hands of our more common mass communications media. In order to remedy this neglect, the proposed alternative system can make itself effective only by allowing itself to be guided by the realities of workers' needs, values, tastes, and styles of life. To neglect these realities is to add to the neglect of the constituencies that the proposed system might be able to serve.

Selected Bibliography and References

1. Barber, Bernard. Social Stratification. New York: Harcourt, Brace and World, 1957.

2. Bendix, Reinhard, and Seymour M. Lipset, eds. Class, Status and Power New York: The Free Press, 1953.

3. Berger, Bennett M. Working-Class Suburb. Berkeley: University of California Press, 1960.

4. Bogart, Leo. "Comic Strips and their Adult Readers." In Mass Culture, edited by B. Rosenberg and D. M. White. New York: The Free Press, 1964.

5. Bradburn, Norman M., and David Caplovitz. Reports on Happiness. Chicago: Aldine, 1965.

6. de Grazia, Sebastian. Of Time, Work, and Leisure. New York: Anchor Books, 1964.

7. Dotson, Floyd. "Patterns of Voluntary Association among Urban Working Class Families," American Sociological Review, Vol. 16 (1951).

8. Endleman, Robert. "Moral Perspectives of Blue-Collar Workers." In Blue-Collar World, edited by Arthur B. Shostak and William Gomberg. Englewood Cliffs, N.J.: Prentice-Hall, 1964.

9. Farber, Bernard. "Types of Family Organization: Child-Oriented, Home-Oriented and Parent-Oriented." In Human Behavior and Social Processes, edited by Arnold Rose. Boston: Houghton Mifflin, 1962.

10. Gans, Herbert. The Urban Villagers. New York: The Free Press, 1962.

11. Gans, Herbert. "Popular Culture in America: Social Problem in a Mass Society or Social Asset in a Pluralist Society." In Social Problems: A Modern Approach, edited by Howard S. Becker. New York: Wiley, 1966.

12. Geiger, Kurt, and Robert Sokol. "Social Norms in Television Watching," American Journal of Sociology, Vol. 65 (September 1959).

13. Glick, Ira O., and Sidney J. Levy. Living with Television. Chicago: Aldine, 1962.

14. Louis Harris and Associates, Inc. "The Viewing of Public Television— 1970." New York, November 1970 (mimeographed).

15. Hausknecht, Murray. "The Blue-Collar Joiner." In Blue-Collar World, edited by Arthur B. Shostak and William Gomberg. Englewood Cliffs, N. J. : Prentice-Hall, 1964.

16. Hodges, Harold M. Social Stratification: Class in America. Cambridge, Mass. : Schenkman, 1964.

17. Lieberman Research Inc. Broadcasting and the Public: 1966. New York: National Association of Broadcasters, August 1966 (offset).

18. Mendelsohn, Harold. "Socio-Psychological Perspectives on the Mass Media and Public Anxiety," Journalism Quarterly, Vol. 4, 1963.

19. Mendelsohn, Harold. "Sociological Perspectives on the Study of Mass Communication." In People, Society, and Mass Communications, edited by L. A. Dexter and D. M. White. New York: The Free Press, 1964.

20. Mendelsohn, Harold. Mass Entertainment. New Haven: College and University Press, 1966.

21. Mendelsohn, Harold. Public and Academic Library Usage in the United States. Denver: Communication Arts Center, University of Denver, 1967 (offset).

22. Mendelsohn, Harold. Radio in Contemporary American Life, Vols. I and II. Denver: Communication Arts Center, University of Denver, 1968 (offset).

23. Mendelsohn, Harold. Edu-Drama, A Mass Communications Technique for Mass Education: A Research Analysis of the Television Series "Cancion de la Raza." Denver: Communication Arts Center, University of Denver, 1969 (offset).

24. Mendelsohn, Harold. "What to Say to Whom in Social Amelioration Programming," Educational Broadcasting Review, December 1969.

25. Miller, S. M. "The Outlook of Working Class Youth." In Blue-Collar World, edited by Arthur B. Shostak and William Gomberg. Englewood Cliffs, N. J. : Prentice-Hall, 1964.

26. Miller, S. M., and Frank Riessman. "The Working-Class Subculture: A New View." In Blue-Collar World, edited by Arthur B. Shostak and William Gomberg. Englewood Cliffs, N. J. : Prentice-Hall, 1964.

27. Rainwater, Lee, R. Coleman, and G. Handel. Workingman's Wife. New York: Oceana Publications, 1959.

28. Roper Research Associates, "A Ten-Year View of Public Attitudes Toward Television and Other Mass Media, 1959-1968." New York: Television Information Office, March 26, 1969.

29. Scammon, Richard M., and Ben J. Wattenberg. The Real Majority. New York: Coward-McCann Inc., 1970.

30. Schatzmann, Leonard, and Anselm Strauss. "Social Class and Modes of Communication," American Journal of Sociology, Vol. 60 (1955).

31. Simpson, Hoke, ed. The Changing American Population. New York: The Institute of Life Insurance, 1962.

32. U.S. Bureau of the Census. Statistical Abstract of the United States 1968. Washington, D.C., 1968.

33. Weiss, Robert S., and David Riesman. "Working and Automation." In Contemporary Social Problems, edited by R. K. Merton and R. A. Nisbet. New York: Harcourt, Brace, and World, 1966.

34. Wilensky, Harold, and Charles N. Lebeaux. Industrial Society and Social Welfare. New York: Russell Sage Foundation, 1965.

COMMUNICATIONS TECHNOLOGY AND COMMUNITY CONTROL: CONFRONTATION AND CHALLENGE

Black self-development and self-determination efforts have consistently emphasized the necessity for control of those public and private institutions that operate within their communities. Pan-Africanists, integrationists, separatists, and black nationalists advocate community control of community institutions. DuBois, Booker T. Washington, Marcus Garvey, Malcolm X, Stokely Carmichael, Elijah Muhammad, and Huey Newton are in agreement on this issue. Furthermore, the increased urbanization and concentration of the black population in the central cities has given additional impetus to this historic movement for community control.

Because of the sophisticated and complex structure of racism and decision making in urban governments, community control has become the dominant theme in the struggle of urban minorities for social justice. Community control challenges white control of those institutions that operate in and serve predominantly black communities. Through these institutions, whites exercise control over the resources needed for local development.

The public school system, poverty programs, unions, police departments, welfare agencies, United Appeal, and every form of urban-based institution and organization controlled by whites are now being challenged. The urban oppressed are demanding jobs and economic benefits as well as a controlling voice in the policy-making functions of those institutions, agencies, and programs operating in their communities. The results thus far are small, but important changes are taking place in the degree and quality of minority participation and control of local institutions, organizations, and programs.

Community leaders and organizations are now faced with a new challenge in their efforts to achieve community control. Cable television, a futuristic communications system ideally suited for community control and local programming, is on the verge of broadscale expansion into the cities and ghetto communities. This development could provide the leverage needed by local communities to achieve a much greater degree of independence and self-determination, or it could seriously weaken the movement. Cable television will have a decided impact one way or the other. Its importance and its potential as a social, cultural, economic, and political force cannot be ignored.

Cable television may be the last communications frontier for the oppressed. Yet, most community leaders and organizations know nothing about this revolutionary communications technology and the plans underway to install sophisticated video systems in homes, schools, hospitals, health centers, courtrooms, police stations, banks, fire stations, supermarkets, and department stores.

A major power struggle is underway among broadcasters, cable operators, the FCC, Congress, the Administration, newspapers, publishers, motion picture producers and allied media interests, and a variety of professional groups and associations. All are jockeying to influence the development, expansion, use, and control of cable systems in the cities. The stakes are high.

Because there is great power and profit potential in the ownership and control of this medium, the oft-repeated "rip-off" by big business interests for private gain at the expense of the public interest is taking place once more. If it succeeds, it will stifle the diversified, highly specialized, local programming potential of CATV and prevent local control and community development. Diversified public and private ownership offers the best assurance that social benefits rather than social disaster will be the end product of this new medium. Concerted action by minority groups can bring positive results.

Among those public groups engaged in the policy debates, most advocate a regulatory arrangement that would guarantee minority groups and individuals public access to one or more channels on a free or minimum-charge basis. A regulatory scheme requiring uniform toll rates similar to the rate system used for long-distance telephone operations has been suggested.

Access is extremely important to minority communities, but it does not go far enough. Access alone will not provide the measure of control required over capital, labor, and technology to stimulate and sustain economic and social development of ghetto areas. Ownership and control must be achieved to meet this objective. A sizable portion of the income and profits from CATV in the major cities will come from minority subscribers, particularly blacks. Unless these systems are controlled by the communities served, the resources urgently needed for development will be lost.

If this proposed "access" policy were applied across the board, there would only be white-owned businesses in every sector: a conclusion not only at odds with the goals of self-determination, but one certain to render blacks and other minorities even more powerless and dependent. If it is adopted as the public policy for minorities in the field of cable communcation, it is certain to increase the power of the white business community, utilizing minority revenues as a subsidy. In other words, ghetto communities will be placed in the position of "paying" for their powerlessness and economic dependency.

Access and community ownership and control are not mutually exclusive or antagonistic. Regulated public access and community ownership and control are equally desirable objectives. In fact, ownership and control may provide the only safe guarantees that access will be accorded to minorities on a non-discriminatory basis.

The continuing oppression and exploitation of blacks and other racial minorities is directly related to their lack of control over indigenous institutions and resources. For example, it is estimated that the annual disposable income of blacks alone is about $40 billion. On the other hand, a current survey of minority businesses by the Bureau of the Census revealed that receipts of all minority businesses in 1969 were less than one percent of total U.S. business receipts—a meager $10.6 billion. Minorities own less than five percent of the total businesses in the country, and most are small retail and service operations with fewer than five employees.

Minority ownership and control of cable television systems could dramatically alter this situation over the next five to eight years. There are approximately twenty-five cities with black populations in excess of 100,000; eight of these have populations in excess of 200,000; five have populations over 500,000; and two have populations over 1 million. After five years a cable system with 10,000 subscribers would generate revenue of approximately $500,000 annually and a system with 20,000 subscribers would generate up to $1,000,000 in revenue annually. Obviously, many of these communities could support several cable systems. Six cable districts have been proposed in Washington, D.C., where the black population is over 500,000.

THE NEW FRONTIER

From a few isolated cable systems in small communities and rural areas of
the United States, this new industry has aroused nearly every power bloc and
organized interest group in the country. Their excitment centers around three
aspects of CATV: (1) profits, (2) the vast signal transmission capacity of
cable, and (3) the imminent expansion of cable television into the cities and
major metropolitan centers.

Cable systems seem to offer unlimited opportunities for making money.
First of all, cable operators have avoided programming costs by retrans-
mitting programs produced by broadcast television. This air piracy was upheld by the U.S. Supreme Court. Accelerated depreciation schedules enable
operators to gain the benefits of a tax loss while increasing the book value of
their investment. The practice has been to depreciate systems over a five-
to eight-year period even though equipment life is actually twenty years. The
results — a guaranteed paper loss during the first three to five years of operation. This loss is allocated on a pro rata basis to investors who then claim
the loss on their individual tax returns. Meanwhile, the cash deductions made
for depreciation before taxes are available for use to expand the system or
purchase new ones. Hence, the assets of the system and the book value of the
stock are increased. When the system is totally depreciated, it may be sold
and the entire process can be started all over again.

The vast signal transmission capacity of cable television is further cause
for excitement among power blocs and organized interest groups. The "economics of scarcity" common to over-the-air broadcasting can be eliminated
by the enormous channel capacity of cable. Early systems provided up to
twelve interference-free channels, and those now under development will offer from 24 to 60 channels. This abundance means the general public can now
afford video programming for a wide range of purposes; for example, education, community meetings, and information programming concerning health,
jobs, and legal matters, to mention a few. Private access, similar to the
telephone system, is possible using rate schedules like those for long-distance
calls.

Cable technology has a potential, however, that goes far beyond increased
channel capacity. Two-way communications, home computer terminals, home
banking and shopping services, transmission of mail, fire and burglar alarm
systems, piped-in music for each home, and other 1984-style communication
services can be provided over the same cable that transmits the video signal.
"Cable television" is a misleading term; "cable communications" more accurately defines the technological parameters of this new medium.

With the highly probable interconnection of systems between cities by domestic satellite within this decade, the communications prospects of cable
technology are genuinely "mind-blowing."

Another, and perhaps the most crucial, factor generating the growing interest and controversy over this medium is the introduction of cable systems
into the major central cities and metropolitan areas.

If the present trends continue, minority communities will be excluded and
disenfranchised. White capitalists who own and control the major print and
electronic media systems in America will own and control the cable communications industry, including the systems that serve black communities. Fifty

percent of the cable industry is already controlled by other media owners. Broadcasters own 36%, newspapers 8%, and telephone companies, advertising agencies, and motion picture companies, 6%. Further, there is a rapidly developing concentration of ownership within these groups. Ten companies now control 52% of the industry. The top ten, in order of rank are: Teleprompter, Cox Cable, American Television and Communications, Tele-Communications, Cypress Communications, Viacom, Cablecom-General, Time-Life Broadcasting, Television Communications, and National Transvideo.

The white middle class that manages and operates major educational, social, and cultural institutions (that is, schools, colleges, universities, foundations, theaters, museums, and churches) is actively vying to dictate public programming policies for cable systems, including those serving black communities.

These two groups — white capitalists and the white middle-class intellectuals, managers and technocrats who have worked so effectively together in controlling and operating everything from the Pentagon to the poverty programs (at a handsome return to each group) — are now moving toward an accommodation of interests in this new communications field. If this act is consummated, the promise and the potential of CATV as an instrument for empowerment and development of underdeveloped ghetto communities will be seriously diminished, subverted, and perhaps entirely lost.

WHAT YOU SEE IS WHAT TO GET

The requisite conditions for community control of resources and development are mass mobilization and unified action.

For example, urban renewal programs provided a significant opportunity for unified action by varied interest groups within urban black communities. Many ghetto communities united to stop these projects because of the insensitive treatment of residents by urban renewal agencies and the disruption of the community for the benefit of white profiteers. Serious attacks were made against the traditional system of planning and decision making from the local to the national level. Blacks demanded and secured important concessions affecting policy making, jobs, and other aspects of the urban renewal process.

Cable television is a better vehicle for achieving sizable gains in community organization, unification, control, and development. Several factors support this statement. First, cable television systems are not presently installed in black communities and central cities. Therefore, no entrenched interest group or power bloc can claim public protection for its investments. Second, franchises are issued by local, municipal governments, and the FCC has recommended the continuation of this process. Third, installation requires the actual stringing of cable on poles or the laying of cable underground along the streets of the ghetto. Individual hookups must then be made from these trunk lines to homes and apartments, and outlets must be installed within these living units. Fourth, black communities are a substantial segment of the urban subscriber market. Fifth, the great potential of cable in technology, economics, and the power of mass influence is ultimately tied to cablecasting or local programming origination. Sixth, cable will be used in a wide variety of applications, apart from entertainment programming. Education, health, welfare, safety, crime prevention, and police operations are a few of the likely uses.

Viewed together, these factors reveal significant opportunities for community participation and the imperative need for community control. Each of the listed factors will require a series of crucial political decisions at the local level. How will CATV be regulated? How many franchises will be awarded? Will a single franchise cover both system management and system programming? What are the qualifications for franchise applicants? How will franchise fees collected from CATV enterprises by the local government be used? Who will own and operate the systems? Who will determine the program content? Who will install the systems? Who will decide on the areas to be served? These are political issues that will be decided with or without community participation, but the options for black communities are still open. How long they remain open will depend on the initiative, ingenuity, and determination of community leaders and organizations.

Whether or not a franchise has been awarded, broad-scale community participation is possible. Early involvement in the franchise process is crucial. Local franchising involves several steps. The commission or city council usually adopts an ordinance giving the political body the legal powers to regulate CATV within the community—including procedures for awarding franchises. The ordinance may stipulate that public hearings must be held prior to franchise awards and that public notice must be given regarding the period established for filing franchise applications by interested parties. The ordinance may further state that multiple franchise awards will be made for various geographical areas within the city or county.

Community organizations should view the entire franchising process as an area of vital interest to their constituencies. Ideally, disenfranchised black communities should be consulted by the local government and included in the discussion and development of the ordinance and all other regulatory aspects of CATV systems for their communities. Needless to say, that is not happening. Local politicians, who are not well informed about CATV, have been selling the "communications birthright" of minority communities to the highest bidder.

Community participation should begin with a systematic, factual determination of the status of the franchising process within the local government. The city attorney or council should be contacted for this information. If no ordinance has been adopted or franchise awarded, action should be taken to establish procedures for community inclusion in the policy-making process.

If an ordinance has been adopted and/or a franchise awarded, a detailed review and evaluation should be made to determine the provisions made for community participation in the monitoring, control, programming, and ownership of the system that serves them.

Most CATV franchises are nonexclusive agreements between the city and the cable operator. Thus, community groups may organize their own company and apply for a franchise covering the same territory as previously awarded.

As a last resort, it may be necessary for the community to exercise its veto power over those CATV projects that disenfranchise blacks. Legal and other forms of protest actions may be required to achieve community participation in the policy-making discussions and the achievement of community control and development objectives.

OPPORTUNITIES FOR COMMUNITY DEVELOPMENT

Cable television provides a substantial opportunity for urban minority communities to develop and control the most powerful cultural and social instrument in their communities. It can also provide a viable economic base and political leverage for power-deficient communities.

A partial listing of the wide variety of program uses will give some idea of the development possibilities:

Educational Uses

Video correspondence courses

Special education programs for unskilled workers, housewives, senior citizens, and handicapped persons

Home instruction for students who are temporarily confined

Adult education programs

Exchange of videotaped educational programs with other schools, for example, science, travel, and cultural programs

Interconnection of school systems to facilitate administration, teacher conferences, and seminars

Greater use of computerized testing and grading — thus giving teachers more time for individual instruction.

Health Uses

Interconnection of medical facilities (private offices, clinics, hospitals) to provide a wider range of consultation services to patients on an emergency or nonemergency basis — especially those without means of transportation

Wide dissemination of preventive medical and dental information to the community

Information programs concerning sanitation, sewage, rat control, garbage control, and similar problems.

Legal and Consumer Uses

Listing of substandard and abandoned housing

Review of leases, agreements, and installment contracts

Discussion of labeling, marking, pricing of food, drug, clothing, automobile, and other consumer products

Establishment of a "hotline" in legal aid and consumer protection agencies to provide immediate notice of fraudulent and exploitative practices

Use of videotaped records and depositions in nonjury cases.

Safety Uses

Installation of fire emergency and burglar alarm systems in every home (these systems can operate over the same cable that brings in video signals)

Automatic gas, water, and electric meter readings

Rumor control

Disaster and emergency warning systems.

Cultural and Entertainment Uses

Minority-owned cable systems in the top 50 television markets alone would provide a major market as well as a distribution system for professionally produced films, plays, concerts, sports events, talk shows, and every other form of artistic, creative, and intellectual expression. There is no shortage of professional talent in the community—only the lack of a mass-based communications and distribution system could have promoted the Ali/Frazier fight. The white promoter of the fight, Jack Kent Cooke, is a major CATV owner.

Production of a black history series from the voluminous materials written by DuBois, Hughes, Malcolm X, Cullen, Woodson, Bennett, and thousands of minority historians, politicians, writers, poets, and leaders who have prepared records of their people's struggle. Such a series could now include an oral history of important historical events by elders of the community.

IS THERE GOLD IN THE GHETTO?

The economic potential of CATV for minority communities should not be minimized or overlooked.

The urban ghettos in America comprise a compact, differentiated, and lucrative market for cable television—a conclusion that is supported by the phenomenal economic success of soul radio stations. This fact has not escaped the attention of Teleprompter, Time-Life, and other white entrepeneurs who are scrambling for ownership and control of cable systems in every large city.

Cable is uniquely suited to serve as a vehicle for economic development, because it is a subscriber-supported system. If an adequate number of households in the community purchases the service, sufficient income can be derived to maintain the system and to produce a profit.

Most community-based enterprises that depend on black customers (with the exception of white-controlled, high-risk, illegal operations like the numbers and narcotics rackets) are small, marginal operations. Cable television is inherently a monopolistic enterprise. Although it is possible, it is highly unlikely that there will ever be more than one cable system serving a given community. Therefore, a black-owned system serving the entire inner city or just the black community would have a captive market—just as a white-owned system serving a ghetto community would have a captive market. (Soul television is not a remote possibility.) The point is that a community-owned system would not have to compete with white-owned systems downtown or in the suburbs as minority-owned grocery stores, restaurants, hotels, motels, clothing stores, drug stores, and so on, must.

SUBSCRIBERS—THE KEY TO CONTROL

As stated earlier, CATV has grown and developed in rural areas and small

towns where no television was available or in areas where signal reception
was poor.

The major incentives to residents of these rural and out-of-the-way communities to purchase cable services have been (1) better reception and (2)
more signals or channels. Similar incentives exist in New York City and a
few other major cities where tall buildings and atmospheric conditions cause
poor signal quality. However, these conditions are not duplicated in the majority of black population centers. In most central cities there is fair to good
signal quality and five to seven broadcast television channels are now available.

Ghetto residents are not likely to subscribe for cable services just to get
better signal quality or more channels. Other reasons must be found. The
emphasis on community control and development can provide some of the necessary incentives. The few soul radio stations owned by blacks and the total
exclusion of blacks from ownership of television stations provide additional
motives for local communities to prevent the continuation of these patterns in
the cable communications industry. Blacks own none of the more than eight
hundred licensed commercial television stations and only about twelve of the
three hundred and fifty soul radio stations. Most black communities do not
have a local newspaper or magazine produced by and for them.

The strongest incentive for local residents to subscribe to a community-
owned and -controlled cable system may well be the opportunity to combat the
insensitive programming of the existing media, the exploitative practices of
soul radio stations, and the discriminatory hiring practices of the radio, television, and print media enterprises.

The National Advisory Commission on Civil Disorders reported that "...
television and newspapers offered black Americans an almost totally white
world; and far too often, acted and talked about blacks as if they neither read
newspapers nor watched television."

As lawyer Donald K. Hill pointed out, however, "...The Commission's
Report only touched upon the tremendous impact which the white-culturally-
oriented media has on the black community. Although black Americans have
the opportunity to fully observe the white world, communication flows in only
one direction; blacks never see themselves as they perceive themselves, nor
does communication flow from blacks to blacks...."

The opportunity for ownership, control, management, and programming of
CATV systems in the cities can be a powerful incentive to powerless minorities. If they are aware of the economic and social potential of this medium
for their neighborhood, ghetto residents may be persuaded to subscribe to a
community-owned system as a matter of enlightened self-interest. Five dollars per month is probably the lowest price they can pay to secure a share in
the wealth and power of the country.

Collective ownership and control of systems will undoubtedly enhance the
incentive to local residents to subscribe. Shares in the enterprise should be
offered to local residents, just as AT&T offers its stock to its employees on
a payroll deduction basis. CATV systems could apply a portion of the monthly
service charge to the purchase of common stock by the subscribers.

Selling cable to urban residents who already have good reception and multiple channels will not be an easy proposition for whites or blacks. It should

not be any more difficult, however, for minority entrepreneurs than for whites. In fact, it may be easier if the strong desire for community control is recognized and if entrepreneurs are willing to include the residents in the ownership.

STRATEGIES FOR DEVELOPING COMMUNITY SYSTEMS

Communications experts estimate the cost of wiring each of the major cities will range from $2 million to $20 million. Where will community-based corporations get the financing to build and operate CATV systems? Financing will not be simple or easy. On the other hand, it's not impossible. Joint ventures with white or nonresident minority investors are one possibility. Such investors are one possibility. Such investors might be banks, national and local church groups, wealthy individuals, insurance companies, savings and loan associations, high-income blacks (such as athletes and entertainers), and local or national industrial concerns.

Available resources within the black community should not be overlooked. In fact, that's where the initial organizing efforts should start. Many black professionals (doctors, dentists, ministers, lawyers, teachers, and businessmen) have relatively high incomes and accumulated savings that can be tapped. The professional class can also provide collateral assets in securing outside financing because of stable employment and high incomes. Black churches and insurance firms are also potential sources of equity capital.

Community development corporations and model cities programs deserve special attention and consideration because of their uniqueness. These groups, operating with both public and private funds, have been established to plan, design, and implement community redevelopment and development projects in the inner-city ghettos. One or the other of these entities, and sometimes both, are presently operating in most of the major cities. These groups usually have a working knowledge of city hall politics, the local financial community, the federal funding structure, and the national philanthropic community. They also have the staff and technical expertise to package a community proposal.

In short, both the community development corporation and the model cities programs are in a position to act as effective brokers for the community in planning, designing, and implementing a program for community control of CATV systems.

There are several approaches that could substantially reduce the financing burden on local communities. One approach that is practical for large cities like Washington, D.C., is to divide the city into four to six cable districts. If a franchise were awarded for each cable district, four to six cable companies with roughly 30 to 40,000 households could be established. Fifty-five percent market saturation in each district would result in a 15 to 20,000 subscriber system. Under this arrangement, each company would have to raise only $1 to $2 million in financing in lieu of one company attempting to raise $10 to $20 million to build a city-wide system.

Tom Atkins of Boston, one of the most knowledgeable public officials in the country on CATV, suggests that municipal governments should "wire up" the entire city and then take bids for system management and operation. This approach would eliminate the big cost of system construction and place community groups in a highly competitive position for franchise awards. This is

an attractive proposition where multiple cable districts are established and
the minority community is not fragmented into several predominantly white
districts. Blacks, in particular, have been disenfranchised by such gerry-
mandering in the past.

A common practice in the CATV industry is the "turnkey" system of construction. Hardware manufacturers have financed, designed and built systems,
turning the completed system over to the owner. Some hardware manufacturers
prefer to enter into joint ventures for turnkey systems, providing from 30 to
50 percent of the financing. Care must be exercised, however, to assure that
community-owned equity in the system is the controlling interest. The community corporation should also secure an option to buy out joint venture partners
on a first-sale-offer basis.

5. POLITICS IN A WIRED NATION

Ithiel de Sola Pool and
Herbert E. Alexander

INTRODUCTION

The purpose of this paper is to propose, not to predict. People often ask, "What will CATV do to American politics?" There is no single answer. What it will do is what we choose to make it do, for CATV is not one thing but a family of things sharing only the physical fact of a wire in common.

Each physical medium — print, cinema, radio, TV — has shaped the message by its material form; but Marshall McLuhan overstates the case. The molding effect of the medium is small. The diversity of uses of which any medium is capable is great. Print can be a learned tome, True Confessions, the telephone directory, or Edward Lear's limericks. Radio can be musical background, Roosevelt's fireside chats, or a device for calling a taxi. So it is with CATV. There is no way to answer what its effect on politics will be. The proper questions to ask are what its effects on politics can be, at what cost, and by what means.

Our subject is the cable medium and politics. Political events change fast. Communication systems change more slowly, and political systems change more slowly still. The period of our concern is the 1980s. Any system proposed must recognize that things will change even within that decade. Furthermore, systems developed for that decade must be so designed as to be hospitable to changes that may come later. Two-way communication systems will be embryonic at best by 1980 but may be very active by 1990. Forty channels may be largely absorbed by entertainment, repeat broadcasts, news, and a few commercial services in 1980. Eighty channels may offer a very different range of opportunities in 1990. Intersystem linkages by satellite or other means may be costly or limited in capacity in 1980 but virtually costless in the 1990s. The CATV systems linked by such means may be an essentially American domestic phenomenon in the 1980s. It may be a worldwide phenomenon twenty years later.

We consider four models of CATV and examine the potential political effects of each system. The first model is commercial CATV with little public control of the use of facilities or of time. This would lead to use of only a few channels, carrying popular, large-audience programming and little or no political or public service material. The second model involves legislative control of CATV for the purpose of protecting existing interests so that there would be only gradual changes in present broadcasting patterns. A contract carrier system, similar to the telephone company, is the third model of CATV. This method would allow maximum immediate use of CATV's capabilities. The fourth model is a compromise among the previous three, with some commercial and some free and paid public service use of CATV time. The FCC's August 1971 letter to the Congress proposes such a mixed system.

With such a variety of options and changing circumstances, to predict clearly makes far less sense than to consider what courses might be wise and what consequences for American politics some of them might have. It is to that task that we now turn. We shall consider how a developed CATV system might handle political material of various kinds, both during election campaigns and in ordinary public affairs programming such as news, documentaries, and public policy discussions.

POLITICAL USES OF CABLE TELEVISION

Public Affairs

Campaigns are brief orgies of citizenship; a healthy community needs a steady diet of involvement in public affairs. What can cable television do week in and week out to keep the public involved in and informed about civic affairs?

One thing it can do is to bring a picture of the government live into the view of the people. The simplest case is a camera continuously focused on the city council, on congressional hearings, on the United Nations Security Council and General Assembly.

The Congress was purposely excluded from that list. Some kinds of legislative sessions can well be caught by live TV and some cannot. The difference lies in the extent to which the real business is transacted on the podium. TV camera coverage of the floor of the Congress as it operates today would present a most misleading picture. To do the job right would require creative reporting with the camera moving from committee sessions to the floor, to offices and back, with summaries of written reports read, as well as camera coverage of speeches made on the floor of the House and Senate. This is the way the Congress works and indeed the way it should work given the vast volume and technical character of its business. In this respect, the Congress is quite different from a court of law where virtually everything that the jury considers must be publicly presented before them. The Congress is also different from a typical city council, which consists of part-time members who come together once a week and really do a substantial part of their business in the meeting. The Congress is also different from the United Nations, which is not a legislative body, but rather an opinion-forming and persuasive body and therefore also one in which the speeches are indeed a substantial part of the business.

Another scene that should not be screened is the courts of law. Considerations of human dignity here prevail over the proper desire of the citizens to be informed. Public as a trial may be, being humiliated before just a roomful of visible people is different from being humiliated before an unknown vast audience whom fantasy can blow up to any size and character.

Yet with all these reservations, some public bodies could well be on the screen to the mutual benefit of the body and the citizens. City councils are a good example. Granted, screening may not be to the benefit of the incumbents who may show up badly. There is experience with councils that have tried going on television and have stopped because they looked too bad. However, most cities require the council to conduct virtually all proceedings in public, and audiences to attend. More attention would likely have a favorable effect on the character of the proceedings. Public hearings are also good material for CATV. Legislative sessions may well be covered at moments of major reports such as gubernatorial or presidential messages.

A fixed camera focused on a debate is the simplest mode of public coverage. Imaginative photographers with lightweight equipment can bring the government to the screen in more vivid ways than the passive shooting of a meeting. The people who live in a housing project, the sanitation crews on the job, the soldiers in Charlie Company, the refugees in a peasant village are all

treated now on TV in snippets running from 30 seconds to one hour. On CATV
they could live their lives and talk their talk for hour after hour, not with a
large continuous audience, but still some audience for long periods of time.
Just as copious radio time has brought the endless "talk show," we must assume that copious CATV will bring the endless "look show."

In addition to shows intended for the general public, there can also be programs using CATV more or less as if it were closed-circuit TV. The City of
New York, for example, is using television in training courses for municipal
employees. A political organization could use CATV to train its workers in
canvassing techniques or in voter registration. CATV can help a police department talk to all its men, or a school system talk to its teachers. These
uses could be open to all to see or else scrambled and perceivable only to the
persons involved.

Today we can only speculate how CATV may come to be used in public affairs. One thing sure is that it will be used and in ways that have not even
been considered. A severe restriction, however, will be money. In whose interest is it to pay for putting public affairs on CATV?

Some programs will be paid for by government as a means of communicating with its citizens. Some programs may be paid for by government in its
public service role, just as government now supports public television. Some
programs will be paid for by civic and pressure organizations with a message.
The costs for at least some kinds of programming can fall drastically. Those
who have studied the matter talk of rates of $20 or less an hour for cable
rental. Production costs, being very much higher, may be the limiting factor.
Use will, of course, be made of expensive, well-produced movies that will be
shown over and over again, justifying their cost. However, what we described
earlier as the endless look show will have to be produced at costs scarcely
above the cost of one man and tape.

Clearly, very few political or civic organizations will want to support a
channel in any one locality at their own expense. Even at one cent per day for
each receiver, in a 20,000-home system, the cost of a channel comes to
$73,000 per year. If subscription fees paid half of that cost, the channel cost
would still be $100 a day. At that level of cost, most civic and pressure
groups would want to buy channel time now and then, though a few private
groups might be able to afford such expense in some locations continuously,
and others could raise funds with the help of CATV itself. Even if most political groups would only be buying occasional hours, the growth of public affairs
channels will require the development of much very cheap and therefore very
simple programming.

Is there an audience for this kind of artistically low-grade though useful
public affairs programming? Will there be audience enough to justify even
$100 a day? The best prediction is a mixed one. The audiences for public af-
fairs channels will undoubtedly be tiny by current television standards. But
that is no tragedy. Consider the city council or the United Nations on the
screen hour after hour, week after week. The citizen who has watched it three
or four times a year will have acquired a better understanding of how his government runs. He does not need to watch every week to learn something. So,
too, an adolescent growing up who has had the opportunity during his forma-
tive years to obtain a clear picture of Congressional hearings has gotten some-

thing valuable even if he does not become a regular viewer. Just as it is important that the library is there even though not everybody is using it all the time, so it is important to have the operation of the government channel there ready to be looked at by the citizen when he wishes. He will know that he can turn it on occasionally, when issues become hot.

Furthermore, among the various tiny audiences of CATV channels there will be the organized ones. In political life one organized person is worth many casual viewers. Suppose 100 members of a conservation group watch a live helicopter survey of smokestacks belching smoke over the city. If these 100 people have responded to the newsletter of an ecology society that told them when to watch, and if they are then ready to write letters or phone their protests to the offending plants, they and that program are a powerful force indeed. Increasingly one may expect organized audiences to be the most extensive consumers of channels, though also the least numerous part of the audience.

How many channels can public affairs programming use? There is no easy answer. The number of channels will probably increase as a growing number of organized groups learn to use public affairs programming in specialized ways. One could imagine as many as four channels supported by government agencies in a major city. One could imagine a similar number being used by public affairs and political groups in a normal evening in a middle-sized and only modestly politicized city. But, the demand for such channels is very price elastic. Our assumption here is of a modest price such as would have to be charged by a contract carrier, rather than either a subsidized price, which would quickly fill many channels, or a price similar to that of current air time (which would lead back to political reliance on short spot announcements and an occasional ration of free public service time). Somewhere in between is a wide range of possible numbers of channels demanded. The estimate is sensitive to the size and character of the city involved, to price, and to the level of understanding of the capabilities of CATV.

News Services

Periodically since 1959, Roper Research Associates have polled the American public about their relative use and dependence on the different media. They have asked from which medium people get most of their news, and if there were conflicting reports on radio, television, magazines, and newspapers, which of these they would believe. By the end of the 1960s television had become the American public's prime news source as reported in answer to those questions.[1] Other questions still establish the primacy of newspapers in the delivery of local news and perhaps in total quantity of news conveyed. Other surveys also showed that the most informed part of the public continues to use newspapers more. But while the media organizations may quibble over question wording, no one can deny that television has become a vital part of the information stream for the American people.

Clearly, if CATV produces any major changes in the way in which news is reported by television, that will be one of its main effects on politics. Is it likely that CATV news reporting will improve over current practice? Or, conversely and more ominously, is it possible that CATV will undermine the news services that the three great networks have now built up?

There is reason to be concerned that the fractionation of the audience will destroy the economic basis of the networks, and that they will no longer be able to support their present expensive national and worldwide live news coverage. The news services, it is often alleged, operate at a loss and are run by the networks to protect their profitable entertainment operations from regulatory assaults. If it is true that news is a propitiatory offering to the gods at FCC, then it could well be abandoned when it no longer served that purpose. What we must examine is this dire prediction that CATV may undermine the American public's most important news source. Is this a real danger or a myth?

Let us start by considering the present economics of the network news services. All three networks decline to reveal any of the essential figures. The authors of this paper wrote the three networks requesting basic information. The only dollar figure given is the statement that the News Department of NBC has an annual budget of roughly $100 million. Our estimates are therefore all secondhand, hearsay, and only moderately reliable.[2]

The CBS news budget is variously estimated at $40 to $50 million. The NBC and CBS figures are, however, not comparable. The NBC budget includes sports, which in CBS is a different department, and the local news budgets of the NBC-owned radio and television stations. In 1968, the ABC news budget was estimated at $33 million. A rough estimate of present news expenditures of the three networks (not including sports or local activities of stations) would come to around $150 million.

Whether the networks make money or lose money on their news operation is virtually incalculable. It depends very much on how one allocates costs, and it is frequently alleged that costs are allocated in a way to assure that the news operations look like losers. Some of the main news programs are clearly moneymakers, however. Broadcasting magazine in 1968 estimated that the Huntley-Brinkley show netted $8 million on a gross of $28 million. The "Today Show" in the same year was estimated to gross $11 million. These were the big moneymakers in the news field for NBC. That $39 million minus commissions may be set against perhaps $40 to $50 million, which in 1968 might be roughly the appropriate part of the total budget that might be called news service. Whether incidental revenue filled the gap or not we cannot say. Our basic conclusion is that news must be close to a break-even operation. The networks are fond of pointing out the opportunity costs that they incur when they put news on instead of more profitable entertainment shows. That, however, is irrelevant to the purpose of assessing whether news as now handled could survive in a CATV environment. Absolute profitability, not opportunity costs, must be estimated, and as far as we can tell, that is marginal.

The manpower used to sustain the network news services is somewhat easier to document than are the budgets. In 1971 the employees of the three news departments numbered: NBC, 1,200; CBS, 800; ABC, 510. (We must repeat that NBC includes more activities in this department than CBS.) ABC gave a careful estimate. About 65 percent of the 510 employees in the department, that is, about 330 people, work on producing the daily news, special events, public affairs, and documentary programming. Applying similar criteria, it would seem that the three network news establishments must employ some 2,000 people.

For the world news, NBC has 17 overseas bureaus, CBS 11, and ABC 9. They employ 100, 120, and 63 persons, respectively. Thus some 280 of the 2,000 news people are posted abroad.

In contrast to print journalism, in which a low-paid newsman with pencil and pad is enough to cover a story, TV picture coverage is very expensive. It costs about $1,000 a week to field a three-man crew, the minimum that can handle the job under present practices. This figure is independent of the costs of film or tape handling and long-distance transmission. Such costs are manageable only because, in the main, TV presents only "the front page." If TV had to fill the newstime of full-time news channels, the total costs would rise markedly.

Unquenchable need exists for something like the present networks to provide daily film coverage of the top visual stories of the moment: wars, floods, sports events, elections, riots, and leading public figures. Only something like the networks could adequately cover a moon landing, the death of a President, a nominating convention, or the returns on election night. The problem is who is to pay.

So far, the facts that we have adduced suggest a very real problem. For adequate national and worldwide news coverage the networks must survive at least as news services. News is the one function for which their ability at live reporting of reality is essential. So if the networks collapse as profitable entertainment peddlers, thanks to CATV and tape, they may be left primarily as news services. There will be a social need for an expanded and more expensive operation to fill the expanded channel time, but it is not clear that even the present level of operation can survive.

Whether one, or two, or three of the present networks survive, and whether they will flourish or retrench depends upon new patterns of revenue. Who will pay and how much? Will advertisers be willing to pay more than they do now? Will cable franchise owners pay to have a news service? What about pay TV? Will the homeowner be willing to pay for his TV news as he does for his newspaper?

In general, our conclusion is that the answer to these questions is a hopeful one. Despite the problems that we have already listed, the economic future of television news is a favorable one. Consider first advertising revenues. We have suggested that the entertainment part of the networks may be in trouble in the cable era. If they are, it will be because in a fractionated audience they can no longer deliver 20 percent of the homes with a single buy. But precisely to the extent that that happens, television news will become a good buy for the national advertiser. News shows are not a bad advertising buy today. If they do not get more advertising it is because there is a still better purchase available. In the long run, news may become the most conveniently available simultaneous national program on which a national advertiser may place his campaign.

Whether cable franchise holders will pay for national and world news and also for processing of local news depends upon the kind of CATV system it is. In a system in which the franchise holder originates some material or sells advertising himself, franchise holders might well wish to have news on their channels so as to win local advertising. If much national advertising were carried, they would want a share of the revenue. Where the equilibrium point

would be in that bargain between the national news services and the local
franchise holder is highly unstable and therefore unpredictable. However, the
franchise holder would wish news to be available because his customers expect it. So if advertising turned out not to carry the full cost of national and
world news services, some revenue could be obtained from franchise holders
themselves.

In a contract carrier system the same interest in providing news would lie
with the channel renter rather than the franchise holder. However, in a contract carrier system there would be a question as to which channel renter
would have access to which news services. Should there be some (perhaps
one or two or three) all-news channels? If so, are other channels free to carry news from the same sources, that is, the network news services? This is
a complex matter comparable to the question of access to the Associated Press
for competitive newspapers. On cable the problem is even more complex, for
the renter of a channel for a particular moment has none of the permanence
and committed investment of even a weak newspaper. If the national and world
news services were to sell film to anyone at any time, it is not clear that
stable well-organized news channels could survive. And without such news
channels, there would be no one to do the job of collecting, editing, and reporting local news either. On the other hand, if only a very few specialized
news channels can get access to news, and the rest are barred by refusals
and copyright, then political freedom of expression is severely restricted.

The most likely outcome would be one in which the news services of the
major networks, which have live current film to dispense, would contract
with all-news channels exclusively for their full live service, while other
channels would receive Teletype text, which announcers could read and on
which they could comment from their own viewpoint. In addition, old background film would be readily available.

This is certainly not a completely satisfactory arrangement, but it does
have some viability. It takes account of what is often not recognized: that an
adequate news processing organization probably cannot exist in a completely
competitive market but requires some oligopolistic scarcity of resources at
some place in the system, so that news organizations large enough to do the
job can develop. In print journalism we typically find only one or two newspapers in a city. If it were a truly competitive market with every viewpoint
having its paper, no newspaper could afford the kind of features and exten-
sity of coverage that we have come to expect. In a contract carrier cable system the element of oligopoly is only in the world news services, not in the
many cable renters. To build effective local news organizations probably requires that the world news services develop partially exclusive relations with
certain local ones, giving the latter the edge that would justify their investing
in local news collection and editing.

While we thus recognize that some element of oligopoly is necessary in the
communications system, we also hope that it does not move toward monopoly.
Every effort should be made to keep all three of the present networks alive.
It is certainly to be expected that a number of specialized news services will
develop. Labor unions, partisan groups, documentary filmmakers, and others
will make material available to the increasingly important and increasingly
hungry screen. The print wire services will probably be glad to serve cable-

casters, too. (Whether they will try to become comprehensive television news services like the networks, we do not know.) But there clearly is a limit to the number of possible full news services. We are not likely to move into a situation in which each tendency with a channel of its own will be able to run a full, though partisan, news service of its own. We would guess that something like the present number of news services with a corresponding number of news channels is a plausible estimate of what the system will support. It is hoped that these channels will develop distinctive stylistic individualities of their own while remaining committed, since their numbers are small, to the principle of objective and balanced news reporting.

Advocacy
Although television has been a growing factor in political campaigning for a generation, we know relatively little about its impact on voting decisions or even on voting turnout. Cable television is a newer phenomenon, and its political usage has barely begun. Whatever the political potential in cable communications, we can do little more than conjecture about the forms that political campaign usage will take or their impact. At the outset, we will make certain hypotheses about the potential impact of cable communications and then discuss them.

1. To the extent that cable permits the reaching of specialized and selected audiences, it may reduce some of the present pressures for political time on commercial over-the-air broadcasters.

2. To the extent that cable fails to reach mass audiences or does so only occasionally, politicians will continue to seek means of reaching large numbers of persons through over-the-air broadcasting, sports or other special events on cable, or wherever the attention of potential voters can be caught. This means continued use of spot announcements.

3. To the extent that cable is used, there will be significant costs for production, promotion, and interconnection, apart from any time costs that may be charged.

4. There have been no known time charges for political uses of cable to this date, and if that precedent were followed as cable incidence and capacity expands, there will develop significant restructuring of political campaigning for certain categories of candidates in a free medium. For them, cable can reduce the need for mail and other means of advertising directed at specific audiences.

This latter point is important at a time when the costs of political campaigning are high, and there is legislative and public ferment about how to cope with the problems of financing politics.

In 1968 political costs at all levels for candidates and committees in primary and general elections were at least $300 million. What the costs will be in a decade or a generation from now, when cable may flourish, cannot be predicted. But campaign techniques may well be different, some old techniques may be at least partially displaced, and the cost components of politics may well vary from the present.

In 1968 political broadcast expenditures were $58.9 million. The outlay represents all network and station charges for both television and radio usage by candidates and supporters at all levels for both primary and general election periods. If average production costs and agency fees of only 20 percent are added to the total broadcast expenditures of $58.9 million, the cost of broadcast advertising was approximately $70 million. To this figure must be added the cost of "tune-in" ads in newspapers and other promotion expenses.

Thus at least $75 million, one-quarter of the estimated total of all political spending, is directly related to over-the-air political broadcasting, making it by far the largest area of political expense. If one were to add in allied costs— travel to the broadcast city, speech-writing, and other such planning and preparation—then a total of 50 percent more than time costs would not be unreasonable, making broadcast-related expenses as much as $90 million.

In 1968 newspaper advertising costs for politics were at least $20 million. This figure represents a projection from a survey limited in size and scope but permitting extrapolation because both the universe and the limitations are known.

Nationwide costs for other political printed matter —for billboards and posters, for mailers, for handouts —are not known. Adding the broadcast and newspaper projections, and other media expenses, at least half and probably a good deal more than half of the $300 million estimated to have been spent in 1968 went into media advertising costs.

Candidates compete against much more than the opposition. Candidates of the same party compete for nomination. Candidates and committees of the same party, at different levels, contend against one another for available funds and for the attention of the electorate. Candidates and parties are also in competition for media time with commercial advertising on a regular basis.

Our system of elections, then, creates a highly competitive political arena inside a universe full of nonpolitical sights and sounds, all seeking attention. But the attention span of the electorate is short; focus on politics is easily distracted and needs fresh and constant stimulation. In this world, politics rates relatively low in interest, and what interest there is tends to be diffused among many levels of candidacy and contention.

Thus political costs are high because political effort must be high for each candidate on each ticket to attract the voter and get him to the polls. At present, the price is rising for those who can afford it —and it tends to drive those who cannot out of the market.

Traditionally, the politician spends money campaigning where he thinks the audience is —this is one reason why television is chosen and is also why broadcasters can charge high rates for political usage. Campaigning on television is often the most economical way to reach large audiences; the cost per hundred may be relatively low. To maximize audience, a spot announcement usually is preferred. About 75 percent of the money spent to purchase air time for politics is spent for spots or network participations.

The wisdom at this advertising strategy is confirmed by broadcast surveys that show audience loss for longer political programs. The broadcasting industry reiterates this point often, and some stations will not sell program time if there is risk of losing the evening's audience. With the exception of a relatively small group of activists, politics does not have a highly motivated audience. If the political span of attention is short, the spot or the five-minute program is preferred to the fifteen-minute or half-hour program, even though the latter may sometimes be cheaper to buy.

These considerations explain why UHF channels and public or educational channels have only rarely caught on as important political media —and the same could apply to cable presentations. UHF and public television today provide selective audiences, and time may be provided free, but politicians are often reluctant to take the time if a substantial audience is not guaranteed. If the money is available or the candidate is willing to go into debt, he may prefer to buy time to reach large audiences rather than receive "free" time that reaches only limited audiences and at significant costs for programming and arrangements.

Yet if free time promises to reach significant audiences, it may be used as it was in 1968, when the general election campaign marked the first time that the messages of both major party Presidential candidates were carried free over cable. The National Cable Television Association had invited the three major candidates to utilize cable when possible. It also urged all cable operators with program origination facilities to make equal time available, at no charge, to all political candidates —national, state, and local. Both the Nixon and Humphrey campaigns made organized efforts to solicit cablecasters to present their candidates. The Nixon campaign reported that 415 systems with a potential audience of 4.7 million people carried the Republican material, while the Humphrey campaign reported that 303 cable systems representing a potential audience of 3.5 million people carried the Democratic material. The Wallace campaign supplied some materials to a smaller but undetermined number of stations.

The Republicans circularized some 500 CATV systems and 65 multiple-system owners, enclosing a post card for response. A letter was also sent to the party county chairmen for each area where there was a CATV system, asking the chairman to urge the system to participate. Follow-up letters were used. A survey indicated that the participating CATV stations showed the Republican program an average of three times, and all showed it at least once the day before the election.

The Democrats reported that their two half-hour programs — "Because It Is Right" and "The New America"—were distributed and shown during the last two and a half weeks before the election. The Democrats reported that their CATV operation costs were as follows:

97 films at $40 including mailing	$3,880.00
113 Ampex Video Tapes (including mailing)	4,745.38
Secretarial salary (estimate)	300.00
	$8,925.38

Expense money was partially provided by the Democratic National Committee, and some funds were raised by individual contributions from some cable operators. Since the two Democratic programs were not specially produced but just copied on film and tape for use on CATV, the actual production costs of the programs are not included in the reported figures. The Republicans did not report the costs of their CATV effort but are known to have spent about three times as much as the Democrats.

Cable operators are not required to keep logs setting forth programming, as air broadcasters are. Accordingly, the FCC's 1970 Survey of Political Broadcasting does not cover CATV, so no national inventory of either paid or sustaining time, apart from the Nixon and Humphrey data, is available. FCC action on logs is pending, and by 1972 FCC questionnaires will surely be sent to CATV operators, and the results incorporated into future surveys.

Surveys of political usages of cable at other levels would reveal the following examples:

In 1968 the systems in Greensboro, North Carolina, and Farmington, New Mexico, made extensive presentations of candidates and incumbents ranging from Congressmen to city officials.

Congressional candidates have used cable in San Diego, California, and in various systems in Pennsylvania.

The cable system in Newport Beach, California, invited 30 candidates to appear live and unrehearsed, and all accepted.

A Honolulu system offered 30 minutes to any candidate running for election in the districts covered and drew more than 50 aspirants for public office. The system met its offer economically by using one black-and-white TV camera and a zoom lens — equipment costing the operator less than $2,000.

A significant advantage of CATV is that it can enable a candidate to address his constituency only and thus avoid the uneconomical expense of multiconstituency media. The problem of multiconstituency media is felt particularly in metropolitan high-cost media areas, such as New York City. The NYC channels cover at least portions of about 40 Congressional districts in parts of New York, New Jersey, and Connecticut. A candidate for the House of Representatives will find it uneconomical to purchase air time to reach audiences most of whom cannot vote for him. A statewide candidate in New Jersey who wants television exposure must buy time on NYC or Philadelphia channels, knowing that he is throwing away 75 cents of every dollar. The NYC audience overlap creates problems for the stations, too, if they want to serve the communities they reach even on a limited basis, say, for statewide offices. In the general election period in 1970, there were eight major U.S. Senatorial campaigns in New York, New Jersey, and Connecticut, and five major gubernatorial campaigns in New York and Connecticut. (In the New York and Connecticut U.S. Senatorial campaigns, there were also major third-party and independent candidates requiring time.)

In varying degrees, the problem exists in other areas: candidates in New Hampshire and Vermont buy time in Boston; candidates in Delaware buy time in Philadelphia or Baltimore; candidates in Maryland and Virginia buy time on Washington, D.C., stations; candidates in Kentucky buy time in Cincinnati, Ohio, or Evansville, Indiana, or in Charleston or Huntington, West Virginia; candidates in Gary, Indiana, buy time in Chicago.

And so it goes. Only in Alaska and Hawaii do the air signals reach only potential voters.

Most American communities do not have their own television broadcast facility, but every community has elections of local public officials. More than 500,000 public offices are filled at regular elections in the United States, and in addition, there are primaries for many of these. The bulk of these positions are in constituencies too small for reasonable use of telecasting. But almost all constituencies could be reached by cablecasting at low cost.

Various means of reaching pinpointed audiences are available, besides cable, including mail or organizational means in door-to-door contact. The candidate may want to combine visual presentation with organizational delivery of audience. For that he would want the services of the local party system, or of a personal following to promote his appearance on cable, for the key to the use of CATV is whether there will be an audience. Attraction of audience is, in fact, exacerbated by the greater diversity of channels.

One way to attract an audience might be to set aside a channel exclusively for politics during a campaign period. The ability to present visually all candidates on a ticket makes a party or neutral political channel functional as a desirable means of giving the many candidates at least minimal access to the electorate. Cable permits audience segmenting, so that the candidates in any given political jurisdiction can be switched in and out in a meaningful way for the viewer, who would see only those candidates for whom he could vote. This ability to structure the presentation the way the viewer's ballot will be structured is a unique feature of the low-cost delivery system that cable provides. The large number of channels that cable will provide will permit a structuring for political presentation that could not be contemplated with the limited number of VHF and UHF channels.

A future potential of advanced CATV systems could be to permit candidates to reach special target groups. In a cable delivery system with selective coding capability, service could be directed to minority and specialized interests in meaningful administrative or legislative districts. It would permit various communities to be reached selectively, giving ingress into homes with selected socioeconomic or demographic characteristics. This would encourage efforts to reach selected segments of the electorate through cable rather than through present methods (including computer mailing; canvassing, whether by personal interview or by telephone; or through organizational means).

Campaigners could computerize information in more sophisticated ways than selecting mailing lists that now reach target groups. Computerized information for coded addressed cable systems will be derived from Census data, from credit bureaus, credit card organizations, magazine subscriptions, and driver's licenses.[3] Applying this information would permit the transmission of specialized appeals to specialized groups. This could cause greater attention to be paid to interests that are essentially local or narrowly specialized.

Whatever means of cable communication are utilized—whether for mass or specialized or selected audiences, whether time is purchased or provided free by cable operators—there will be the added costs of production and promotion for the candidate or party. Once a master is produced, duplicate spot announcements can be produced for between $2 and $5 apiece, and half-hour tapes for between $30 and $100 each. These costs seem low but, in a state having scores of cable operators, many films or tapes are needed if an effort is made to present material simultaneously across the state. Distribution of the films is also a consideration. Although hand delivery is costly and mail delivery less expensive, the latter may present time and scheduling problems.

CATV is presently characterized by a large number of small operators. The economics of the small systems may mean that many would find it hard to support their own origination and programming. The FCC has taken note of this problem in requiring origination of only those systems with 3,500 or more subscribers. The large number of operators also means that there are interconnection problems for national, statewide, even Congressional district campaigns, and in major cities, for citywide campaigns.

A single live program can be interconnected, as the Nixon telecasts were on a statewide or regional basis in the 1968 campaign, but there again there are production as well as interconnection costs. Nixon had ten live telecasts; costs for each of these, not counting time charges, varied from $11,000 to $27,000 and consisted mostly of expenses in set-building (such as theaters-in-the-round) and in interconnecting the various stations. These figures indicate the kinds of costs related to cable where extensive production and/or interconnection are involved.

Another example of production and conncection costs is derived from an experience in 1970, when Florida pioneered in a "politithon '70." On the night of October 28, a statewide network of educational television stations carried live a four-and-a-half-hour presentation of the candidates for Governor, U.S. Senator, the State Cabinet, and the Public Service Commission. A $25,000 grant from the State Department of Education paid for the costs, mainly for interconnection, and the candidates got prime exposure free. This suggests potential usage for cable and also for government subsidization or assistance.

Government assistance in production may also be available in other forms. The U.S. Senate and the House of Representatives each provide studios for producing video tapes, films, and recordings. These facilities are used mainly to prepare programs in the form of weekly reports for incumbents to send to radio and television studios in their constituencies. Members sometimes use the facilities to record endorsements of nonincumbents and thereby assist others on their ticket. But the facilities are used mostly by incumbents, giving them an advantage even if not used during campaign times in that they help the incumbent become better known. In recent years, several Congressmen have taped programs for distribution to CATV systems in their districts. State governments and city halls could provide similar facilities.

With the potential of numerous channels available, consideration could be given to providing a channel to each of the two major political parties. The channels could be available only during specific prenomination or general election periods, or they could be available permanently. In the latter case, party channels could substitute for the party press, which is found in other democratic countries but never developed in America beyond the early years of the Republic when all newspapers were party- or faction-oriented. The cost of running a channel on a year-round basis probably would be prohibitive for the parties and would require imagination and resourcefulness they rarely evidence. It has been estimated that delivery service would cost about one cent per day per household, and in addition there would be origination costs. Assuming a party committee sought to reach one million households for a two-month period, say from Labor Day until election day, the cost of carrier service alone would be $600,000. Unless cable presentations displaced all other types of campaign advertising and became the basic campaigning medium for all candidates on a given state or local ticket, such high costs would hardly be feasible. However, buying a few hours of such service now and then and several hours just before election day would be relatively inexpensive, particularly if the cablecast presented more than a single candidate or more than a few candidates at the top of the ticket.

Providing a channel to each of the major parties would raise questions about treatment of minor parties or candidates and how to satisfy the campaign informational needs of independent voters who would have to keep switching from channel to channel to get a proper balance of information on which to base a voting decision. Of course, if sufficient channels were available, the minor parties and independent candidates could share yet a third channel. Various definitions of "minor party" exist in federal and state laws, and proposals regarding air broadcast time could be applied to the cable situation.[4]

Still another possibility is to telecast hours for given offices so that viewers could switch from party to party to get information about candidates for specific offices. A more promising possibility exists in providing only one political channel and asking the League of Women Voters or some other independent group (such as Citizens Union in New York City or the Committee of 70 in Philadelphia) to supervise the use of the channel to assure fairness to all major and minor parties and candidates.

The states of Oregon and Washington mail to registered voters elaborate voter's publicity pamphlets, which carry information about the records and programs of the candidates. In Oregon, there are 27 regional pamphlet editions that coincide with the ballot faced by voters in each area. A number of other states send voters information pamphlets concerning ballot issues, constitutional amendments, referenda, and so on. Any such communication can be adapted for visual presentation over cable. Similarly, states that now provide sample ballots to voters or information about when and where to vote, could provide this information to cable operators for visual presentation.

Cable today is primarily a substitute for bad reception in rural areas. Much can be made of the future of wired cities, but from a political point of view, there is substantial difference between a city the size of New York or Los Angeles and a city ten or twenty times smaller. The largest cities could use perhaps, profitably, 30 or 40 channels; but in a city of 25,000 or 100,000, that many channels would be divided among such small selective audiences that it might not be in the candidates' interest to prepare programs for seldom-watched channels — even if these were provided free of charge or at nominal cost. Without a substantial investment in promotion, politicians might well think that certain cable channels will not provide a large-enough audience to justify the expenditure of their time and energy.

The mass audience of television may become atomized and fragmented if the wired city produces an audience pattern like that on radio, in which there are some stations specializing in news, some in talk shows, some in symphonic music, some in soul music, and so on. It is certain that promotion will be needed to point out to potential audiences what political programs will be presented when. In a multichannel system, there would probably be a di-

rectory channel, which would carry information on a daily basis as to what
information or programs will be available on what channels at what time.
Political broadcasts will be listed along with other channel offerings, but because of low-motivation problems, politicians would likely desire newspaper
or other types of promotion; these could add another five to ten percent to the
bill.

    If cable produces a more fractionated audience, it could lead to more local-
ization of political focus, a topic that will be discussed in more detail later.
If the local cable system serves to demarcate neighborhoods by giving a sense
of community to a section of a city or a suburb now mainly dependent on the
central-city media, then politics could become more decentralized, with less
attention to the nation and state and more to the local.

    Specialization and localization and the needs for promotion could give stim-
ulus to the revival of local party organization and to grassroots organization
of interest groups. Local party organization would have several functions. One
would be to organize local cable listeners by promoting the cable cast in various ways, by putting up posters or by phoning neighbors to listen in. Party
organization might also have an important role in refereeing time purchases
by the various national, state, and local candidates all vying for time on local
channels. The development of the equivalent of a party press over cable is
another potential local party function. A significant effect of localism could be
the recruitment of talent into local politics, both as candidates and activists.
A kind of participatory democracy could develop around the cable screen, and
the party could play a key role. Because there would be more diversity at the
local level and more personal interaction at the studio or on the streets, less
blandness on the air might result. There might be an accompanying decline in
responsibility.

    With abundant time for political discussion available, campaigns can do bet-
ter than provide the "cuing of the party label, or established voting tradition,
or the consensus of a community." [5] The abundance of channels may offer the
media a new opportunity for discussion that could improve the quality of politics. The opportunity exists for "reasoned argument" to prevail where party,
class, and community cues are relaxed. Political communication over cable,
with its wide potential and its potential for diversity, can offer reasoned argument in ways that over-the-air broadcasting cannot afford to. The American
public should ponder whether it can afford to miss this opportunity for improved political dialogue and participation.

PROBLEMS AND POSSIBILITIES

Two-way Communication

CATV in the 1980s will undoubtedly still be largely one-way communication.
CATV in the 1990s will begin to be two-way communication. That change is as
important a revolution as CATV itself. Communications applications in which
the individual will no longer be limited to receiving a set of messages chosen
by a sender but can instead choose the information that is sent him from a
vast library of materials are possible. This kind of communication, which
may be called "on-demand communication" in contrast to "mass communication" is illustrated by information retrieval systems. In some hypothetical fu-

turistic information retrieval system, the audience member will address a computer and ask for the current information about whatever is on his mind, be it today's news about the Middle East, the latest FCC ruling, a recipe for cheese fondue, or the price of lamb chops at a particular market.

Bidirectional systems would permit more than the scheduled presentation of propaganda, sample ballots, facsimile mail, or newspaper advertising. In the 1990s such systems could create a political demand media, providing the set-owner with the opportunity to request whatever information he wants about a candidate or the upcoming election at his convenience. With advanced computerized information systems, the viewer could request the candidate's views on a given subject and receive them immediately on demand in his home. If the system were plugged into a current events service or a political research center (whether or not operated by the parties or newspapers or encyclopedia companies), the viewer could seek out information he wanted about past speeches or events in the candidates's life. He could learn what the candidate promised two or four years ago. But these are prospects for the distant future.

The creation of widespread systems of demand communication for the ordinary citizen lies beyond the era of our present attention. Nonetheless, the urge for achieving feedback is great, and various limited forms of feedback from the audience can be expected to begin to appear, taking advantage of the technology of CATV. We postulate that for the first decade of CATV we can dismiss consideration of fully switched systems or of video feedback (as with a picturephone) from many sources, or even of audio feedback by telephone from large numbers of people talking to each other.

The feedback that we anticipate as possible during the next twenty years will be of four kinds: the mails; oral telephone responses from viewers but only from a sufficiently small number of viewers so that the calls can be handled by a limited bank of answerers at the other end; video feedback from one, two, or three locations in the field; and digital feedback. All of these kinds of feedback have considerable political potential.

Digital feedback means that the viewer can push a button on his set or tap a code on his touch-tone telephone and thereby record some simple coded message into a computer at the other end. The message may be a vote on a public opinion poll, an order for a product that has been shown on the screen, a pledge of a campaign contribution, or any such simple record—with or without identification of whose button was pushed. The message can go back from the viewer either over the same cable that brings the TV picture or over the home telephone. The extent to which digital feedback becomes a common feature of the CATV system will make a great deal of difference in the ways CATV is used in politics. The most common science fiction notion about the use of digital feedback in politics is that of the instant referendum. The notion is that the ancient dream of direct democracy, in which the people themselves vote on the issues instead of merely periodically choosing representatives, can at last be made a reality. This is sheer fantasy. It rests upon a total misunderstanding of the legislative process.

The essence of the decision-making process, whether in the Congress or in private life, is division of labor and allocation of time. The Congress considers thousands of bills a year. Almost all of them are highly complex with consequences that turn not only on the broad statement of purpose but on the

details of verbiage and punctuation and on the precise administrative mechanisms provided. On most bills the crucial vote is not the final vote for or
against the bill (for on more issues than not there is a broad consensus on the
need for some action) but the prior votes (often in committee) on matters of
detail never covered in the press, yet decisive in determining the social consequences of the action. The process is far too complex even for full-time
Congressmen to keep track of it, much less a citizen in his leisure time.
What Congressmen do is specialize. Any one Congressman is an expert on a
small range of subject matters. For the rest he casts his votes by following
the guidance of others whose general philosophy he shares and who have in the
past proved trustworthy guides to him.

Once one has given thought to the legislative process it becomes immediately obvious that it cannot be replaced by a referendum process. The chance
collection of citizens who happen to look at their TV screens at a particular
moment could not conceivably consider even one issue a night seriously, much
less scores or hundreds. Furthermore they won't want to; they may be happy
to answer a quick poll question as to whether their economic condition is better or worse than last year, but most certainly don't want to learn what the
issues are in the alternative welfare plans. Clearly, any instant referendum
scheme is so destructive as to be inconceivable.

Referendum proposals that have some validity are those that might allow
for one or two major issues to be submitted to the electorate after extensive
prior analysis and discussion. We are not taking a stand against referenda.
The kinds of referenda that do make sense, however, have little to do with
CATV. They are well prepared. The issues put up for vote are important
ones, so voter attention can be obtained, but costly campaigns pro and con are
required. Voting is a major event, and so there is no particular reason why
it should be done by pressing a button in one's home.

The fact that CATV is not likely to be used for official referenda does not
preclude its being used for public opinion polls. It will not be used for serious polls because the self-selected audience is not a good sample. However,
one can well imagine a pressure group organizing a "people's referendum,"
using digital feedback on CATV for the purpose. Also, a nonpartisan public
affairs program might end up by encouraging people to express themselves by
pushing buttons to indicate how they lined up. It is our belief, however, that
these more or less inevitable uses of CATV will turn out to be rather minor
entertainment gimmicks without much impact.

Propagandists could, undoubtedly, use digital feedback to create the image
of a groundswell of opinion. For example, the mere demonstrated fact of
having a high rating for a public affairs program could have some persuasive
impact on a politician, just as it does today in the rare instances when it happens. Also, a politician might be influenced if after a genuine and apparently
fair presentation of the issues on both sides, he were able to ask his constituents which side they agreed with. One can imagine Congressmen putting on
programs of that kind to establish rapport with their constituents.

While CATV is not a useful device for the conduct of public opinion polls,
it is a superb device for copy testing. It lends itself well to learning how the
public reacts to various ideas, themes, or issue presentations. One may expect that the national committees and other political organizations will try

speeches and spot commercials on some sample population via CATV and will ask for some kind of feedback, for example, by having people vote among alternative presentations. CATV is already being used commercially for copy testing by splitting a community onto two cables and presenting different ads on each. The same kind of capability will be useful to politicians.

Digital feedback will also be a useful device for politicians who wish to create a sense of personal relation to their audience. One can well imagine a Congressman talking to his district and saying something like, "We took up three issues this week: the SST, rising prices, and the war in Vietnam. Which one would you like to hear me talk about now? Push button one, two, or three and let me know." Conceivably, skillful use of such interaction could turn a speech into something more like a dialogue. On the more conventional side, a Congressman could say, "If you want a copy of my speech on George Washington, push the little red button and I will send you one."

If one moves beyond mere digital feedback, the possibilities of personalization expand. Coffee parties could be simultaneously arranged at five, ten, or twenty homes in the community, each with an open phone line to the candidate in the studio. He could talk, answer questions (addressing the questioner by name), listen to comments and complaints, and in general try to behave as he would if he were at the coffee in person. Similarly, a mayor on the screen could carry on conversation with several city officials with the general public as the audience. The show's effectiveness might be enhanced if there were a TV pickup at one or two of the remote office sites, so the public could see both ends of the interaction.

In the 1968 Nixon campaign, 88,000 persons were invited to "speak to Nixon-Agnew" through tape recordings about the problems uppermost in their minds. At major campaign headquarters or at rallies, a citizen could speak into a microphone to record a tape, which was funneled to the Response Center Coordinator in the Division of Participating Politics in the Nixon-Agnew organization. There a robot-typed "personal letter of reply" from one of the candidates was triggered. At major campaign headquarters, an audio response with the voice of Nixon or Agnew was given immediately to any of a number of precoded questions a citizen could seek an answer to. Cable could execute programs such as this.

The Nixon campaign in 1968 also operated another program with great potential over cable. This was an attempt of the United Citizens for Nixon-Agnew to recruit five million "commitments" of volunteers to help elect Nixon. "Commitment cards" were distributed, but similar appeals for volunteers could be carried over cable to millions of homes; with bidirectional signals, the viewer could respond immediately from his home, and his commitment would be recorded.

More immediate usage may be to communicate with party or campaign workers, to serve as a pep rally, to kick off a registration or fund-raising drive, or to send down instructions for election day activity. In many ways cable's potential seems great as an inexpensive substitute for closed-circuit television. There are any number of political uses of closed-circuit, such as those linking various cities for simultaneous fund-raising dinners or events, events such as the McCarthy Day rallies in 1968 or inspirational hookups to various committee or headquarters staffs across the country or around the state. Of course, the opposition could tune in, so strategy sessions would not be held openly, except with the use of scramblers.

Another important campaign activity is the registration of voters. With a bidirectional system, registration could be brought into the living room, and the formalities accomplished at once, if the law allowed it. This could be accomplished on a channel allocated to the city government or on any channel in a common carrier system.

The advent of the 18-year-old vote may lead to new modes of campaigning for that vote. It may become desirable to reach students on campus. Since universities are planning to have internal wired communication systems, cable may be a means of conveying political messages to students in auditoriums, at meetings, at fraternity houses, or in their rooms. Cable could also be a means to instruct students on registration procedures, requirements for absentee voting, and so on.

There is potential here for uses other than by candidates and political parties. Interest groups, too, could attempt to reach their members and sympathizers through the use of cable. Labor union political committees, trade association committees, ethnic group committees, peace groups, and gun lobbies could all reach homes in given areas where concentrations of their members reside.

Conceivably, the biggest effect of having digital two-way communication on CATV might be on campaign financing and on financing political movements in general. An effective pitchman saying "We need money; push the button once for every dollar you are willing to give," might raise an extraordinary amount of money. Impulse giving is far more generous than what people will give if they have a chance to cool off. There is no way, at this time, to estimate how much money that kind of appeal might produce. However, if channel time is priced reasonably in terms of cost, there is no reason to believe that a pitch will not bring in substantially more than it costs to put it on. If so, the screen may well fill up with fund appeals. Political parties may find themselves more dependent on charismatic pitchmen than on a few rich men.

The registering of instantaneous reactions made possible through bidirectional cable brings some disadvantages. Immediate feedback gives the candidate the opportunity to recast strategy, speeches, or appeals according to the responses given. Such responses, however, may tend to be impulsive, not based on considered and balanced thinking about the subject. There is some danger that the amount of serious and reflective political discussion might be reduced by instant feedback or if politicians play to the more easily swayed, volatile segments of the population.

Feedback can give the citizen a better sense of real participation combined with a feeling of efficacy. The citizen could be better informed because visual, personalized channels would be available for his receiving in depth the selected information he wants or may need to perform his duties as a citizen or as an activist. The public official or the politician could be better informed of his constituencies' opinions and therefore might be more responsive. But the potential disadvantage also must be weighed.

Localized Audiences
In its cable franchises the City of New York requires that by 1974 Manhattan's

cable systems be capable of transmitting simultaneously discrete, isolated signals to at least ten subdistricts in each franchise district. Each franchisee is further required to develop a plan "to divide the District into the greatest number of sub-districts possible, which may be variously combined so as to constitute neighborhood communities, school districts, Congressional districts, State Senate and Assembly districts, and the like... ."

The most striking political effect of CATV will be to make television an economic medium for reaching small subcommunities. This political advantage results from a technological limitation, namely, the relatively short distance that can separate the receiver from the head end of the cable. That makes CATV a community medium. As we have already noted, broadcast television today is an uneconomic buy for use in small constituencies since the boundaries of the audience reached do not correspond to the boundaries of the constituency. CATV, on the contrary, will be able to provide service for ethnic neighborhoods, Congressional districts, legislative districts, and various municipal districts.

CATV will be a valuable instrument for communities within a city, for example, a "Model City." (Bedford-Stuyvesant wants to control its CATV franchise.) CATV could help achieve the restructuring of city government in the direction of decentralization.

A particularly important type of community cablecasting is ghetto cablecasting. It is an unfortunate fact of American life that the most distinctive and self-contained neighborhoods are the ethnically defined ones, black ghettos in particular. Cable channels can give such communities a sense of identity and can provide an outlet for their distinctive cultural products. In black ghettos the demand may be not only for their own channels but for their own franchise. A franchise would not only serve to provide media of expression for the community but would also help foster black capitalism. Either with a separate franchise or without, it is clearly necessary for ethnic communities to have their own channels. Cablecasting on such channels will undoubtedly accelerate the development of community consciousness and community organization.

Indeed, the net effect of CATV may well be the localization of American politics. At present, no medium does a good job of focusing attention on local community problems. TV stations have to serve their total market area, which is, typically, a metropolitan area or more than one city. Radio can afford to specialize more, but its coverage is geographically just as broad or broader. The major newspapers are also set up on a metropolitan area basis. CATV may be the first medium to favor community groups. With CATV it will be practical for persons with local concerns to talk only to their respective neighborhoods, possibly at very low prices.

A neighborhood will not be reached even on CATV without organization. A group opposing a zoning variance, for example, may buy time, but unless they organize their audience no one will be listening to them. Unlike entertainment TV today, there will be no natural mass audience for specialized local cablecasts. However, local audiences can be organized. Local community CATV is not an anonymous mass medium like commercial TV. It is an auxiliary to political and community organization.

Special-interest audiences that are geographically dispersed are not as easily reached on CATV. It is technically possible to provide a channel to physicians or to members of a particular union. If channels are cheap enough, the material for a specialized audience could be cablecast throughout the metropolitan area. If one wished to prevent nonmembers from viewing, this could be accomplished by the same technology that permits pay TV. However, that procedure would be less economical than one that reached a geographically defined group.

At first glance one can only applaud anything that will advance civic participation in community affairs. There may be however, an unanticipated consequence — the disintegrative effect of community organization on national politics. There are so many forces in the modern world pressing toward nationalization and centralization that it may be hard to find a reverse tendency credible. However, CATV, depending upon its structure, could have disintegrative effects. We must look at the factors working in each direction.

A pessimist might predict as follows: CATV, plus the use of the inexpensive hand camera, cassettes, and video tape, will fragment the audience, depriving the networks of their captive mass audience. The networks will be supplanted by a pluralism of competitive producers. Associated with that development will be a de-emphasis on real-time simultaneity of broadcasts. This will mean that the President of the United States will no longer be able to command a third or more of the viewing public as an audience simply by pre-empting prime network time. Now, when he broadcasts, he displaces a popular program. In the future, when he appears on one of the many CATV channels he will be competing with popular programs. It will become more difficult for him and for leaders in Congress to catalyze national awareness and national sentiment around the great issues of national politics. An already somewhat rudderless nation will become even harder to mobilize and to govern.

At the same time CATV will be breathing new life into local politics, which are far less principled and ideological than are national. Local politics are mainly the pragmatic clash of vested interests. Local politics entail a bargaining process largely unrelieved by broader considerations of social costs and benefits. America lived through the consequences of localism in the era of machine politics. Once again local politicians, because they have the ear of their constituencies, will become powerful in American political life at the same time as national leadership is weakened. The Congress and the President will increasingly have to operate by pork barrel concessions to local leaders to achieve anything at all.

To that prediction of doom, the optimist would reply that the revitalization of local politics will bring able young people into the political process earlier, for opportunities in local politics are open to them in a way that they are not in national politics. Out of that reinvigorated school of local politics may well come a more talented generation of intelligent political leadership than we have known recently.

The optimists also question whether the rise in attention to local politics necessarily implies the decline of national political coverage. Under certain circumstances the President might retain his ability to speak to the nation. The law, for example, might give him the right to pre-empt all or many channels in some situations. Furthermore, the optimists argue, whatever else may happen to the networks they will retain their function as news services. Indeed, with more or less continuous news coverage on some channels, the demand for their news services will grow. Their live coverage of national and world news can continue to keep the citizen aware of the great issues.

Both the optimistic and pessimistic predictions are in a limited sense valid. Both describe, perhaps in caricature, the tendencies that are let loose by the shift from a few channels in over-the-air broadcasting to a plenitude of channels on CATV. The outcome among these tendencies is not to be predicted but to be chosen, and it depends on the design of the cable system.

It probably makes little sense, except in national crisis, to permit the President to pre-empt enough of the CATV channels to give him an effective means of monopolizing public attention. In our free and competitive political system such an advantage for the President would be resented by his partisan opponents, by the local interests he pre-empted, and by the citizens whose freedom of choice he restricted. That clearly violates American concepts of free politics and free speech.

What makes a lot more sense is to provide the Federal Government with at least one full-time channel on all CATV systems. Time would be divided according to law, with members of the Congress having a large proportion of the time, shared between spokesmen of the two parties. The time allocated to the Executive branch would be largely devoted to explaining public programs in health, welfare, agriculture, and similar fields that are largely "nonpolitical." The President, as always, plays a mixed role between executive and advocate, and his time could properly be used for mobilization on behalf of national policies.

Such a channel or channels could, in what we have described as a compromise system, be among the public service channels that the franchise holder would have to provide free or at low cost. In a contract carrier system, it would probably be undesirable to give the federal government such a price advantage over the other users any more than we do on the telephone system. The federal government should pay the low rates involved.

Objections will be raised to what will be called a federal propaganda outlet. These objections stem from an image of CATV more like that of over-the-air broadcasting than like printing; that image is inappropriate. With few channels and scarcity of access it would indeed bias the political process to give the federal government a channel of its own. But with many channels, objecting to one that is a federal voice would be rather like objecting to the Government Printing Office. An enormous amount of printed material emanates from both sides of the Congress. Much of it derives from floor debate, hearings, etc., and is technical in nature, but the government must produce it if it is to have a workable relation to its citizenry. So, too, in an era of electronic communication, with plentiful channels, the government can cablecast various of these materials.

To maintain democratic citizen involvement in national and international affairs requires that governmental as well as private institutions have a voice to the public. The process of government is two-sided. The government is not simply a passive reflector of public opinion, nor are the media simply reflectors or critics of the government. Issues are made and national viewpoints defined in Presidential statements, diplomatic negotiations, commissions of enquiry, Congressional hearings, and Congressional debate, all of which are reported in official publications. Issues are also defined in journalistic coverage of these, and in the private voices of criticism and support reported by the media about those public actions. A healthy national dialogue requires that both the voices of the heads of government and of the national opinion media be heard.

How to maintain the health of the national news media in an era of CATV is a matter that we have discussed in a separate section. Suffice it here to note that it is a matter of cardinal importance. Local news handling based on text wire services is not enough to keep public identification with national and international issues in a video era. Live worldwide video coverage is essential. Both for live video news and for a federal channel, it may be useful, as we shall note in the next section, to have linkage of CATV systems by satellite. To make it economic and practical to maintain the national political dialogue at present levels or better, this could be a measure of considerable importance.

Smaller Cities, Suburbs, Rural and Low-Income Areas
CATV may in some respects increase rather than decrease the disparity between the service provided the top 100 markets where 80 percent of the public lives and the service provided for the rest of the population. Perhaps the comparison should be between the top 50 markets and the rest of the population, since substantial amounts of serious local programming are likely to occur almost entirely in the larger cities.

The usual assumption is that importing remote signals via CATV will give the smaller markets a quality of service comparable to that in the main metropolitan areas. In regard to entertainment that is probably correct. In regard to politics and public affairs, however, the reverse is probably the case. The kind of novel community-oriented public affairs programming that we anticipate will be local in its reference and interest and will not be easily exported or imported.

Suburbia is by definition linked to a city or metropolitan area. The news fare is generally dominated by the central-city broadcasters and newspapers, and what local media exist may not be much more than shopping news throwaways. Yet suburbs contain inhabitants having higher levels of education and income, and these are characteristics that should make subscribers receptive to serious political cablecasting. Thus CATV could help fill a void in news distribution to a politically crucial segment of the population. Reapportionment is providing suburbs with increasing representation in both the House of Representatives and in state legislatures, and the hot suburban political battleground can receive the kind of local programming that the central-city mass media could never provide. Moreover, suburban income levels make wide subscribership possible.

Penetration of cable systems into low-income areas is a special problem. In terms of politics, it should be stressed that the lowest rates of political participation are found in low-income and low-education areas, and these are where both educational information about how to register and vote and political information about who is running on what programs are most urgently needed. Insofar as poverty areas get organized by community organizations, cable television offers them an effective new instrument for action.

There are many technologies whereby local cable systems can be linked:

landlines, mailed tapes, satellites, over-the-air broadcasts. Each has assets and liabilities in regard to cost, simultaneity, capacity, and so forth. These technical questions need not be considered here, but we shall assume that local cable systems will be linked by a mixture of these technologies.

Politically, characteristics of the linkage other than its technology are important. If it is inexpensive to link up all cable systems nationally to hear a presidential candidate live, then campaigning will be conducted quite differently than if it is just as cheap or cheaper to mail tapes around for local origination or if it is cheaper to put local citizens on the screen speaking on behalf of their candidate.

CATV creates the opportunity for local community coverage and interaction on the screen, which, by reason of the shortage of channels, over-the-air television does not allow. CATV, however, does not reduce, and may even increase, the options for filling some channels with uniform, nationwide, network material.

Indeed, there is reason to expect some reductions in costs that will be favorable to national campaigns. Bell System long-distance landlines currently are expensive, and sending tapes of spots or speeches to local stations is also expensive in practice, because the reusable tape often is not sent back to the politician. Satellites promise to lower the costs of live long-distance communication. Also, there are means whereby national campaigners and other national communicators can put their taped material in the hands of the cablecasters around the country on a daily basis, even if not live. Since the cable systems will be contiguous to one another in densely populated areas it would be entirely feasible to link the cables already laid for the service, thereby creating a long lines grid. Between 1 AM and 5 AM, when many channels will be unused, material could be sent out to be recorded on the cablecaster's tape or cassettes (instead of tapes provided by the politician). The tapes could then be cablecast locally during the day.

Community Power

Not since the trolley car has there been a technological innovation with as much impact on local machine politics as CATV. It is a franchise in which the holder of the franchise may become a major influence on elections and politics. The temptations are enormous.

Evidences of corruption are already in the courts. It would be strange if it were otherwise. Characteristically, television station owners are powerful political figures in American cities. Station owners have the power to make it hard or easy to acquire time for political broadcasts. Their public service programming can help or hinder anyone in public life. But television station owners are not chosen locally. They receive their franchise in Washington. Now local CATV franchise seekers appear who want to compete with and perhaps injure the local station owners. Depending on the rules the FCC adopts for CATV, new cable franchise holders may acquire similar power in local politics as the television station owners. If CATV franchise holders are allowed or required to originate programs, they may become the local political powers of the next era. In addition, if the system is not a common carrier, the CATV franchise may be a gold mine.

The CATV franchise holder will be even more deeply involved in local poli-

tics than the typical television station owner now is. He is dependent on the
goodwill of local authorities for his franchise. Since the CATV franchise
holder must dig up the streets in order to lay cables, or use the poles existing, he knows that a friendly attitude on the part of the utilities and the phone
company is a virtual necessity. He is likely to be selling channel time to a
variety of conflicting and dissident community groups. The television station
owner occasionally has to make difficult decisions about which partisan groups
he will put on the air in accordance with the fairness doctrine, and for how
much time and in what way. Often there are complaints, but the total amount
of such conflict over the airwaves is small since the shortage of channels
stifles most of the presentation of this kind of material. With growth by an
order of magnitude or two in the demand for channel time for controversial
uses, the CATV franchise holder will most likely find himself in a constant
controversy over terms of access, responsibility, and rates, unless he is
operating under a contract carrier system with uniform treatment of all
comers.

There are two approaches to reducing the undesirable involvements of the
franchise owner in local machine politics. One is to remove the franchising
authority to a higher level of government; the other is to put the franchise
owner in the position of a contract carrier and bar him from corporate parti-
sanship. The first of these measures, depriving the local community of the
authority to choose the recipients of franchises for local services, runs counter to American traditions and attitudes on local self-government. The other
approach, restricting the franchise holder to a strictly carrier function,
seems more acceptable.

Who May Hold a Franchise

The rules that govern persons who may be granted a cable franchise are matters of great political significance. Franchise holders are likely to be important political forces in their communities. In anything but a contract carrier
system these men of great local social and political power will also have substantial influence on programming policy and content.

As a matter of franchising policy, several kinds of persons might be de-
barred or looked on with scepticism as franchise holders. Among these are
partisan political groups. While a channel might be rented to a political party
or political interest group, the person who has the channels available for rent
should be presumed to be nonpartisan and ready and willing to serve all parties evenhandedly.

It is a serious question whether newspapers or broadcasters should be assigned CATV franchises. The objections are far less serious in a common
carrier system, where the cablecaster presumably has no influence on content. A producer of content for dissemination clearly has a conflict of interest, however, if he also controls other media. The possible facsimile delivery
of newspapers complicates the problem because newspapers claim the right to
own the equipment necessary for their production and delivery.

There is some question, however, whether a cable franchise needs to be
or should be monopolistic. The technical and regulatory issues determining
whether more than one cable service can be efficiently and economically offered to the same homeowners obviously have profound political implications.

We shall proceed here on the less hopeful assumption that within any one geographic area the cablecaster will be a monopolist.

The largest problem is to assure that the cable system offers diversity of
programming and more points of view than are presented by the local broadcasters or newspapers. With clear evidence of substantial concentrations of
media power residing with dominant broadcasters or with single-ownership
newspapers in a community, the great potential and good of CATV is its promise of greater diversity than most localities now have. Pressures to exploit
cable politically will most likely come most strongly, not from the centrists
in the two major parties, but from representatives of the right and the left
who see an opportunity for their voices to be heard, and certain civil liber-
tarians who see an opportunity for everyone's voice to be heard. The two major parties may be among the last to recognize the potential for exposure that
exists and the last to bring pressure to bear to ensure that wide political uses
are assured. There is little in the history of the two major parties to suggest
that they will be enthusiastic about programming that gives equal exposure to
all or many minority voices. Candidates of the major parties often seem to
prefer to pay their own way rather than obtain free time in ways that produce
prominent minority exposure.

Moreover, major party politicians now seem to be playing off the broadcasters against the cablecasters. This will enable politicians or their representatives to extort campaign funds from both groups wanting to protect their
own interests. In a competition for funds the broadcasters are wealthier and
have a tradition of influence at the federal level. At the local level where franchising occurs, the situation may be different.

Cable television is not without political resources. A Political Action Committee of Cable Television (PACCT) was organized in 1969, and in that year it
reported receipts of $33,982 and expenditures of $8,983. In 1970 PACCT reported receipts of $17,058 and expenditures of $30,002. The committee made
contributions mainly to members of the U.S. House and Senate Judiciary and
Commerce Committees, committees that deal with legislation of concern to
the industry.

The Organization of the Industry

Cablecasting is a business that must cover its costs. There are various ways
it can be organized. Let us consider three different "pure" models and one
compromise and the different political consequences of each.

In the first model, the economics of monopoly runs its course unchecked.
In the second, regulation is used to protect broadcasting as it exists today.
The third is a contract carrier system. The fourth is a plausible kind of compromise, responding in part to the potential of new technology while at the
same time protecting vested interests that exist.

Politics Without Policy. As a base line, let us imagine CATV as it might function if no federal policies deliberately shaped it to public service. What would
then emerge would be determined very largely by what would be most profitable to the monopolistic local owners of cable systems. They would start out
simply relaying broadcast programs but would eventually go beyond that when
penetration grew to the point where they could supplant the broadcasters.

Cable operators would become engaged in program distribution, because
that is where big profits lie. They would not develop the system beyond perhaps twenty channels, since additional channels would not add to the total audience but would fragment it, raising costs and probably reducing revenues.
Of the channels, perhaps five or six would carry extremely popular programs
such as sporting events and major theatrical performances, as pay TV. These
major events would be broadcast more than once to suit people on different
schedules and thus maximize the audience. A similar number of channels
would provide commercially sponsored entertainment free including old films
and music and variety shows so as to win subscribers. These programs would
have to be cheap enough to make money with smaller audiences than the networks now require. There would be one or two channels devoted to weather,
public announcements, news, TV schedules, and talk shows, also supported
by advertising. In ways characteristic of monopolies everywhere, the entrepreneur would be inclined to restrict production, in this instance, channels,
so as to maximize revenue. In this instance he would maximize revenue by
getting as much as possible of the audience to watch a few pay-TV channels.

In such an unregulated world, the present three great networks would lose
most of their functions to various special-purpose producing and distributing
corporations. In such a system there would be little public service broadcasting. There would be news, and there would be spot political commercials
scattered around wherever a substantial audience could be found. There would
be little else of political interest.

Minimum Change   The CATV system just described, in which the FCC and the
Congress are passive and monopoly economics holds sway, is highly unlikely;
it is too adverse to the spirit of American politics. Far more likely is a system in which government regulates cable, with the main thrust being to protect
existing interests in broadcasting. Such an objective would require severe restrictions on pay TV, on importation of distant signals, and on origination of
programs by franchise holders or other nonbroadcasters. Such a highly restricted system is unstable in the long run. Sooner or later the economy,
power, and flexibility that advancing technology makes possible will overwhelm
established ways of doing things. For as much as a decade or two, however,
CATV could be held back by unfriendly regulation. There would, furthermore,
be long-term effects of such enforced delay. For example, cassettes might
take over functions that CATV would otherwise perform. Such regulation would
mean that for a decade or two CATV would produce minimum change in present
broadcasting patterns. That would be true in politics as well as other fields.

A Contract Carrier System   In the two systems just described, the potential
for CATV is restricted, in the one case by monopolistic practice, in the other
case by bureaucratic practice. Let us now consider the opposite pole, a system designed to take full advantage of the potential of CATV. That would be a
contract carrier system.

By a contract carrier system we mean something very like a common carrier system, of which the phone company is the best example. We choose the
term "contract carrier" to avoid automatically assuming all the legal implica-

tions that go with the well-defined concept of "common carrier," though most
of them we do mean to adopt.

In a pure contract carrier system, the franchise holder would do no origination of programs and would be prohibited from having financial interests in
his clients or in producing corporations. He would simply sell time to all
comers on the basis of a publicly announced and nondiscriminatory rate card.
We leave open the issue of how far and how rates would be regulated. We assume, however, that the franchise holder may levy a connection charge and
may and will collect a monthly subscription fee. We assume further that extra
charges may be assessed against both sender and receiver for special services, such as pay TV, or for reaching highly specialized audience groups,
like physicians, or for extra services, such as burglar alarm systems.

In general, in a contract carrier system the franchise holder makes his
income by expanding the number of subscribers and of channel-time buyers.
He has no vested interest in what gets transmitted, only in the volume of traffic. The franchise holder has no personal interest in keeping ratings on any
one or few channels high. He prefers to see many channels in use, even if each
has a small audience. The result is, therefore, likely to be a copious system,
with a great variety of material being made available to a highly fractionated
audience.

A Mixed System   More likely to emerge, perhaps, than any of the three pure
types of CATV system just described, is a mixed system, born of political
compromises and intended to achieve most of the benefits while avoiding most
of the drawbacks of each pure system. That intent is clear, but whether it
would be realized, or whether compromise would produce only a camel, is
far from clear.

Let us consider a likely kind of compromise system, indeed the kind of
compromise adopted by the FCC. It will be highly regulated, for only regulation can enforce the kind of unnatural alliance among diverse elements that a
compromise implies. It will, we assume, be a relatively copious system, for
the regulatory agencies will insist on that. It will also be a system designed
to preserve the present broadcasting stations in a new and competitive environment. It will, therefore, sharply restrict the amount or character of competing entertainment material that the cablecaster will be permitted to provide.
The cablecaster will have to carry all over the air channels. The cablecaster
will be allowed at the same time, to support some local origination channels
by advertising. At least in places where local channels are few, one or two
channels will be used for importing remote signals. Perhaps there will be one
or two additional channels that use a news, weather, and service announcement format.

Some channels will be for pay TV. We would expect that at most two or three
of these would be allowed, the restriction being to protect free TV. For the
same reasons, the kinds of content carried over pay TV might be restricted.
Some of the channels might be reserved for special services, such as stock
quotations, burglar alarms, etc.

While regulation will require the franchisee to make channels available on
a lease basis to private groups and on a lease or free basis to schools and gov-

vernmental jurisdictions, such proliferation of channels will be checked by
the franchisee as far as possible. His main weapon against such channels is
charging high rates for them and giving them poor service. However, regulation may require him to provide channels for such material; if so, he will
want small audiences on them. What is most profitable for the entrepreneur
is originating and distributing commercial and pay programs to concentrated
audiences. The franchise holder, if he is also a producer or distributer, has
no reason to compete with his own offerings, fractionating his clientele. He
has but little motive to sell channel time. He is in a conflict of interest situation, which he can most profitably resolve by not offering too much variety on
the cable and by not merchandising noncommercial channels aggressively.
From the cablecaster's point of view the ideal public service channels have a
tiny audience, pay their own way, and fill excess capacity by repeat programming, for which discount rates might well be offered.

A cablecaster might even be happy to dole out free broadcast time to civic
groups who would otherwise want channels of their own and thus press him for
greater channel capacity. He might be particularly happy to give free time to
such broadcasters if they could be counted on to put on dull programs.

In short, a mixed system that looks attractive at first glance turns out to
suffer the consequences of creating deep conflicts of interest for its operators.
In the early years in which we now are, franchisees may seek out any kind of
low-cost public service role that will draw support to them. In the long run
the franchisee who is in the entertainment business has motives for not wanting to do much for developing the social uses of cable.

Pricing
If some or all channels are made available for lease to politicians, then the
rate that the franchise holder charges them will become an important consideration in the planning of political campaigns.

In current discussions of mixed CATV systems, it is often proposed that
channels be provided free for political campaigns and for public service activities. On June 24, 1970, for example, the FCC proposed an allocation by
CATV systems of channels, including at least one for local origination, at
least one for free use by local governments and candidates, channels for free
use by local citizen groups, leased channels for commercial use, and instructional channels.[6]

Until now, the usual practice of cable franchise holders has been to provide
time free to candidates and officeholders, though that might be interpreted as
merely a come-on in a period in which cable penetration is not yet high enough
to attract many political buyers.

In a mixed system in which the franchiser makes considerable income from
advertising and selling of various services, it makes considerable sense to
require him to provide public service channels free even though that immediately puts him in the position of rationer and distributor of favors. It reopens
all the problems of the fairness doctrine and its administration. However, in
an era in which campaign costs are skyrocketing and in which various proposals for public support of campaigning are being seriously considered, to levy
an appropriate charge on the holder of a very profitable public franchise makes
much sense.

However, in a contract carrier system the return to the franchise holder is much lower. In such a system the burden of the argument would seem to favor charges to all users, with political and public service users entitled to the "lowest unit rate."

The return to a cable franchise holder for the lease of a contract line ought to be adequate to create an incentive to multiply such lines. Any other pricing policy creates incentives to distort the operation of the system in undesirable ways. If the cable franchise holder makes money from origination or royalties on popular commercial or pay-television programs, he has an incentive to keep the audience concentrated on those programs. This would reduce the competing channels to relatively dull, esoteric fare for infinitesimal audiences. He also has an incentive to keep the number of channels down.

We would predict, though admittedly without any solid evidence, that the demand for political channel time will prove highly price elastic. It is true that politicians with large area constituencies, like other advertisers, will often prefer to purchase the most expensive mass media buys — such as spots on popular entertainment programs. But that is because they deliver large audiences at a low cost per thousand. That will be true also in a CATV environment. If interconnection costs between CATV systems are high, or if the campaign has to assume the cost of negotiating separate buys from each system, CATV might not be attractive to the manager of a statewide political campaign.

At low prices, on the other hand, vast amounts of political time would be bought. At a zero price or something near it, there is almost no limit to the amount of channel space that would be requested for politics in major metropolitan areas.

If cable costs are low, a group should be able to raise more money broadcasting fund appeals than it spent to buy the cable time. Here is one of the reasons for expecting the demand for time to be highly elastic. At low prices, renting cable time pays because it permits one to raise funds. Under such circumstances there is hardly any reason for not going on cable frequently (especially in wealthy and sympathetic communities) and using material for which production costs are low.

In the absence of experience, we cannot prove our contention that the demand for political time would be very large in some places at low prices. However, we proceed on the assumption of a highly elastic demand, which would result in very different patterns of political programming depending on whether the price for channels is high or low.

Fortunately, all the evidence seems to be that mature cable systems will be able to lease channels at a profit at prices that are very low by broadcasting standards. We assume that subscribers will pay an original installation charge and perhaps $5 a month as a subscription charge. A 20,000-home system with each home paying $5 a month would collect $1.2 million per year in subscription charges. Assuming a 16-hour day there are 5,840 hours available on a channel. Each 20 channels that would average $10 per hour in rental over the 5,840 hours would produce just about the same amount again ($1.16 million). Costs for a year, using the figure of one cent per set per day, for a system of that size would be just under $1.5 million. It seems clear that once

CATV takes hold relatively low hourly charges will produce quite adequate revenues.

Minute Groups: Telemeetings

At $10 or $20 an hour, or even at somewhat higher prices, the demand to lease channels for political broadcasts can become very great indeed. In fact, all kinds of special-interest groups will find that cable communication can become as significant and useful a medium to them as the printing press and the mimeograph machine. For numerous organizations inexpensive CATV offers an opportunity to bring people together in front of their respective screens at the functional equivalent of a meeting. There are various ways to use CATV to gain many of the advantages of a meeting, though not all of them; indeed attention-getting devices must be used or the mild sympathizer in the passivity and anonymity of his home will not bother to watch such unexciting fare. However, if activists telephone their friends to tune in, if contests and rewards are used to capture people's attention, it should not be hard to reach audiences in the hundreds for many activities that today draw small attendance.

One problem is that the success of small telemeetings requires at least enough feedback to establish that the audience is there. It is one thing to talk to 50 known and identified individuals. It is a very different and less satisfying thing to talk into a faceless microphone in the uncertain hope that somewhere in the big city 50 people may be listening. Another problem will be that all groups will want time between eight and ten on weekday evenings. The market mechanism will hold proliferation of channels down below the level that could be filled at prime time only. Higher prices at prime time will drive the less passionate users off the screen and the more passionate users into the undesired off-hours. But passionate users do exist. Militants on behalf of many causes may be expected to operate on a more or less marathon basis, at least in major centers at hours when prices are low. The majority of groups will be hard-put to do one telemeeting a week, but not so the peace movement, or women's lib, or Jehovah's Witnesses. A number of militant movements may very well try to carry on almost steady programming even if only one person is viewing. In New York or Los Angeles, one could easily imagine dozens of such dedicated and TV-oriented groups going almost all the time.

It is undoubtedly true that a telemeeting loses some of the interpersonal impact of face-to-face contact. One will not replace the other. But considering the difficulty of getting people to meetings, the telemeeting will undoubtedly meet a real need.

One particularly interesting and frightening possible use of CATV with digital feedback is to keep an automatic record of just which sets were tuned into a program. Mailing lists can be made up of the households that tuned into particular programs, enabling the politician to follow up with appeals for funds, or votes, or memberships. Such uses depend, of course, on such invasions of privacy not being barred by regulation.

By the standards of contemporary TV, with audiences in the millions, telemeetings with audiences in the hundreds may seem mere trivia. But a few such nonconforming programs in each of several thousand localities add up to a significant breaking-out of the present shell of limited channels. The opportunity

for the emergence of many small but active and autonomous audiences is a
matter of considerable political significance.

There is no reason to assume that the organized groups that may be interested in holding telemeetings will be of any one or limited number of viewpoints. The more active sponsors of such political expression, however, will
be disproportionately drawn from the more extreme and intensely committed
groups. It is that way with ordinary meetings; it will be that way with telemeetings. There is one difference, however. Telemeetings are readily accessible to eavesdropping by the general public. This may have many consequences. Will indignant viewers, shocked by outrageous statements or conduct,
be less tolerant of free speech when a meeting comes right into their homes
than they are when the same things happen out of their sight? Conversely, how
will political activists respond to this situation? Will they moderate their message when they know the audience watching represents a broad range of political persuasions?

The Hand Camera and Amateur Production

"Cable television is not the technological change that will revolutionize American politics. It is the cheap hand camera." That is what one of our interview
respondents said. It may be an overstatement, but it is true that the two things
interact in a very important way. When channels become available, at least at
certain times of day, for $10 or $20 an hour, production costs and production
quality, not distribution costs, become the bottleneck.

Perhaps amateur film-making will be as natural to the next generation of
young people as the writing of stories and essays has been for the creative
young for the last century. Will the inner-city black who now might write a
leaflet or make a film protesting ghetto housing, someday televise his work?
Will the Congressman who now sends a form letter to his constituents prefer
instead to let a cameraman follow him through his daily routine?

If this optimistic suggestion has even a shred of validity, then there will be
another kind of time buyer of comparable importance to the political organization—namely, the creative group that wants its stage. At the costs we are
projecting, there will be hundreds of film-making groups that will raise the
small fee required to screen their product on the tube. Some of the films will
be political. Some of them will also be good. Experience may be the school
for a large group of producers of visual communication. Cinéma verité or
some successor to it could become a major form of political expression.

Equal Opportunity and the Fairness Doctrine

There no longer is question of federal jurisdiction through statute or FCC regulation of intrastate cable systems. At present Section 315 of the Federal Communications Act, the Fairness Doctrine, and sponsor identification regulations
do apply and are crucial in consideration of political cablecasting. Should any
question of federal jurisdiction arise, it is most likely to affect transmission
when local cable functions as an originator of programs. A distinction must be
made between programs the cable system originates and the programs of other
broadcasters whose signals it picks up. In the latter case, presumably the on-
air broadcaster would be observing the "equal opportunity" and fairness pro-

visions, so the cable operator would carry these programs on the same "equal time" or "as purchased" basis if he was not unfairly selective in what he carried. (Where distant signals were being picked up, the candidates may be in another constituency or even another state, so there may be little local interest.)

Popularly called the "equal time" law, Section 315 is more accurately termed the "equal opportunities" provision. It requires that the licensee giving or selling time to one candidate must provide similar free access, or the opportunity to buy similar time, for any other candidate seeking the same office or nomination.

The Fairness Doctrine is a set of supplemental regulations, which the Federal Communications Commission applies to politics as well as to other areas of controversy. It specifies that a station presenting a viewpoint on an issue of public importance must provide an opportunity for the expression of opposing viewpoints. However, the Fairness Doctrine allows the station substantial discretion in permitting the appropriate response to the viewpoint, and it does not require equal time, only that adequate reply time be allowed.

The Fairness Doctrine is supplemented by the FCC's personal attack rules. These require that a person being personally attacked must be so informed by the station and furnished with a transcript or a synopsis of what was said and time to reply.

One further rule applies: When a station editorially endorses a candidate, it must inform the opponent within seven days and provide equal time.

If time for campaign purposes is sold on CATV, then opponents would need to be offered an equivalent opportunity. If time is given free, then it would be under an access formula determined by government — at present, "equal opportunity."

Despite the present application of the fairness rule to the presentation of political viewpoints, the question is whether fairness is appropriate if there is no channel scarcity, as in on-air broadcasting, but an abundance, as is the potential in cablecasting. What fairness requirements are appropriate depend on various aspects of the system. Is there enough channel space to provide access to all who wish it? How reasonable are the prices? Are channels controlled by partisan interests, and if so, are those channels de facto dominant ones, or are there many channels on a par? Until it is clear how the system operates in such respects, it is not possible to recommend any particular fairness procedures as optimal. Further, while federal regulation of cable prevails, licensing is presently a local matter. Elements of "equal opportunity" or the Fairness Doctrine may or may not be specifically written into local franchises in the form of certain requirements demanded by the locality.

The larger question of fairness is at issue in terms of an abundance of channels because, given a large number, there should be more opportunity for overall balance and rejoinder than where the numbers of channels and amounts of time are limited. The more channels, the more diversity, the more inexpensive the time, the greater the opportunity for more voices to be heard. However, on anything but a pure contract carrier system, without application of something like Section 315 or fairness, there would be no assurance that any given voice would ever be heard. On the other hand, if cablecasting were given contract carrier status, enforcement of fairness would, presumably, not be necessary. Channels and time would be available in abundance to anyone or any group capable of paying the minimal charges.

Yet the alternative of applying some standards of fairness or reply ought to be explored even if contract carrier status is achieved. Without standards, one side with funds could dominate a campaign or issue while the other side without funds could be blacked out. The problem is not only the costs involved but also the availability of an organized voice on the other side of a given issue; say, a voice representing the consumer, the voter, the radical right or left, or even a voice representing the corporation or the government policy being attacked. It is not likely that a state government would purchase time to reply to every local attack. The closest parallel is not on-air broadcasting but newspaper or magazine publishing. In these latter media fairness is not guaranteed, but under some circumstances very similar practices seem to be necessary. Letters to the editor may be published by a newspaper or space for a formal reply provided, particularly in media such as newspapers that de facto dominate a market (even if, in principle, competition could come in).

In cablecasting the greatest need for protection would be in the right to reply to a personal attack. Just as government could require right-of-reply, it could also require an electronic equivalent to letters to the editor, a kind of public rebuttal at a given time on a given channel or a given time on a single channel permitting reply to any program cablecast on any channel in the system. Such a weekly feature might attract regular audiences while providing an electronic soapbox with meaningful potential.

Certainly the protection of candidates for political office is a warranted use of governmental power; one wonders, however, what protection minority candidates might receive in some states and localities. Surely some extension of the "equal time" provision or some new standard such as "differential equality of access"[7] is desirable on a national basis. A strong national policy can assure that cable is used fairly. Such a policy can also assure cable's use as a great national educational resource.

One further suggestion is warranted if the operator is required to provide political time free. Unless special provision is made, the pattern would tend to follow that of on-air broadcasters, each of whom is free to decide for himself for which contests for office he will offer free time. If we are to take advantage of the new media potential of cable and to attempt to restructure some elements of political campaigning, then all elective offices on the ballot in the range of the cable system should be presented—not just the most visible or interesting ones selected by the cable operator. The alternative would be to require the cable system to turn the time over to the political parties for decision as to which candidates at which levels are to be presented. Earlier, we suggested one function of the local parties may be to referee cable-time allocations to various candidates on the ticket, and of course, if there were party channels or some equivalent of a party press over cable, then the party and not the cable operator would be making the allocation of time decisions.

As noted earlier, party-controlled newspapers are common in other countries, but not in the United States. In Italy, to get anything like a fair view one needs to buy at least two papers, for each paper tends to bury the activities of other parties in silence. Newspapers in the United States, while less balanced and bland than the electronic media, do try to do a full reporting job on both sides.

Under cable television, partisan groups (parties or pressure groups) that have cable channels might well become engaged in newscasting. A party or other partisan group with a channel of its own might well find news announcements and commentaries to be one of the cheapest and easiest ways to fill time. It is far easier to set up partisan newscasts than to start a party paper. A party paper must compete for the reader's purchase with a general newspaper if it is to be seen by him at all. The party newscast, on the other hand, is available in the home, requiring only a flick of the switch to get at least a few minutes of attention. There is, therefore, reason to expect the emergence in this country of partisan dissemination of news. However, the consequences of such party participation in cable newscasting may be less biasing than a party press, for all partisan channels presumably will be present in every home.

Copyright and Libel

Abundance of channels will permit the screening of a wide diversity of tendencies and views. That is one of the blessings of abundance. No longer need the viewer be limited to the circumscribed moderation of the mainstream, which now necessarily receives the bulk of the limited air time available for politics.

But every blessing has its price. One of the prices is an increase in the amount of irresponsible and even illegal material transmitted. There are certain extremists and inadequately-funded irresponsible groups who, in the case of channel abundance, may properly claim some portion of the ample time available. These individuals may be more likely than established network executives to overstep the bounds which are set by law, even under the First Amendment.

In a truly copious cable system, in which each of 2,000 to 6,000 local operators supplies input to perhaps 40 channels of material 16 hours or more a day, no one will be able to monitor violations of copyright or libel laws.

Obscenity is a different matter. Material that offends much of the community will, unless controlled by scramblers, cause a backlash from irate citizens. The person who is injured by copyright violations or libel is an isolated individual who rarely will happen to be watching at the moment of the offense.

Print media leave behind an indelible trace of the violation. Electronic media do not. Cablecasters could be required to keep an audio tape of all output for a specified period of time. The cost of reusable tapes would not be burdensome if the storage period were reasonably limited.

The locus of responsibility for violations of law is a thorny policy matter. A contract carrier should have an obligation to serve all under nondiscriminatory terms and, correspondingly, should have no liability for what the cable leaser sends out. Unlike the telephone company, which serves millions of senders, the operator of a system with 40 or even 80 channels could be required to keep a record of the responsible person to whom he leased time. Such a record might be useful in enforcing regulations about campaign spending limitations.

Free Speech

While free speech is a guiding principle for the organization of any communications medium in a democracy, it is not a simple guideline. The technology of every medium limits who uses it, when, and how. The restrictions are of three kinds: rationing, prices, and conditions.

Pricing or rationing occurs with any scarce medium, whenever there is not enough of it to meet the demand. But as anyone who has studied economics knows, demand is not a number but a curve. The amount demanded is a function of the price. What, for example, is a copious system of cable television? It is a system with one more channel available than the number demanded. But at what price? If the price is unregulated it will rise to eliminate excess demand. If the price that society judges fair and proceeds to enforce is below the current equilibrium, some kind of rationing will be inevitable.

By conditions for the use of a medium (as distinct from pricing and rationing) we mean those restrictions that do not control quantities used but specify who may do what, where, and when. For example, all free speech in our society is conditional on certain laws of libel and copyright. There are also conditions on particular media. Use of the streets by pickets or paraders is confined to certain areas. In the following paragraphs, we shall consider, not such universal conditions as the law of libel that apply to all media alike, but rather those conditions arising from the technology of the particular medium.

The ideal image of free speech is of a medium that is unrationed, unconditional, and free of charge. This is the state of affairs for personal influence. No price or special constraints restrict us from talking to each other; but there are few media that are free of such restrictions.

Mail and the print media are generally unrationed, unconditional, but priced. The use of the streets on the other hand is unrationed and free, but it is conditional. In contemporary television broadcasting outside of campaign periods, political time is not sold so speech is both rationed and conditional, but without cost. During campaigns speech is rationed, conditional, and priced—the most restrictive situation of all—although the rationing tends to be less severe than between campaigns. Radio, on the other hand, has reached the point of copiousness of stations at which, at least at current prices, there is no real rationing of material during campaigns; the candidates can get whatever time they want to buy. What can and should the situation be for CATV?

With all its abundance of channels, a CATV system cannot be like personal influence—unrationed, unconditional, and unpriced. Perhaps communities exist somewhere in the heart of mid-America where this would be possible. Those are communities where political life is quite dead, where, if 40 channels were introduced, they could be paid for by advertising and by pay television, but where the cablecaster could not fill 10 channels with local origination even if free time were offered. Perhaps there are places where this would be true; it is not generally the case.

Consider rather what the situation might be in New York, Los Angeles, Chicago, Washington, and San Francisco. Let us speculate about what the demand might be if channels were made available at no cost and unrationed to any group meeting the condition of providing at least five hours a day of public service programming. (We leave it to wiser men to define that criterion.) A speculative census of applicants might look something like the following: several political parties (including the Republicans, reform and regular Democrats, and at least one party to the left and one to the right of the majors); spokesmen of ethnic groups (in New York perhaps five black community groups, three Puerto Rican, and five miscellaneous); government agencies (the police, the state and city governments); protest groups (with their notable factionalism one might expect perhaps three on the left and three on the right); several women's and peace groups; special-interest and civic organizations, perhaps half a dozen (including internationalist, conservation, and some forum groups) that would debate a wide range of issues; and church groups (perhaps a dozen when one allows for the various denominations and evangelical movements). That adds up to about 50 channels without branching off from public affairs to educational uses.

We have no illusion that this is a reliable prediction. It would be drastically low if the culture of amateur film-making takes off as a new wave in a world of modest production and distribution costs. It could be much too high if people learn from experience that they are talking to no one. But that is the best guess we can make. It seems to us that the ideal of an unpriced, unrationed system accessible to all is, in fact, neither realistic nor ideal.

There are, as we've noted, alternative patterns of access to television depending on the type of cable system. On a mixed CATV system having substantial advertising and box-office revenues, the franchise holder might provide unpaid public service channels, which he would have to ration. On a contract carrier system, given the relatively low cost of cablecasting, it would seem that political time, like other time, should be unrationed, unconditional, but priced. At a modest price, use of the cables would become more responsible than if free, and rationing could be dispensed with. If all users were expected to pay the basic costs of transmission we believe there would be no reason to try to formulate fairness doctrines or to decide who could get on when. The revenue generated would justify expansion of the system till it was indeed copious at that price level. On such a system we would urge that there be no restriction beyond the general laws of libel, obscenity, and copyright.

Some differences from the print media are inherent in the monopolistic character of a cable franchise. There must be a regulated ceiling on rates. It should be a flexible ceiling, so variable rates by time of day can smooth out demand. We would recommend setting rate levels to yield a public utility type of return and letting the market determine how much of each kind of content would be transmitted at that price. At low rates political time may not need to be subsidized.

Privacy

Cable usage raises questions of invasion of privacy. If cable transmits individualized messages, it is not inconceivable that government or the cable operator could learn more than many would think was justified about a citizen's political proclivities. The situation would be particularly troublesome if two-way signals were in use and the citizen were registering, contributing money, responding to a candidate's speech, copy testing, or in some way participating politically through his home set. Presumably, the cable operator could record individual responses in such a way that would be an invasion of the citizen's rights of privacy and of secrecy of the ballot or of political preferences.

For example, a two-way system enables an observer at the head-end terminal to determine what each subscriber is watching, which individual subscriber is responding, and how he responds. The potential for political and for market intelligence is immense. The transmitter could thus identify his interested audience. At the extreme, the potential even exists for a political opponent to block out or deflect certain votes or contributions. To overcome these problems would require strong laws and alert voters or contributors. Technology provides means of scrambling as well as unscrambling, and laws or regulations can prohibit both invasions of privacy and interferences. The dangers are no doubt greater than those now existing when the mailman or the bank clerk or the registrar of voters may learn something of an individual's preferences. These dangers exist and must be faced up to, lest electronic fraud be added to other voter frauds.

CONCLUSIONS

In summary, let us emphasize three features that CATV seems most likely to have and draw four central policy conclusions.

CATV with a multiplicity of channels will be addressing a highly fragmented audience. Much political material will go to small audiences; viewers will receive this material in a form resembling meetings on closed-circuit television, a form very different from today's mass-media programming. Most of this type of cablecasting will be supported by the political groups themselves because CATV can be low in cost, and virtually nobody is going to pay to receive political material.

In this fragmented environment, audiences will have to be organized. Political grass-roots organizations, which have atrophied in the era of mass media campaigning and propaganda, may revive in order to fulfill these new organizational and promotional functions.

American politics could become increasingly localized. Local community organizations and political machines may grow, perhaps at the expense of national ones. Public attention may increasingly focus on local problems that CATV will cover well. Conceivably, it might be harder under such circumstances to mobilize the country on behalf of shared national goals.

To guard against any such tendency to weaken public attention on the operations of the federal government, there should be a federal CATV channel as part of all CATV systems.

Although it has some drawbacks, the best system of CATV control is, in the long run, a contract carrier system. Under any other system the franchise holder acquires undesirable political influence and conflicts of interest.

Under a mixed system, in which the franchise holder earns advertising and pay-television revenue, he might be required to provide free time for politics and public affairs. All charges under a contract carrier system should be regulated to keep them modest. If they are as low as they should be, no special structure of rates is required for politics. Political users could and should pay these low rates.

In a developed contract carrier cable system there need be no rationing of time. Users should have as much time as they choose to buy. Under these conditions, balance and fairness doctrine matters are minimized, if not completely removed.

Notes

1. In November 1968, 44 percent said they would believe television in case of a conflict, and 21 percent said newspapers. In mentioning main sources of news, 59 percent included television; only 49 percent included newspapers.

2. We have recently found an unpublished doctoral disseration, NBC News Division: A Study of the Costs, the Revenues and the Benefits of Broadcast News by Alan Pearce, Indiana University, 1971, which contains many useful figures, generally in line with those offered here. For instance, from his 1970 budget figures NBC news costs without sports or radio or owned stations would come to about $45 million. On the "Nightly News" they made $10 million on a gross of $28 million; on the "Today Show," $10 million on a gross of $15 1/2 million; and on "Meet the Press," $375,000 on a gross of $1 million. The rest of the news programs lost money. Total net revenues (after commissions and station compensation) were probably $40 million in what was a bad year competitively for NBC News.

3. Herbert Goldhamer, ed., The Social Effects of Communication Technology (Santa Monica, Cal.: The Rand Corporation, 1970), p. 14.

4. See, for example, Roscoe L. Barrow, "The Equal Opportunities and Fairness Doctrine in Broadcasting: Pillars in the Forum of Democracy," University of Cincinnati Law Review, Vol. 37, No. 3 (Summer 1968), pp. 447-549; and Voters' Time, Report of the Twentieth Century Fund Commission on Campaign Costs in the Electronic Era (New York: The Twentieth Century Fund, 1969).

5. Robert E. Lane, "Alienation, Protest and Rootless Politics in the Seventies," in Ray Hiebert et al., The Political Image Merchants: Strategies in the New Politics (Washington, D.C.: Acropolis, 1971).

6. On March 31, 1972, rules were adopted for such an allocation in the top 100 markets with a free channel for local government and one for education.

7. See Barrow.

6. PROBLEMS OF COMMUNICATION IN LARGE CITIES   William T. Knox

COMMUNICATION-BASED PROBLEMS IN U.S. CITIES

Life in the typical U.S. city of 1970 is full of problems. Decent housing is scarce, criminals prowl, the air is polluted, streets are dirty, health care is expensive, public and private transportation is erratic and irritating, and a large, frequently unresponsive city bureaucracy steadily grows larger. This list could be doubled easily, but the focus of this report is on the problem of inadequate communication. Communication among people, between people and organizations, and among organizations — much of it is distorted, uneven, or wrongly focused.

Communication affects all the cities' problems, sometimes for better and sometimes for worse. As an example, take the extreme problem of riots. Some urban riots have been provoked by rumors or distortions in communication, other have been aggravated, and still others have been avoided or ameliorated through constructive uses of communication. Lack of simple radio-communication devices among transportation employees recently left New York subway and commuter passengers hurt and suffocating. Attempts to use the telephone system for bus or train information are guaranteed to result in frustration and annoyance. Newspapers not only tell news and goad the city's administration, but they also litter the city's streets. Cheap portable radios are a delight to some people, and have added an immense communication capability, but they are a public nuisance when used indiscriminately in the city.

In an age notable for the diversity, ubiquity, high quality, and inexpensiveness of communication techniques and devices, it is remarkable that communication is still a problem. Yet after every major upset to the city's life there are cries for better communication. An active concern for the capability and use of the total city communication "system" would help greatly in making cities better places in which to live. Since deterioration and distortion of the communication system have aggravated many of the most serious city problems, a proper concern for communication must be an integral part of any attempt to solve these problems. With a communication system connecting people at the neighborhood level and with their diverse pluralistic communication needs satisfied, many social problems would be solved at the neighborhood level and would not fester until city, state, or national action is required.

Constructive action has been impeded largely because we rely too much on mass communication in a pluralistic society. Our pluralism is especially evident in the cities.

Mass communication has the potential for keeping us all informed about some important national and local issues, but it is completely unsatisfactory for handling the numerous small local issues that affect neighborhood groups and individuals. The mass media thrust information at the individual but allow him essentially no access to the media. It is no accident that the young and not-so-young rioters have shouted that their actions were the only way they could communicate, "the only language they will understand." The surprising ignorance of our people about major national issues and personalities testifies that most people prefer to get their own information about matters of personal interest to them, when it is of interest to them, and that they do not turn to the mass media information suppliers.

Constructive action has not been taken on the critical problems of the city because the vital role of communication has been underestimated. People as-

sume that communication will take place, and they are often surprised to find
that it hasn't. The "substantive" problems of criminal justice, garbage pickup, the welfare cases, et cetera, get the attention. The communication system is not recognized as a problem area that is susceptible to study and improvement.

Only in the case of certain telecommunication devices — the telephone, telegraph, and cable TV — has the city asserted its right to regulate the development and exploitation of communication technologies. Even these rights have
been challenged by state and federal authorities. The city must be more its
own master in the use within its boundaries of the vast array of modern communication and information-handling technologies. A stable city society at all
levels and through all devices is critically dependent on an effective, adequate
communication system.

Psychological Aspects

The human need to interact and communicate with other human beings is beyond question. Cities are, in fact, a primary physical means of satisfying this
need. Specialized interest groups can be easily formed, broken, and reformed
out of the hundreds of thousands or millions of people within the city, although
the ease of accomplishing this depends heavily on individual wealth and mobility. Deprived of human interaction and communication, most people undergo
an undesirable change in personality and behavior. Just as people want to communicate their thoughts to others, they also want feedback of information from
others. A friend has "bad manners" if he fails to express thanks for a favor —
he may have been grateful, but he has to express it, and express it properly.

We take pride in the lessened influence of political machines and patronage
in our cities. But there is a pervasive feeling that the present system does not
respond, that it is inaccessible, that its rules and decisions are inflexible,
and that the individual cannot communicate with or act through the system. If
the frustration at not being heard, now affecting a few people, spreads to a
larger minority, the "legitimacy of government" — the willing acceptance of
authority — will be further endangered.

It is an inescapable fact of city life that overcrowding results in almost continuous overload of person-to-person interactions. People act hurriedly and
save their time and energy for events of high priority to themselves. Screening devices, such as secretaries, doormen, and unlisted phone numbers, are
used to block off undesired inputs. The sidewalk drunk's tragic situation is
ignored. Changes in social organizations and functions within cities are made
to relieve the increased stresses on individuals. People in some U.S. cities
now must have exact bus fares; organized social services take care of the
poor, the sick, and the homeless; and people must place zip codes on mail.

These adaptations, necessary as they are, limit the individual's close
personal interaction with minority interests other than his own. The total communication system within the city, therefore, should not only lessen the individual's burden of stress-provoking information, but it should in addition give
him easy accessibility to the subjects and areas of importance to his fellow
city dwellers.

Essential Communication Needs of Individuals

Individual alienation and social deterioration in large cities are both primarily related to communication needs, and neither can be solved by purely economic measures or physical renewal. There are, for many city dwellers, essentially no communications that build or reinforce their own self-esteem. Yet an individual must have a good self-image in order to be a stable, productive member of society. One of the easiest ways to get this is to identify himself with his neighborhood or with other people with similar interests and background—a similar culture. This has never been a problem for those members of the dominant, affluent U.S. groups. Both their face-to-face contacts and the mass media have reinforced their self-esteem.

However, it has been a major problem for members of minority cultures, especially those who have been clustered in cities. The dominant means of communication—the mass media—has until recently shunned minority interests and has held up as the American standard the values and images of the dominant groups. The printed news media—newspapers, mass audience books, and magazines—and the broadcasting news media have generally been aimed at the "average American." City schools have traditionally ignored teaching minority languages, dialects, and cultures. Although several dozen minority-interest radio stations have been established in the last decade, there are only two minority TV stations. A youth absorbed in the rock music of his portable radio as he walks among the symbols and presence of another "establishment" world is shielding the image he has of himself.

In this chapter "minority" is used to mean _any_ minority. In fact, minorities according to interests are more important, more meaningful, than racial minorities. Our pluralism is a pluralism of human interests, and the communication channels must accommodate those interests.

Structures or mechanisms must be available—and everyone must know about them—that offer help to individuals in both crisis and noncrisis situations. The very recent adaptation of the telephone in some cities to accept emergency calls from pay stations without first inserting a dime is a step forward. But the anonymity of the city also makes it possible for people to scream for help and still get murdered, without assistance from nearby people. Then, too, there are many noncrisis situations in which people simply want to know about something for the fun of knowing, or in order to buy or sell something, or to get something repaired, or to improve one's health, or to get a job, or to carry out a transaction. The modern city has a number of such mechanisms (the yellow pages of the telephone directory, tourist centers, municipal clinics, et cetera), but there is no attempt to create an effective _system_ out of this set of distinct and puzzling informational aids. Nor do the city's schools know about or instruct youngsters or adults in the use of these services.

The individual must further have constructive, sympathetic, and meaningful response from the institutions against whom he is thrown. The size and complexity of our modern cities makes it almost impossible to have the personal contact with neighbors, proprietors, servicemen, and city officials that used to exist. Yet there is no excuse for the endless waits, the circular referrals to someone else, the alleged inability to change the system, the "your problem is really unimportant" attitude on the part of many _civil_ servants and company employees.

Finally, the individual needs to be educated and re-educated, trained and retrained throughout his lifetime — an intensive communication process. Cities frequently provide a large array of formal educational services, but it is now clear that these services have for years been inadequate for large numbers of the minority groups within the city. Not only has communication been poorly utilized to provide education per se, but what was actually happening in the city schools — the gap between promise and results — had been imperfectly communicated to those who had the organizational and economic power to change it for the better.

DESIRABLE QUALITIES FOR A COMMUNICATION SYSTEM

With the benefit of years of experience with many kinds of communication devices and mechanisms, and by using some of the new and powerful concepts of system engineering, it should be possible to specify those characteristics of a communication system that are desirable for satisfying human existence. Some of these qualities are outlined, especially as applicable to city living, in the following tenets.

The system should have a network structure that is similar to the nervous system or the telephone network. There should be alternative pathways within one medium and among different media for communicating if one way fails. If there is a newspaper shutdown, the radio and TV stations and the phone network should fill the gap quickly and effectively. Messages moving in one medium should be easily accessible via another medium. Major news is treated this way; other messages are not handled so well. In a recent death-filled fire and riot, the firefighters could not communicate directly via radio with their police coworkers at the scene. Allocation of radio frequencies to different groups had given inadequate priority to the need for systems operation. Present-day communication devices and systems are designed to do one job well, but they were never intended to be parts of an interrelated set of networks.

The communication system should automatically respond to user success or failure in using the system. There should be enough feedback from the system itself to educate the user on how to successfully use the system. Most present systems don't. A phone connection uncompleted is lost effort: "Please dial again." An individual who loses his argument is never told how he could have communicated with greater success. Some computer systems, however, educate the novice in correct computer use. When the computer is confused as to what the user wants it to do, or when there is an obvious error in the user's command, the computer will automatically instruct the user on what to do next to clear up the confusion. When people are involved as elements in the communication process, they should act as cooperatively and responsively as a well-programmed computer.

The system should provide the right amount of communication. This does not mean the most communication. People, groups of people, and organizations have limits as to the amount of information they can utilize in a given length of time. The mass media typically operate continuously, thus saturating the individual's environment with information of all sorts, without giving him adequate means for selecting what he wants. This is not good system design.[1]

The system itself and the content of the messages going over the system should be under the direct influence and control of the users of the system. Under these conditions the users of the system will optimize the allocation of resources among the various parts of the communication network, instead of, as at present, letting the system operators do the optimization. In general, the common carrier telephone network has this desirable quality.

The communication system should provide for resolution of conflict among groups. This is not to say that the system should inhibit conflict, but rather that the system should provide for effective, peaceable resolution of conflicts. It is inherent in the nature of people and their organizations that there will be conflicts. Our democratic society revels in its ability to foster and tolerate diversity of viewpoints, with the knowledge that through such techniques the greatest good is done for the greatest number of people. But there is a real danger that our present communication practices will exacerbate conflict rather than ameliorate it.

The mass media, as well as some of the special-audience media, depending as they do on money for advertising, have an incentive to attract large audiences, and they frequently use stories about conflict to get such audiences. There should be no governmental control over such media practice, although private groups and individuals should be encouraged to express their views on the subject, and the media operators should be urged to develop better measures of their impact on the social environment. The remedy for overdosage on conflict items would be information systems that give individuals or small groups the information and entertainment they want. Face-to-face communication, for example, is a better way both to recognize the conflict potential and to resolve it, or at least to keep it under control, than TV news on-the-spot reporting is.

The communication network should, by its very operations, lead to change in social institutions when change is required. Automatic logging and classifying of messages could provide city executives with information about the most pressing city problems as a prelude to action. Groups in power typically operate within a different set of communication systems and get different messages than do the out-of-power groups. The "intelligentsia" deliberately avoided the use of TV for years after the average city dweller had become a TV addict. With such rapid, ubiquitous communication as we now have, such isolation is dangerous to society.

The system should act to unify rather than to splinter the city. It should utilize the well-known local spots as rallying points for groups of people and should satisfy divergent groups. Museums, taverns, and laundromats are communication devices, and they could serve as rallying points for groups of people if they were deliberately employed in this way.

The communication system should tell what the present state of the city is as well as the state of the system itself. It should be possible for city administrators to find out easily and quickly and inexpensively what they need to know about the current city situation. The system should also have internal monitors that flash telltale warnings when some parts of the system are not functioning well.

The communication system should be able to start small and be capable of growing incrementally. These conditions, fortunately, are met by almost all of our communication subsystems, even the very large ones now operating.

The communication system should have sufficient alternatives so that the failure of one part can be counterbalanced by extra use of some other part.

A good communication system will transmit affective as well as cognitive information. TV is superlative here: You "get the message" when you see and hear the confrontations that appear to have become a part of our communication system.

People responsible for the effective operation of the communication system should consider on a cost/effectiveness basis trade-offs among the communication system, the transportation system, and physical facilities designed for communication purposes. For example, would it be better (that is, more effective, and perhaps cheaper) to have most meetings held over the conference Picturephone? Do people have to be brought into close physical proximity in order to transact most of the business and the services that take place within today's city? What are the relative merits of building a new museum or library, compared to giving each of the residents in a specific area his own set of colored slides, audio tapes, and microfilmed books?

EVALUATION OF PRESENT SYSTEMS
The main reasons that television suffers follow:

There are not enough channels, even in our very largest cities, for the future demands that are expected (in spite of over 100 presently unused ETV channels).

Diverse groups cannot get access to the TV system.

Even when access is possible, it must be on the system operator's terms.

The radio system is considerably better than TV, primarily because there are far more radio channels available. Within the last ten to fifteen years, a number of minority group radio stations have begun to flourish in the very largest cities. However, the systems operator still controls the content of the messages and the terms under which diverse groups gain access to the radio system.

The telephone network is probably the closest approximation to an ideal communication system that we now have in cities. However, there are many informational and action resources within a typical city that most people do not know about and with which the telephone system does not easily connect one. Although telephone charges are relatively modest (about the price of two or three movies per month), credit restrictions keep telephones out of precisely those areas that are in need of access to information and action sources. The cities' poor normally have no phones.

The press has traditionally suffered from an inability to provide adequate coverage of information that is of interest to small groups. It is a typical mass medium. Modern techniques of photocomposition, offset reproduction, and other print technologies have greatly lowered the production cost of printed material. However, these technologies have not created a low-cost means for distributing messages to small, widely scattered groups. It is time to consider distribution schemes other than the traditional individually delivered copy.

Film is a medium especially appealing to the young and to the disaffected in our cities. New film technologies — for example, screens that are usable in bright sunlight and automatic cameras — have swung wide a door for expression and communication. Parks and empty city lots can supplement the theater and the auditorium for film showings. And these new possibilities offer an outlet for neighborhood interest and amateur films. Such productions would help people come to grips with the real world, to an understanding of societal norms, and they would offer them a new channel for communication.

The present face-to-face communication mechanisms also generally provide inadequate communication. This applies to churches, schools, town meetings, professional societies, unions, civic organizations, fraternal clubs, and other types of groups. Part of the increased difficulties are due to the increased demands of an increasingly complex and diversified society. Perhaps the individual is already suffering an information overload and really doesn't care about additional communication with or from these organizations. However, it is equally possible that those in control of these organizations are inadequately educated to recognize poor communication when it exists and are equally inadequately trained to do something about it even if it is recognized. The ward boss has disappeared, and the organizations mentioned earlier have not filled his role as a personal intermediary with "the establishment" or as a source of information and help.

Another major city communication system consists of libraries and museums. They operate more or less unrelated to the rest of the overall communication system. They employ, in general, late 19th-century technology, and they are moving to automate this antiquated technology rather than to devise and utilize completely new technologies of communication. Even the radio guide or pretaped telemessage in museums, helpful as it is, does not allow the user to ask questions.

A very recent study on public libraries indicates that their administrators are concerned about the failure of libraries to adapt to today's needs. They are concerned about the possibility that other institutions may be created to replace libraries, but they are, nevertheless, uncertain how to make the change. The library system is almost a classic case of a system without minimal feedback. Library administrators insist that library schools produce people with technical skills that are adequate for present functions. People have very little chance to tell administrators what they really want out of a library, and library schools thus have slight chance of producing the right kinds of library managers. However, providing an information service for the community should be the major public library function, and all sources — not only to traditional library schools — should be drawn on for the necessary competence.

Museums are useful means for establishing an individual's self-image and the common culture of a subgroup. However, the traditional, big museum is obsolete for serving this need. People must be convinced that what they see in museums is a part of their world — past or present. What is required is a system of museums placed in civic building lobbies and in natural group-meeting places such as laundromats. Some experiments along this line have been extremely successful.

Public agencies normally consider providing an information service to the public as a low-priority service. The self-image of city employees is mostly tied to their professional function — as policemen, accountants, or sanitation workers — rather than to their role as communicators with the public. Yet they are the people seen, heard, and talked to by the public — not the personnel of some centralized public information agency. All city employees should understand the importance of their role as communication links between the city administration and the public.

TOWARD BETTER CITY COMMUNICATION
Better communication will begin to function within the city when the knowledge is fully appreciated that our pluralistic city society needs a pluralistic communication system. This system must be one that is composed of many different kinds of communication devices, mechanisms, and techniques, and it must be orchestrated to serve the ultimate user in a systematic way.

Better communication will begin to come when the process and function of communication is specifically recognized and funded as an integral part of every new solution formulated to ease the city's social problems.

Better communication will come when people are specifically trained to enhance their skills in communication — not only in sending messages but in receiving them, not only in the printed-word medium but in all information media. A society so critically dependent on effective communication as ours is can no longer sidestep the development of communication skills in all of its citizens.

New technologies for communication offer much promise. There is a much-talked-about but not-yet-attained technical feasibility that each home and apartment in the city will become an information terminal — one capable of receiving text, pictures, and sound on a TV screen and speakers; making cheap copies of the TV images; requesting specific information or programs via teletype or phone; interacting with a computer via electronic pencil or tablet, with picture playback from home scanners; and capable of handling audio and video tape records. Further, these individual terminals are to be linked via multichannel cable or electromagnetic waves into networks, with flexibility to change the size and the membership of the networks.

The 2000-odd CATV networks are the forerunners of these electronic information networks. There has been much speculation about the uses to which such networks can be put, and a major foundation has recently funded yet another large study of cable communication. The Federal Communications Commission, the courts, and the Congress have been heavily involved with legal and policy issues in CATV. Much pro and con publicity was given to New York City's recent study and award of 20-year franchises for CATV in segments of the city. There is, in summary, little that has not already been said, printed, and reported over TV and radio about CATV.

Since cable communication will dominate much of our total national communication in the coming decades, such systems must be regulated and used to serve the pluralistic needs of our pluralistic society. They must be simple extensions of the broadcast mass media.

Cable communication systems should be installed with enough flexibility to make, break, and remake different groupings of terminals into subsystems. It

is especially important that these systems make possible the grouping into
subsystems of all city residents with like interests or culture; ethnic racial,
and other culture-dependent groups should not be divided by franchise boundaries. Police and other municipal-employee groups should also be aggregable
by these systems.

CATV systems should operate perhaps as many as half their channels as
public service common carriers, open to all groups and individuals. As Publishers Weekly stated recently, "Denial of access to (communication) technology is a most insidious kind of censorship." Requests for access should be
in terms of audience size and interest, and, on the same basis, several channels should be available on a first come-first served basis. It becomes vital
to poll audience response quickly and quantitatively.

Two-way communication, providing transmission to as well as from the
central station, is the long-term, essential need for full development of this
communication system. Systems given franchises today should be required to
have enough flexibility to permit eventual conversion to two-way systems.

Copyright laws and regulations that make it easy for cable operators to obtain distant signals for local retransmission will aid the expansion of cable
systems. The more program variety offered, the more likely it is that people
will subscribe to the cable system, and the faster these systems will expand.

Technical standards are a crucial issue. Inadequate or too-costly-to-
change systems would preclude many of the new communication services that
offer major improvements in the quality of urban life. At the same time, in-
sistence on uniformly high technical standards will limit the ability of community groups or individuals to send amateur messages through the system.

In the meantime, there are many things that can be done with existing know-
how and with devices that will help immediately. Without any pretense at priorities or completeness, and with compliments to those who have pioneered in
the application of present communication knowledge, the following ideas are
suggested as worthy of consideration and action by city governments and by
private firms in the communication business.

The telephone is an indispensable instrument for communication and should
be more available to the cities' poor. Commonly, large deposits are required
for phone service in poor areas. This is because on the average an excessive
number of users in poor areas fail to pay phone bills. The phone system itself, coupled with computer technology, should be able to treat customers as
individual credit risks, thus overcoming the barrier of the large deposit for
some city residents.

An information and call-routing service should be an especially fruitful way
of using the phone network. Many people are unsure about whom or what to call;
the 911 policy emergency number in New York is used for many nonemergency
calls. Calling without getting results is time-consuming, frustrating, and
costly (from pay phones). A network of information services going beyond the
separate "Dial-a-_____" services should be constructed in a city to involve

a wide range of public and private organizations, interested people, and groups of all sorts. The incoming call would be listened to until the need became clear, then switched to another information source until a responsive reply could be given. The call would be routed for the caller, and the router would explain the need to the next-in-line. The service should be available from all phones without payment. City-government services that are set up to serve the citizen should be the first group of organizations to be placed in the network. Citizens' questions about health care may be a top-priority category to be provided for.

Greater use of good, recorded phone announcements and phone-message receiving units should be encouraged. No one minds a recorded announcement saying (and repeating periodically) that the reservation, order, or information clerks are busy but will accept one's phone call shortly, or that one's message has been received and will be answered at the earliest possible moment. More widespread use of these devices would relieve much tension and frustration.

Consumers need to know what a product is really like, what it will and will not do, and what it requires from the user in sophisticated use and maintenance. Although neighbors, acquaintances, and trustworthy dealers supply much of this information, there is much more available from the manufacturer and from independent testing organizations — including government agencies. Such information should be easily accessible to the public; a phone call should suffice. Governmental agencies may possess the statutory authority to create such an information service.

Consumers also need to know where a product can be bought for the lowest price and about dealers of good reputation. Present media (newspapers, word-of-mouth, and comparative shopping) furnish cost information fairly well. Dealer reputations are probably to be found in the combined files of the Better Business Bureau, governmental bureaus of Consumer Protection, and the Neighborhood Legal Services Office. Computer technology would make access to such information easy; legal problems associated with public release of such information would be more difficult to handle but could be resolved.

Consumers want to know how to use government agencies — established for their benefit, presumably. Similarly, consumers want a direct, open channel into a business firm so that they can directly express their reactions to that firm's products, practices, plans, et cetera.

Consumers, finally, want and need to know about remedies that are available to the victim of illegal practices. While there are still problems with the legislatures and the courts to provide adequate consumer weapons, the major deficiency is inadequate knowledge on the consumer's part about the many present remedies and helpful agencies. Should not information about these sources be included in every phone directory, every city tax bill, and be distributed at neighborhood gathering spots?

Small, frequently-changing neighborhood libraries and museums, building on local resources and interest, are more important at this time in promoting individual self-esteem and group self-esteem among minority groups than growth of large central facilities. Using normal community gathering places as sites for libraries and museums rather than building and manning separate, distinct structures has much to recommend it. Rented or donated space in taverns, churches, office buildings, and transportation terminals and stations are possibilities.

Libraries should also work toward becoming more of an information center for the community; worth considering is a free phone in each neighborhood library, connected directly to a central information service. Also worth considering is having the library act as a barter exchange for books, et cetera, that are owned by local residents.

The increasing availability of relatively low-cost media allows small communities and "subcultures" to set up their own communication channels. The "underground press," for example, is possible because of new, cheap methods for page composition and offset printing. More community papers should be encouraged. Other technologies are also proliferating — cassettes for audio and video use, Polaroid and Super-8 cameras and film, and microphotography, for examples. A portable, rugged TV camera and tape deck, operated by one person, is now cheap enough (ca. $1500) for many neighborhood and like-interest groups to produce material for broadcast or cable transmission — or for small group viewings. The widespread exploitation of this new technology could do much to create group self-esteem and tolerance for other cultures. At present, such devices do not meet FCC technical standards for broadcasting. The city administration has, however, the power to open up cable channels for such amateur productions.

The success of the TV show "The Opportunity Line" in matching job-hunters with employers is encouraging. Cities not now exploiting TV for this purpose — utilizing prime TV time — may find this investment well worthwhile. The operation should, of course, be coupled with the local government and private employment organizations. The power of computer technology to match job requirements with applicants' interests and expertise should also be utilized.

Allied with the employment situation is that increasingly-common city problem of getting prompt, reliable, economical repairs and service for oneself and for one's property. Modern communication and information-handling technologies have the potential for creating a community fast-response service system, matching servicemen wanting jobs with those needing service on a "real-time" basis. The scale of operations might be large enough to increase productivity and lower prices. Faster service would also result, with a resultant minimizing of travel for the serviceman and waiting time for the citizen.

The consumer also needs to know how well service suppliers have served customers in the past. A computer-based registry of servicemen coupled with a consumer-marked update card after each service call — with the summary results available from a central bureau by telephone inquiry — would offer, with all of its possible defects, more information for consumer choice than at present.

In the realm of education, communication failures are perhaps more noticeable than in most other areas. Instructional materials other than the on-site textbooks are rare in many city public schools. The enhanced learning that is associated with films, monographs and fiction, photographs, games, and maps is difficult to achieve; the materials are usually not available locally and take months or years to get via the central-purchasing system. More locally accessible instructional materials are an easy application of communication technologies. Establishing a computer-managed inventory and scheduling of this material, coupled with phone inquiry and response, is a logical, proven application of computer and technologies communication. Flexibility and efficiency could both be achieved.

More difficult will be the inversion of the present top-down system of communication to the advanced student to the necessary bottom-up system, wherein the student (with no more guidance than necessary) selects much of the curriculum and instructional materials that he uses. Modern communication and information-handling technologies make the individualized, bottom-up system technically feasible for large numbers of students with widely varying interests and abilities.

Note

1. In another article, "The Pathology of Information," Book Production Industry (June 1971), pp. 43-46, William Knox has drawn some drastic and controversial conclusions on this point. The abundance of information being disseminated, he argues, overloads the human and social processing mechanisms. "Our laws and adult practices are also based on the twin assumptions that (1) there is a scarcity of information and of communication mechanisms, and (2) a good society results from a maximum flow of information of all kinds. The result is public subsidy of the communication pushers via free grants of the limited radio and television spectrum, franchised rights-of-way for communication lines, and below-cost postal service for magazines and newspapers.... We have not changed our attitudes, our practices, and our laws to conform to the new environment — to an over-abundance of information and undreamed-of capabilities for communication.... Laws framed to promote maximum communication in a time of information scarcity must be reformulated to recognize the individual's plight of overload, and to give him greater control over the communication environment in which he must live."

Public safety figures prominently in speculation about the future of cable communications. Among the predicted services cited by the Federal Communications Commission are "municipal surveillance of public areas for protection against crime, fire detection, control of air pollution and traffic." New York Mayor John Lindsay's Advisory Task Force on CATV and Telecommunications declared that the medium would "permit our municipal agencies to scan hundreds of public areas for protection against crime and violence, for the detection of fire, for the control of fire, for the control of air pollution... ."

The Ad Hoc Committee of the Industrial Electronics Division predicted annual savings of $3 billion in the area of police protection and $1 billion in fire protection through the application of broadband communications technology.

In Liberal, Kansas, the local cable company has designated Channel 3 of its 12-channel system an "Emergency Alert" channel. The company installed a special line from the head end to the police chief's office, which enables the chief to originate programs from his office. The office is equipped with a microphone and camera supplied by the cable franchisee.

Channel 3 ordinarily broadcasts a map of the area to the accompaniment of background music. In the event of the existence of tornado, dangerous road conditions, or other emergency, the system can broadcast information about the emergency by audio and video on all channels. The usual procedure for routine announcements is to employ a tone generator to signal all channels. A brief announcement is made, and viewers are advised to turn to Channel 3 for fuller details.

A favorite use of "Emergency Alert" in this city of 14,000 is the locating of lost and runaway children. Police have on hand the photographs of youngsters furnished by the schools. The photograph is broadcast when a parent reports a child missing.

Prior to 1965, all fire alarm calls in Weston, West Virginia, a community of 10,000 were routed through the telephone operator, who notified the fire station. A siren was sounded to summon the city's volunteer fireman, who make up the bulk of the department. The volunteers responded to the siren either by calling the operator to find the location of the fire or by rushing to the fire station. In response to a request from the fire department to develop a more efficient notification system, the local CATV company devised a closed circuit public address system utilizing the company's cable network.

Fire alarm calls are now routed directly from the fire station to the head end by telephone line. The calls are converted at the head end to an FM frequency and placed on the cable, preceded by a ringing sound to alert listeners. All 60 of the fire volunteers were provided FM radios at cost by the cable company management. By attaching the radios to the cable and keeping them continuously tuned to the designated FM frequency at their homes and business offices, the volunteers hear each alarm as the fire is actually reported. The firemen note the reported address and proceed directly to the fire scene.

Allband Cablevision of Olean, New York, a division of TeleVision Communications Corp., initiated a closed-circuit street surveillance system in downtown Olean, a city of 22,000, in September 1968. The system consisted of eight cameras mounted on city light poles. The cameras covered 75 percent of the city's busiest street and were tied by cable to monitors in the police station. Police at headquarters were able to pan the cameras right and left, move them up and down, zoom and adjust for focus and exposure.

Allband Cablevision installed "virgin" cable for the street surveillance operation. Michael Arnold, manager of the company, said he chose separate cable because the company did not have cable in the downtown area to serve subscribers. Mr. Arnold maintains that if the company's regular cable had been laid in the area, he would have serviced the surveillance operation as part of his CATV system by placing the pictures generated by the street cameras in the "midband" of his 12-channel system. In most systems, this would require two-way capability—a means of moving the picture back over the cable to the head end for retransmission. Mr. Arnold believes local CATV companies are logical installers of surveillance systems even when the pictures are not carried on their CATV cable because of their cable know-how and ability to provide service.

The Olean system was developed by TeleVigil Systems, Inc., a special division of TeleVision Communications. The company invested approximately $250,000 in the project to demonstrate the feasibility of TV surveillance. Mr. Arnold estimates the same system could now be installed for $30,000-$40,000.

No charge was made the city of Olean for the installation. The arrangement with the city called for a five-year contract at $6,500 a year if and when the city agreed to assume responsibility for the operation. A new contract was to be renegotiated at the end of the five-year period.

The system was initiated in September 1968 and terminated in December 1969. Allband Cablevision dismantled it following criticism of the surveillance network during the municipal elections the preceding fall. William O. Smith won election as mayor after describing the system as an "eye in the sky" and charging it smacked of invasion of privacy. Mr. Smith also contended it would be too expensive and that it did not function effectively at night. A resolution was introduced in the city council calling for removal of the cameras, but the system was taken out without formal council action.

The expense charge referred to the possible ultimate costs to the city. No city money actually was expended for the experimental program. The lack of effective night pictures was conceded, but the introduction of new low-light-level cameras was expected to provide 24-hour surveillance.

Olean's 16-month experience with video street surveillance presents a mixed picture. The police chief believes the presence of the cameras reduced burglaries and freed manpower for other duties. The cameras reportedly were useful in calling police attention to traffic jams, a malfunctioning railroad crossing gate, panhandlers, and congregating youngsters. Mr. Arnold attributes the hostility of Mayor-elect Smith to Mr. Smith's employment by the local telephone company. But concern about the privacy aspects of TV surveillance, its effectiveness and ultimate costs, evidently was substantial enough for Mr. Smith to campaign on the issue and to succeed in removing the cameras.

Less publicized was a burglar alarm surveillance system installed as a part of the street surveillance setup in Olean. Commercial customers were offered the opportunity for cameras to be placed on their premises and connected by the same cable to police headquarters. The bank alarm consisted of several devices in the bank for manually signaling police headquarters. When an employee activated one of the devices, a bulb lit in headquarters. Police responded by switching on the TV camera, observing the interior of the bank on a TV monitor at headquarters and making a videotape of the picture. No breakins or holdups occurred during the time the surveillance alarm was in operation, but one alarm was inadvertently signaled. Police turned on the camera, saw nothing was happening and were spared responding to a false alarm.

In the process of installation is a burglar alarm-fire alarm-"panic" button system designed by the Advanced Research Corp. of Atlanta, Georgia. The corporation reports that a cable TV company in Pensacola, Florida, has contracted with it for a pilot installation in 250 subscriber homes.

"Our idea," states the company, "is to provide a relatively low cost central station alarm monitoring system that could be applied on a mass basis throughout the community." The company's system consists of sensing devices in homes and businesses linked to transponders which are continuously monitored from a central station. Two-way communication to operate the system is provided by the CATV cable and a second cable stranded along with the main cable. The company cites as the principal advantages of multiple cable low cost and reliability. It asserts that a relatively small extra cable can provide about 1,000 channels for data transmission for a sizable community without the use of repeater amplifiers.

The modest applications of cable technology for public safety purposes make it evident that considerable innovating and technological development will be necessary for some of the glowing predictions to be realized. When the Electronic Industries Association was asked for specific detail on its projected annual savings of $4 billion for police and fire protection services, John Sodolski, staff vice president of the Industrial Electronics Division, replied, "Those projections were a consensus of the best judgments of a number of men in our industry. There is no documentation behind these numbers... ."

Television surveillance, one of the most often cited public safety applications of CATV, provides a good illustration of the uncertainties. This application usually is described in terms of surveillance of high-crime neighborhoods. These neighborhoods almost always are in low-income, ghetto areas. Would residents of these areas tolerate the close police monitoring of their activities associated with electronic surveillance?

The President's Commission on Law Enforcement and Administration of Justice noted that most crimes against the person occur on the streets or in other public premises. Heavy police patrolling appears to be associated with a decline in crimes against citizens walking the streets. Street surveillance by camera can be regarded as nothing more than the presence of an officer at a fixed point 24 hours a day. But many persons are likely to resent intensive police surveillance, human or mechanical. While the President's Commission conceded that massive police presence probably would reduce the incidence of street crime, it concluded that "few Americans would tolerate living under police scrutiny that intense." Roger Reinke, assistant director of the Professional Standards Division of the International Association of Chiefs of Police, believes citizen concern over the "Big Brother" aspects of TV surveillance will prevent its acceptance.

Some law enforcement specialists question the utility of street TV surveil-

lance on strictly law enforcement grounds. Edwin Shriver, police programs
specialist in the U. S. Justice Department's Law Enforcement Assistance Administration, regards TV as having "very limited application" chiefly because
of its lack of mobility. He notes that a suspect need only turn a corner or duck
into an alley to be out of camera range. He considers street surveillance to be
of minimal value to business establishments because most attacks against
businesses occur at side or rear exits. He believes, however, that video could
be useful in observing traffic, especially on bridges, in tunnels and other critical places, and when employed flexibly, for investigative surveillance of particular dwellings or locations.

The New York City Police Department uses television for surveillance both
by helicopter and on the ground at several fixed points —the United Nations,
City Hall, Bryant Park, and outside the Court House during a controversial
trial of Black Panthers. The helicopter TV pictures are relayed by microwave. The other video pictures are transmitted by coaxial cable furnished by
the telephone company. Several of the cameras are manned by police, others
are operated by remote means from police headquarters. The sites of the
fixed cameras were chosen primarily because they are areas where large
numbers of persons congregate and are potential trouble spots.

New York police now must wait several days for the telephone company to
lay cable whenever police desire surveillance at a particular site. Deputy
Chief Inspector William J. Kanz, one of the New York department's top communications men, visualizes tapping into the cable system at will wherever
CATV cable is located in the city.

The St. Louis Police Department is one of the few to make a major commitment to television. The department operates its own television station on an
Instructional Television Fixed Service Channel, an over-the-air channel ordinarily reserved for educational institutions. The daily lineup of arrested
persons is videotaped and broadcast to all officers at rollcall. Videotape is
also used to record crime scenes for rebroadcast. Officers gathered at the
district stations for rollcall are provided up-to-date video information on
wanted subjects, missing persons, stolen property, runaways, labor strikes,
and community unrest situations. The department has used TV against coun-
terfeiting by videotaping blowups of circulating bills and instructions on how
to spot them and to apprehend suspects.

A major use of the St. Louis police channel is for training films and other
in-service training programs. Taping of lectures makes it possible for experts to instruct officers on all three shifts through a single lecture session.
In the case of live lectures, a talk-back device in each district close to the
TV receiver permits officers to ask questions and make comments. Approximately 200 St. Louis officers attend classes at three junior college campuses.
Identical classes are scheduled during day and evening hours to mesh with the
shifts worked by officers. The department plans to eliminate the duplication
by taping class sessions and telecasting them immediately preceding or following tours of duty.

The St. Louis channel operates on a frequency that cannot be received on
the ordinary home receiver. The ability of "ham TV" operators to intercept
programs limits the use of certain wanted persons and other information sent
over the channel. Scott Hovey, the department's consultant on communications,

believes that coaxial cable would provide greater security while serving other needs of the department in addition to performing the same tasks as the existing over-the-air channel.

Among the extra advantages of cable cited by Mr. Hovey are the high-speed transmission of computerized crime information in digital form and the transmission in image form of police reports that are too lengthy and costly to digitalize. An example of the latter is a police theft report. These reports are now handwritten, reproduced, and sent within the department by mail. Mr. Hovey is confident that video pictures made into hard copy will become the accepted method for sending such reports.

He believes that by going to cable, the St. Louis department could have attached street call boxes to the cable at less cost than it now takes to lease telephone lines for the boxes. The cable could at the same time provide police with multiple video channels and high-speed facsimile capability.

New York's police department also has access to over-the-air television in the form of Channel 31, the city-owned and -operated television station. It is possible to achieve privacy by scrambling. Police use of the channel is restricted to certain time slots, and the channel is regarded as having limited value.

Chief Kanz is an enthusiastic supporter of cable as more versatile and useful than over-the-air TV. He is confident that once the city becomes wired by the CATV companies, it will be cheaper and more advantageous to tap into the systems for necessary surveillance than to rely on telephone company-laid cable.

New York police ran an experiment for one year in transmitting fingerprints on TV cable and having hard copies of the prints made from the video tube at the receiving end. The recipient office then returned criminal records by facsimile. There were problems of picture resolution and machine reliability in the present generation of equipment, but the concept seemed good. Chief Kanz believes the same operation could be performed on CATV, provided there is suitable security and that utilizing the cable for facsimile reproduction would result in more rapid transmission.

Chief Kanz agrees with Scott Hovey's assessment of the potential value of cable for law enforcement. It is noteworthy that both men, serving departments that have sophisticated communications networks and access to over-the-air television, regard cable as an important public safety resource.

What may be good for New York and St. Louis law enforcement agencies is not necessarily beneficial for other areas. The nation's law enforcement system is fragmented into upwards of 30,000 police forces, more than 80 percent of them with fewer than ten full-time officers. Police departments range in size from a single man operating out of a patrol car to 30,000 men directed from a central headquarters and numerous neighborhood precincts. Community public safety problems cover a similar wide range. Any discussion of ways cable communications may aid public safety necessarily will have greater relevance to some communities than others.

The London Metropolitan Police Force has long recognized the usefulness of television by employing TV to provide the public with advice on crime prevention, to solicit help in obtaining information about specific cases, to present the problems with which police are faced, and to secure public cooperation. CATV's potential for directing programs to particular neighborhoods and audiences makes the public information role of CATV of special significance. Regular or special programs could be employed to inform a neighborhood or community about crime reports, to warn of patterns of criminal conduct, to instruct in crime prevention, and to publicize pictures of wanted persons. The Liberal, Kansas, system of broadcasting emergency notices by audio and video an all channels could be adapted to make possible extremely rapid dissemination of information about wanted persons, vehicle license numbers, and related data to aid apprehension of suspects.

The growing use of motion picture cameras to record holdups and police use of videotape could facilitate identification, especially when the films and tapes are widely broadcast. A recent study of the efficacy of various means of identification concluded that still photographs are less reliable than pictures portraying a moving subject from various angles. The study showed that "video tapes and color photography are statistically superior to black-and-white photography in facilitating the identification of suspects. ... There is reason to believe that the video medium is superior to color photography as well as to black-and-white photography."

The most potent police use of CATV may well be as a medium for improving police-community understanding. As the Task Force Report on the Police of the President's Commission on Law Enforcement and Administration of Justice warned:

"Police-community relationships have a direct bearing on the character of life in our cities, and on the community's ability to maintain stability and to solve its problems. ... The police department's capacity to deal with crime depends to a large extent upon its relationship with the citizenry. ... No lasting improvement in law enforcement is likely in this country unless police-community relations are substantially improved."

The commission's studies showed "serious problems of Negro hostility to the police in virtually all medium and large cities." The Task Force concluded that police departments "must become increasingly aware that isolation from the neighborhoods they protect can interfere with good policing as well as good police-community relations."

The Task Force endorsed formation of citizen adivsory committees on city-wide, neighborhood, and precinct levels to conduct police-community dialogues, air grievances, elicit citizen views of police practices, dispel rumors, and explain police procedure. Formation of ad hoc committees representing specific minority groups was also urged. Such committees would have value chiefly for the relatively few persons who participated in the sessions. Televising of citizen advisory committee meetings, including the beaming of neighborhood and precinct committee sessions to the affected localities, could substantially enhance their effectiveness.

The Task Force observed that the "modern urban police department needs closer citizen contacts to maximize its integration into neighborhood life." The neighborhood meeting, coupled with citizen participation through telephoned comments and questions, broadcast throughout the neighborhood offers a partial means of overcoming the isolation of citizens from police. Videotaping and rebroadcasting of pertinent parts to police could be helpful in making police aware of the nature of citizen complaints.

A prime purpose of police-community relations programs is to alert police to tension-breeding situations. CATV could be an important resource for police-community relations units to defuse explosive local conditions.

Citizen negligence is a major contributing factor to crime. In the District of Columbia, 20 percent of break-ins were through unlocked windows or doors. Forty-two percent of all stolen cars in the United States had the ignition unlocked or the keys visible.

Insufficient lighting, inadequate locks, and breakable windows are associated with a high percentage of business burglaries.

In Des Moines, Iowa, police have conducted a crime prevention course to educate businessmen to cope with robberies, larcenies, and bad checks. In Oakland, California, police have distributed weekly bulletins, including information and pictures of bad-check artists. These and other crime prevention measures readily lend themselves to video presentation. In the words of the President's Commission on Law Enforcement and Administration of Justice's Task Force on Police, "Public education to alert citizen and businessmen on how to avoid becoming victims of crime can be a valuable adjunct to a crime control program."

The National Laboratory of Urban Communications, an applicant at one time for a CATV franchise in Kansas City, Missouri, proposed broad-scale use of cable broadcasting for public safety purposes. Its plan included programs to publicize the "Crime Alert" telephone number, to inform viewers of the advantages of anonymous crime reporting, to provide a rumor control service, to further police recruitment, to report the disposition of complaints against police, and to serve as a "community forum."

Police in several cities attempt to alert merchants to crime situations through pyramid telephone warning systems. Upon receiving reports of check fraud, shoplifting, confidence game, or other offenses likely to occur in series, police telephone liquor stores, grocery chains, gas stations, clothing stores, and other appropriate businesses. These merchants in turn telephone others, passing along descriptions when available. In some places merchants have set up alarms to notify businesses within a radius of several blocks. The frequency with which certain crimes are committed in series has resulted in the apprehension of appreciable numbers of suspects through merchant warning systems.

The closed-circuit public address system in Weston, West Virginia, is readily suited to merchant alerts. Merchants need only have an FM radio attached to the cable and tuned to a prescribed frequency to receive all crime warning information over the network. The information could be sent over the closed-circuit channel directly by the merchants who call in the reports, in the same way that fire alarms are broadcast directly in Weston, or be relayed by police who receive the calls and choose appropriate reports to be put out over the merchant alert channel. In either case, use of the FM system would be more rapid and save police time. It would avoid the necessity of making multiple telephone calls and the risk of encountering busy signals.

Coaxial cable is richly endowed with FM channels. The cable is capable of accommodating about 100 separate FM broadcasting frequencies per video channel. Perhaps as many as 50 can be received without interference. It is possible for external devices to assure voice security and to enable signals to be received while the FM set is off to advise the owner to tune in for a message. (In Weston, the volunteer firemen, utility companies, and other FM users on the cable keep their sets on around the clock.)

The multiplicity of FM channels on the cable makes it useful for maintaining contact with auxiliary police, off-duty fire and police, snow removal crews, and other emergency service personnel. Many police departments are unable to contact off-duty officers except by telephoning them individually. A closed-circuit public address system could be used to summon them for emergencies and for interdepartmental announcements. A department could broadcast its police radio calls on the FM channel for off-duty officers to keep abreast of situations in their neighborhoods and to respond to them. This would be in keeping with the practice in several cities of increasing police presence by equipping off-duty police officers with police cars.

High police visibility is a deterrent to crime. When New York City assigned uniformed patrolmen to every train during the late night hours, crime in the subways dropped 36 percent. The President's Commission on Law Enforcement and Administration of Justice concluded, "Large numbers of visible policemen are needed on the streets." Yet many police man-hours are spent not on patrol but in processing prisoners and in court to testify.

A substantial portion of the time spent in court is in connection with speeding and other moving motor vehicle violation cases. The large number of motorists who ignore or neglect to respond to traffic tickets on the specified dates while police are on hand to testify frequently results in the police appearance being a total waste of police time.

The New York City Budget Bureau has estimated that it takes 9 1/2 hours of a police officer's time to process a prisoner through the arraignment stage each time an arrrest is made on a felony or serious misdemeanor charge. The Budget Bureau has proposed employing closed-circuit television to demonstrate the feasibility of speeding the process and eliminating unnecessary detention.

The proposed demonstration would link a designated precinct house and the Criminal Court by TV cable. The cable would provide the means for conferring to draft the complaint, to conduct the interview to determine eligibility for release without bail, and for arraigning the accused. Fingerprints would be transmitted to the Identification Bureau via photographic facsimile or television, and mug shots would be reproduced from the arrestee's television image. The image would be transmitted to the Identification Bureau and automatically converted to hard copy. The Budget Bureau proposes moving gradually from arraignment via TV to use of the system for transmitting direct police testimony in nonjail misdemeanor and traffic cases.

Authorities in Dade County, Florida, have applied for a federal grant to establish closed-circuit TV among four to six precinct stations and magistrate chambers at the Dade County jail. The system would be used to conduct pretrial release interviews, arraignments, and preliminary hearings. The object of the proposed Florida program is to weed out an estimated 18 percent of "bad cases" promptly and to save police time.

New York Budget Bureau officials believe that CATV could be the means for expanding use of the system if the proposed single-precinct demonstration showed expansion to be warranted.

In the case of police testimony in traffic cases, CATV origination points in precinct stations, firehouses, and other places convenient to police on the beat could make it possible for police to be notified by radio to proceed to the nearest point to testify via cable without the necessity of time-consuming waiting and trips to court. Such a procedure would necessitate two-way audio and video communication. It also would be necessary to establish that testimony by television satisfies the constitutional right of confrontation by one's accusers and does not adversely affect cross-examination.

Facsimile reproduction of legal documents by cable provides another method of maximizing police time on the beat and speeding the apprehension of suspects. Search warrants and arrest warrants could be reproduced and picked up for execution at station houses and other fixed points. It may be possible for an officer to "appear" before a magistrate by cable TV, present his warrant request, and receive the warrant and conduct the search or make an arrest in a matter of minutes.

Rapid arrival of police at the scene of a crime correlates closely with likelihood of arrest. A study of emergency calls in Los Angeles found that the time between receipt of a call and police arrival averaged 6.3 minutes in offenses subsequently not cleared by arrest. In cases where police were able to make an arrest, the average response time was only 4.1 minutes. The arrest rate was 62 percent when arrival time was one minute, 57 percent for two minutes and 49 percent when the delay was six minutes. Almost 36 percent of all arrests were made within a half-hour of the commission of the offense.

A call from a citizen usually initiates police action. The more rapidly a call for aid can be made, the smaller the time lag in police arrival and the greater the likelihood of apprehension. The Commission on Law Enforcement and Administration of Justice's Task Force on Science and Technology cited widespread availability of street alarms as one method of enabling citizens to summon aid quickly. The Task Force noted that if fixed alarms were located 40 to the mile in high-crime areas, the maximum distance to the nearest alarm would be 1/80th of a mile. It would take no more than 11 seconds to cover this distance even if it was walked at a pace of only four miles an hour.

New York City is now in the process of facilitating street emergency calls by converting street fire signal boxes for use by citizens to summon fire and police aid by voice communication. In cities without existing signal or voice boxes, it may be more practical to construct a street emergency call box system using the CATV's coaxial cable. The Advanced Research Corp. of Atlanta, Georgia, reports it has developed a voice call box system geared to operate in neighborhoods served by CATV, provided a separate cable is laid for two-way communication. A study by the company for the city of La Habra, California, concluded it would be less costly to install its system for street fire alarms than to expand the existing system, which uses telephone wire pairs. The company states the communications capacity of the cable would enable the call box to be used for a variety of purposes, including tie-ins with numerous fire and burglar alarm sensors in neighborhood dwellings.

Most calls for fire, police, and other emergency aid are by telephone. The confusing array of public safety organizations and telephone numbers, especially in metropolitan areas, is one of the roadblocks to reducing response time.

An American Telephone and Telegraph study found that about 40 percent of

emergency calls are placed each day by dialing zero (Operator). The operator determines the nature of the emergency and calls the appropriate agency.
The 40,000 emergency calls to operators are a fraction of 1 percent of the 14
million calls handled daily. AT & T Vice-President H. L. Kertz has declared:
"Reliance on telephone company operators acting as an emergency service
bureau has inherent limitations which make it less than ideal as a general system for modern metropolitan areas. Operators are necessarily chosen and
trained to handle ordinary telephone traffic, such as person to person long
distance calls. ... Telephone company operators cannot be expected to have
the same background and training in handling emergency calls as attendants
at a specialized emergency switchboard might have. ... A second call by an
operator to an emergency agency ... involves unavoidable delay. Fluctuations
in the volume of traffic handled by operators are such that calls to the operators are sometimes subject to longer delay than is acceptable in emergency
situations."

A single emergency number has been proposed to eliminate the confusion
of multiple numbers and to increase speed of response. The Bell System in
1968 announced the availability of "911" for use as a universal emergency
number. To date the number has been employed in few communities. A major
stumbling block is the fact that telephone exchange boundaries frequently do
not coincide with the jurisdictions of local police, sheriffs and fire departments.

CATV's potential for placing emergency signaling devices in each subscriber's home makes it a possible means of providing a fast and simple call
for aid. The Versacom system developed by the Advanced Research Corp.
features an emergency button that, when activated, records on a display monitor showing the location of the alarm. The alarm can then be relayed or
switched to a responding agency. It presumably would be possible for a subscriber to be furnished separate emergency buttons for police, fire, and other
aid.

Authorities are notified of an emergency either by a witness to an event or
by sensors that detect the presence of an intruder or fire and automatically
signal an alarm. "Silent" alarms are serviced by private alarm companies.
The alarms are either received by central stations maintained by the companies and relayed to police and fire departments or are sent directly to police
and fire stations.

The value of alarms in reducing police response time and apprehending
burglars is illustrated by the experience in Cedar Rapids, Iowa. Intrusion
alarms were placed in 350 businesses in 1969. A federal grant made it possible to install them mostly in gas stations, grocery stores, taverns, and
other small businesses that usually cannot afford alarm systems. The businesses were selected on the basis of crime statistics that made them appear
to be likely criminal targets. The alarms are connected directly to the police
station by telephone wire pairs leased from the phone company.

In the first 18 months of the project, Cedar Rapids police caught 40 persons in the act of burglarizing the establishments. This is more burglary ap-
prehensions than in the previous four years combined. Police have succeeded
in reducing their response time to the alarms to two minutes. In addition to
the burglars caught on the premises, police found evidence of break-ins but

nothing stolen in about 200 places equipped with the alarms. Cedar Rapids police theorize that in many of these cases burglars fled after being tipped off by accomplices who spotted arriving police.

A study by the Small Business Administration shows that small business-men bear the brunt of the losses from burglary. In 1967-1968, the loss from burglary for all American businesses totalled $985 million. Small businesses — defined as firms with annual receipts of less than $1 million — accounted for $677 million, or 71 percent.

The Senate Select Committee on Small Business declared in 1969, "Within any community location ... a substantial number of all businesses and of retail establishments go without protective service of one kind or another.... A substantial majority of businesses go without any major form of protection such as a central alarm or a protective service."

An alarm costs between $200 and $500 a year to install and maintain. Almost all alarm systems operate on leased telephone lines. Telephone line charges account for 15 to 20 percent of the total annual charge. In Cedar Rapids, the phone company initially submitted a charge of $8.50 a month for telephone lines per alarm. The company subsequently reduced the charge to $6.50 monthly, plus $10 for each hookup. Many phone companies charge by distance. Even though the alarm may be located across the street from the police or commercial central station, the distance is figured from the alarm site to the monitoring station by way of the telephone exchange. Cedar Rapids Police Chief George Matias has commented, "The problem of telephone line charge is a major stumbling block in the way of large-scale use of alarms since it is a recurring charge. Even the simplest alarm system would require a basic cost of $66 a year for line charges. Methods of reducing or eliminating the cost should be investigated."

The Small Business Administration study noted several drawbacks to re-liance on telephone lines, including the increasing cost of line charges and the prohibitive costs to businesses located far from the monitoring station. In examining alternatives to use of telephone lines, the SBA report stated: "If co-axial cable were to become the general method of communications, the implications become far-reaching. The single carrier going into each business, home or other establishment could carry telephone, television, fire alarms, civil defense alarms, and of course, crime alarms. The multiuse economics would increase competitiveness. In effect, then, alarm systems could become public utilities, privately owned and operated commercial central stations, with a billing of the approximate order of magnitude of that for electricity or telephone."

More widespread use of alarms could substantially decrease the loss from burglary. This would be especially true if coupled with more rapid police re-sponse to alarms. The 1969 Small Business Administration study concluded that it was possible within ten years to cut the average police response time of five minutes to sixty seconds or less in high crime areas. A response time of a minute or less would produce estimated on-site capture rates of as much as 90 percent. The Small Business Administration study states that cable could be one of the major contributors to achieving that goal both by providing the means for distributing alarms and facilitating communications from central stations.

High false alarm rates make widespread installation of alarms a two-edged sword. False alarm rates of 90 to 95 percent are not uncommon. Enormous amounts of police time are wasted responding to false signals. The problem has caused several cities to consider imposing fines on commercial central station operators for each false alarm. The likelihood of an alarm being false has prompted Los Angeles police to give lower priority to burglary and robbery alarm calls.

Alarms tied to TV cameras that permit video surveillance of the premises and audio alarms that transmit the sounds actually made by an intruder are among ways of checking on an alarm and partially coping with the false alarm problem. Both types of alarms lend themselves to employment in cable systems.

Early detection and reporting of fire is as important in safeguarding life and property from fire as early detection and reporting of crime is in apprehending criminals. The movement nationwide is toward ultimately requiring the installation of fire detection systems in all multiple dwellings, and detection systems coupled with extinguishing systems in all high-rise and commercial buildings.

A recent study of the fire protection needs of Des Moines, Iowa, by Gage-Babcock and Associates, a leading fire protection consulting firm, illustrates the stress being placed on private protection devices. The report recommended that "consideration be given to requiring all except single and two family residential property be provided with either automatic detection or automatic extinguishing systems, arranged so that actuation of the system will automatically transmit an alarm to the fire department. ...It is recommended that an educational program, strongly encouraging all homeowners to install home fire detection systems, be developed."

Cable systems could provide the means for much wider dissemination of fire detectors and alarms. Fire alarms are more reliable than burglar alarms and create less of a false alarm problem. The data-carrying capacity of cable may make it possible to inform fire authorities of the nature and extent of fires, rather than just the bare fact of the existence of fire. The New York City Bureau of the Budget has under consideration the possible use of sensors in connection with cable systems to record and transmit the amount of heat and other indicators of the size of the blaze to enable the department to make a judgment on the equipment to dispatch. The New York Budget Bureau is also considering cable as the means for communicating the call for equipment from the central station.

A principal cause of police delay in arrival at the scene of a crime is the lack of ability to pinpoint the location of police cars. The result is that a car is sometimes dispatched when a nearer one is available. The Task Force Report on Science and Technology for the President's Commission on Law Enforcement and Administration of Justice recommended research on two types of car-locator systems. In one of the proposed systems, each car would emit an identifying signal, which would be picked up by sensors and sent back to the police communication center by land lines. The Hazeltine Corporation has done field tests on a similar vehicle-locator system in New York City. The company has noted that the fixed stations needed to receive the signal from the vehicle and to relay the information back to a central station may be provided by video cable. Similar fixed stations, or "signposts" on the cable, have been proposed to determine the location of appropriately equipped individuals who trigger personal distress alarms.

Rehabilitating offenders is at least as important as apprehending them. A census of the nation's jails conducted March 15, 1970, for the Law Enforcement Assistance Administration showed that on that date 160,863 persons were incarcerated in 4,037 city and county institutions. The total city and county jail population accounted for nearly half of all adults in correctional institutions. About 65,000 of the local jail inmates were serving sentences, 10,000 for a year or more. Yet most jails are devoid of even minimal rehabilitation programs.

The census found that 85 percent of the 3,300 jails in large cities have no recreational or educational facilities of any kind. Richard W. Velde, associate administrator of the Law Enforcement Assistance Administration, noted, "Most prisons have at least some correction and education programs. It is clear that many jails don't even have that." An earlier study by the Task Force on Corrections of the President's Commission on Law Enforcement and Administration of Justice found that only 3 percent of total jail staff in the United States perform rehabilitative duties, and that on the average there is one psychologist for each 4,300 jail inmates and one academic teacher for each 1,300 inmates.

The channel capacity of cable television makes it possible for some of it to be used to reach inside jails to provide access to rehabilitation services. This could be in the form of programs geared to the special needs of jail inmates, spelling out the services available. Community agencies that supply welfare, medical care, family counseling, vocational training, legal aid, drug therapy, and other services could describe them in detail and request interested inmates to sign up for visits by agency representatives. The programs could be videotaped and repeated periodically. Television could also be the means for providing academic instruction and up-to-date information about job openings for inmates being released on bail or by discharge.

Current thinking in corrections stresses the importance of treating institutionalized offenders in community-based institutions rather than in remote prisons. CATV could be of increasing importance as these institutions are developed. Penitentiaries with access to CATV may also find it practical to use the systems for programs for inmates, in-service training, and possibly surveillance.

Technological and economic uncertainties make predictions about the ultimate dimensions of cable's contributions to public safety hazardous. It is evident, however, that the medium holds substantial promise.

The attempts to date to realize the potential have been few and isolated. No central clearing house exists to record successful applications, to suggest approaches, or to exchange ideas.

The development of CATV to serve the public's safety needs is likely to continue to be inadequate, piecemeal, and fragmented in the absence of stimulus to encourage the industry to develop programs and services in the public safety sector. Systematic efforts to foster experimentation through demonstration grants, to promote information exchange, and to alert community agencies to possible public safety applications are necessary in order to achieve a reason-

able level of use of the medium to serve law enforcement, fire protection,
criminal justice, and related public safety goals.

If cost were no object, the technology of telecommunications would permit construction of virtually any interactive two-way facility that one might specify. But any rational discussion of cable as a means toward a more responsive society must start with the engineering economics of the existent technologies along with whatever indications there are as to what future improvements and new approaches appear most cost effective. Pay TV is one of the new options.

What exists today is a technology that provides each area with three to seven channels of broadcast television out of twelve channels reserved for VHF nationally. What also exists are cable systems that provide as many as twelve active channels of television. The first step into the future, beginning to be taken in a few locations, is the provision of 20 or more channels on cable. The technical problem produced by this expansion is interference. So the first part of Ward's first paper deals with the problem of interference and how to beat it.

The next step into the future is to provide two-way transmission of signals. Alternative ways of doing that, and their costs, are treated by both Ward and Lemelshtrich.

The main rival approaches for dealing with multiplying channels as demand grows and also for providing upstream channels are "tree networks" or "hub systems." A tree network (the generally used system in this country) has main trunks running from the head end, branching into feeder lines, branching into drop lines that enter the individual households. The cables are all coaxial and carry 20 or more channels on a single cable into the home, where a selector switch is used to choose one of them. Hub systems are more like the telephone system. A switching exchange for a neighborhood receives all the programs from the head end (whether on coaxial cable, wire, or other medium). Signals from the home activate a switch there at the switching center. From there the single chosen channel is sent into the home, perhaps on wires or on cable.

There are profound differences in costs, services, advantages, and social consequences of each approach. Ward's first paper helps us to understand, as far as the present stage of development allows, the technology and implications of each. His paper consists of materials originally prepared for the Sloan Commission. The Sloan Commission report, On the Cable, published an extensive appendix containing the summary material of Ward's report to them but not the more technical material. The heretofore unpublished portions are important for serious students of the development of cable. Along with them we have included a small part of what already appeared in the previous volume to help provide context for the material that is now published for the first time.

Ward's paper reviews state-of-the-art candidate technologies. Lemelshtrich's paper, on the other hand, suggests some new approaches intended to make powerful two-way interaction available economically. These approaches are new, but not in the sense of being here invented. They utilize devices that exist, but whose possible application to home television uses has rarely been noted.

Theodore S. Ledbetter, Jr., and
Susan C. Greene

WHAT IS PAY TELEVISION?

Pay or subscription television (STV) is a means of providing additional and specialized programming for a fee. The FCC has never adopted an official definition of subscription television by cable but has stated that pay cable is "akin to" a per-program or per-channel charge above the normal monthly cable TV subscription fee.

Pay cable has been tried only in a few experimental situations. But when CATV systems are built in the nation's major metropolitan markets, pay cable will offer subscribers a diversity of communication services and provide entrepreneurs with a lucrative new field.

Initially, subscription television by cable will provide greater options in entertainment—particularly first-run films and sports events. Later it will offer specialized programming, such as refresher workshops for doctors; ultimately it can offer a wide range of services like monitoring security alarms and remote control of household equipment.

Briefly, this is how the pay-cable TV system works. Scrambled signals are sent from the central head end (transmitter) to all subscribers. The subscriber's TV set is equipped with a converter, which decodes the frequencies of channels carrying special programming. A different code is used for each pay channel or program. If the subscriber has ordered a particular program or channel, the scrambled signal will be decoded by the converter, and the subscriber will receive the program. If the subscriber has not ordered the program or channel or tries to watch it without authorization, the converter will not unscramble the signals.

PAY CABLE—PAST AND PRESENT

For twenty years broadcasters, film studios, and theater owners have been pressuring Congress to severely restrict or completely prohibit pay television. In 1968-1969 the issue burst forth into a bitter public struggle when theater owners across the country put slogans on their marquees and passed out petitions to "save free TV." One animated short showing the specter of pay TV taking advantage of the unsuspecting public was widely circulated.

Pay television has been tried in two or three experimental projects on over-the-air stations and on cable. All have failed. Pay television has yet to be instituted as a regular service by a cable system anywhere in the country. But the technology needed for such a system already exists. More than likely it is only a matter of time before the first system is installed.

The arguments against subscription television, whether over-the-air or cable, are much the same. The National Association of Broadcasters (NAB) and National Association of Theater Owners (NATO) have spent twenty years fighting STV. Theater owners predict that STV will diminish their already dwindling audience. On the other hand, film distributors believe STV will provide a new market for their product; but until pay TV is established, they cannot afford to alienate over-the-air broadcasters, their current biggest buyer. The NAB contends that pay television in any form will "siphon off" programming from "free" TV. For example, if Flip Wilson were offered more money by the owners of pay TV than by a network, he might take his program off the network stations and onto pay TV. Then only viewers who paid a special fee could receive "The Flip Wilson Show." If enough performers followed suit,

viewers would end up paying special fees for TV programs they used to see
for no charge. The NAB further argues that the greatest financial burden will
fall on those viewers least able to pay, the low-income groups.

STV opponents want Congress (rather than the FCC) to abolish or at least
severely limit the scope of operation of all forms of pay TV. Proponents of
subscription TV state that their function is to provide viewers with the option
of watching superior or special programming for which they would pay a per-
program fee. This might include special college courses, first-run movies,
or a refresher course for professionals or special sports events, just to name
a few alternatives.

Thus in the STV controversy, the FCC has been caught in the middle. The
FCC took no action until February 1955, when it issued a Notice of Proposed
Rulemaking[1] requesting the views of interested persons on the question of
whether the commission had the power to authorize subscription television and,
if so, whether the exercise of such authority would be in the public interest.
In October 1957, after reviewing the comments filed, the commission issued
a first report on subscription television,[2] concluding that the Communications
Act of 1934 empowered the FCC to authorize pay TV. It also concluded that an
extensive trial operation should be initiated to determine whether subscription
television would serve the public interest.

The idea met with strong opposition from broadcasters and theater owners.
Several pieces of prohibitory legislation were introduced in Congress, and the
House Subcommittee on Communications adopted a resolution against the au-
thorization of STV.

In a second report[3] issued in 1958, the FCC postponed processing applications for trial authorizations until Congress considered public policy questions
on subscription television. A third report[4] issued in 1959 announced the commission's readiness to consider applications for trial authorizations.

The first application for an STV license came in 1951 when the Zenith Radio
Corporation tried to persuade the FCC to authorize an STV experiment. The
FCC took no specific action until 1962, when it finally authorized a seven-
year pay TV experiment in Hartford, Connecticut. The FCC's power to au-
thorize this experiment was upheld by the Court of Appeals in March 1962.
The Supreme Court declined to review the decision.[5] WHCT was licensed to
RKO General, Ind., which ran the experiment in cooperation with Zenith Radio
Corporation and Teco, Inc., its patent licensee. Its sole purpose was to offer
"programs of box office quality... which are not now available on free TV."
The experiment was a financial failure. It also offered less cultural programming and more general entertainment than had been generally expected. In addition to the normal rental and installation charges, the average per-program
cost was slightly over one dollar per showing regardless of the number of viewers for each set.

In its first two years, the station offered 599 different programs; most were
repeated, so that the station actually provided 1,776 separate presentations,
which fell into the following categories:

Movies	86.5%
Sports	4.5
Education	3.5
Special entertainment (ballet, opera, Broadway plays)	5.5

An experiment in Bartlesville, Oklahoma, in the early 1950s, which consisted of sending pay TV by cable, was also a financial failure. The only other
experimental system of pay TV by cable was stymied before it ever got started.
In the early 1960s, about the same time Hartford was trying STV using over-
the-air signals, Sylvester (Pat) Weaver and associates set up a three-channel
wired pay TV system in California. Theater owners were able to get enough
signatures to put the project up for referendum on the ballot. The public was
told it would have to pay for what it was presently seeing for free, and the
project was defeated at the polls. Later the Supreme Court of California found
the referendum unconstitutional,[6] but Weaver had already gone broke.
Despite these financial failures, the FCC decided in 1969 that "subscription
television provides a beneficial supplement to the television program choices
now available to the public"[7] and gave its official blessing to STV.

To ensure that subscription television would not be furnished at the expense
of free television either by siphoning off the types of programs now available
free or by reducing the amount of free programming available, the FCC adopted
a series of rules in 1969 governing the content of pay TV. These rules prohibit pay systems from carrying serial programs (such as Mission Impossible)
or sports events (such as The Super Bowl), which have been carried live on
conventional television in the two years preceding the start of STV in a television market. Special events, such as the Olympics that have previously been
televised on conventional TV are also prohibited from STV. The FCC further
stipulates that all films on STV must be first-run and less than two years old.
In addition, STV must devote 10% of its programming to offerings other than
sports and film.

TECHNOLOGY OF PAY CABLE
Cable operators could offer two types of subscription systems — subscription
television and multipurpose systems. Subscription television would provide
additional programming choices; multipurpose systems would provide a variety
of services and eventually would give the subscriber the capability to originate
his own audio and video transmission. Both operate on the same premise —
that specialized programming can be restricted to only those viewers who have
paid for it.

All subscription systems include a scrambler or encoder, which scrambles
the program at the transmission end, and a converter or decoder, which re-
assembles the program at the subscriber's end. The converter, which may be
rented or bought, is attached to the subscriber's television set. A multipurpose
system also includes a transponder, a mechanism which can switch any piece
of equipment in the subscriber's household that has an on/off control.

SUBSCRIPTION TELEVISION SYSTEMS
Subscription systems have not yet been marketed for the homeowner; however,
several versions are currently being installed in hotels and motels. One such
system uses the hotel's master antenna (MATV) to transmit three TV signals
at frequencies below standard channels. One of these channels, which all
guests can receive, carries free program information. If the guest decides to
watch one of the pay programs (usually a first-run movie), he calls a special
hotel clerk who sends the desired program to the room on one of the two re-

maining available channels. The fee is automatically added to the guest's hotel
bill. This system is currently being installed in hotels throughout the country
by Columbia Pictures, Inc.

A home system, marketed by Optical Systems, Inc., was tested in early
1972. It utilizes a black box, which would rest on top of the subscriber's
TV set, and operates electronically when a plastic card the size of a credit
card is inserted into the box. One card might serve as a season ticket for pro
basketball telecasts, another would decode a channel for current movies, a
third for children's programming. The subscriber is offered a variety of
services. He can select as many or as few as he pleases and is charged a fee
according to the number of services selected.

MULTIPURPOSE SYSTEMS

Multipurpose systems all consist of a computer-controlled central station located at the CATV system hub, which controls the flow of digitally coded messages to and from the remote subscriber locations. The central station transmits a series of digital messages, each preceded by a unique address code.
For each message transmitted, only the subscriber corresponding to the address code transmitted actually receives the message. A response is immediately transmitted from the subscriber's terminal to the central station. The
process is repeated with every station. The response message from each station can contain information requested by the central station, by the subscriber, or between subscribers. Even though the subscriber terminal can transmit
only when interrogated, the rate of questioning is sufficiently fast (several
times per second) to give the impression that the subscriber is actually initiating the transmission of information. Because of the constant monitoring,
these multipurpose systems can offer a variety of "watchdog" services, including monitoring alarm circuits, remote control of home appliances, data
transmission and opinion polling.

With its high-speed central computer the system monitors each transponder
every five seconds, or 17,000 times every 24 hours. With slight modifications
it can provide interactive teaching and game playing, central message and
wake-up services, industrial lighting control, and a host of additional services.
This system is not unique. All two-way cable systems have this potential, and
several versions are currently on the market.

Multipurpose systems are currently in use on all Boeing 747 airplanes. The
passenger entertainment system for the entire plane is run by an onboard computer that monitors the channel selection system of every seat as well as controlling the lights and oxygen supply for every seat.

ECONOMICS — WILL IT PAY?

Subscription service via cable will be a fact of life in the near future. Cities
are installing cable systems, the FCC is accepting the responsibility of regu-
lating them, and the technology for more advanced systems is presently at
hand. From the subscriber's standpoint pay cable is economically attractive;
from the owner's standpoint it is economically lucrative.

In the home subscription system marketed by Optical Systems, Inc., the
subscriber pays approximately $1 per month for the black box and $1.25 for
each first-run film viewed. If the same individual took his family to see the

same film in a theater, the cost of a babysitter, parking, tickets, and a snack could easily run $15. So while the TV screen is smaller, the price is right. Any number of viewers can watch for the $1.25 fee. And there is the comfort and convenience of watching at home.

Subscriber television allows the home viewer to watch events he would ordinarily be excluded from. Take sports. Season tickets for the Milwaukee Bucks have not been available since Kareem Abdul Jabbar (Lew Alcindor) joined the team. Subscription television could make the games available to home viewers at a moderate fee. The Bucks would profit by having a larger audience available and by sharing the subscriber service's revenues from the home viewer. The same principle can be applied to symphonies, operas, extension courses, and numerous other events. Unlike advertiser-supported programming, pay cable reflects the subscriber's viewing preferences. His program charge pays for the program directly.

The key to the financial success of pay TV lies in the success of marketing sports and entertainment programs. For example, the Muhammed Ali-Joe Frazier fight held in March 1971 took in $15 million on closed-circuit television, at anywhere from $10-$30 per ticket. If the next fight is offered on cable at $3 per home and if 75% of the 5 million cable homes watch,[8] the gross income would almost double. In the future, when most of the country is wired, income will soar. If half today's 61 million television homes had watched a similar fight, the gross income would have exceeded $60 million. Cable does not presently have this kind of penetration, but investors estimate that 25-30 million homes will be wired by 1980.[9]

Investors are so certain of the widespread development of cable, Time-Life has begun shifting its emphasis in electronic communications from broadcasting to cable and cassette television and in December 1971 acquired an equity position in Computer Cinema. West Coast multimillionaire Jack Kent Cooke, one of the promoters of the Ali-Frazier fight, owns 500,000 shares of TelePrompTer, the nation's largest cable system.

Movie theater owners and film studios will closely watch the success of subscription entertainment. Despite their fight to prevent subscription television, they can view it as their potential savior. Every film studio is showing deficits. Audiences have diminished 66% since 1956.[10] Admission prices have risen 300% in the same period. A recent study showed that the median age of a person attending a film was just over twenty. There is a potential audience of 96 million adults thirty and over who do not now attend films.

Sidney Dean, Chairman of Americans for Democratic Action and spokesman for the American Civil Liberties Union, estimated in a hearing before Congress that the volume of communications services and products marketable on pay systems, especially on cable, will range between $40-$60 billion a year within five years of the establishment of a national cable system.[11] This compares to the current total revenue of the television industry of about $4 billion a year.

PROGRAMMING
Ultimately, content is the name of the game. It will determine whether a cable system can serve the nation's diverse needs. The answers to what will be of-

fered on pay cable and when it will come will be determined by the success or
failure of the initial subscription systems.

Interest in subscription services is broadening. Recently several media
organizations have considered developing pay television. The first indicator
occurred in September 1971, when Sterling Cable, Inc., a subsidiary of Time-
Life, Inc., asked the FCC for an interpretive ruling on a dispute between
Sterling, which has the cable franchise for the southern half of Manhattan, and
the New York City Department of Franchises. The dispute was whether Sterling
could charge extra for special programs, a practice that ran counter to its
franchise agreement. In a major policy decision the FCC indicated that any
cable system could make per-program charges regardless of the contents of
the local franchise agreement.[12] Since then Sterling has stated that it is "actively exploring" the possibilities of subscription programming.

Since January 1972 three organizations have announced plans for home subscription systems. Sterling Cable of Hicksville and Plainview, Long Island,
is now giving its customers a choice of two services. For a $3 monthly fee a
subscriber receives 12 stations, a weather channel, and a channel programmed by the Board of Education. For an additional $3 he receives five more
stations, a UPI news channel, a channel for sporting events from Madison
Square Garden, and a channel for films.

Telepremier International, headed by Dore Schary, executive vice-president
of MGM & RKO, has announced "Theatrevision" for home movies. And Optical
Systems is introducing its own system in the San Diego, Santa Barbara, and
Bakersfield CATV systems. Television stations in Chicago (WCFL), Boston
(WQTV), and Detroit (WXON) are also considering subscription systems.
Gridtronics, Inc., a subsidiary of Television Communications Corp. (TVC),
a cable TV company serving 70,000 subscribers in ten states, is considering
marketing a converter that will provide four additional programmed channels
to standard TV sets. One will be an entertainment channel showing uninter-
rupted films; another, an instructional channel to teach all levels; a third, an
informational channel on contemporary subjects; and, the fourth, a professional channel that only doctors could subscribe to.

Later, when multipurpose systems are in general use, the emphasis will
be on services, rather than entertainment. These services will alter the economic, political, social, and cultural habits of the nation. Among the services
that will be commonly available are in-home shopping, two-way education,
reservations, credit checking, checkless banking, centralized bookkeeping,
library services, audience polling, highway surveillance, household management, and patient monitoring.

A possible medical service frequently discussed is a medical vest a person
at home puts on. It instantly reads body temperature, blood pressure, respira-
tion, heartbeat, and other statistics and sends this information to diagnosti-
cians in the hospital clinic who can then advise the patient. Such a device would
be a considerable boon for the elderly.

ISSUES AT HAND

Every new technology has the potential for exploitations as well as public benefit. One of the biggest issues concerning cable television will be the invasion
of privacy. Every two-way (multipurpose) cable system enables the cable oper-

ator to monitor individual sets. A subscriber's program choices and services used can be easily surveyed without his knowledge.

Large cable systems will have data banks with information about all its subscribers. While it is not yet a practice of cable firms, there is nothing to prevent these companies from selling or trading lists of subscribers and pertinent personal data without the knowledge or permission of the subscriber. In fact, this is already a common occurrence in other industries. Publishers sell subscription lists, certain states sell motor vehicle lists, and the Post Office freely hands out antiobscenity lists.

Obscenity is another issue. Should pornographic material be allowed over cable television? The general response from the industry and regulatory agencies is a tentative yes, provided that the cable system includes some method of positive subscriber control where the subscriber takes some positive technical action (turns a key, opens a lock, or dials a combination) to receive the program. In the home subscription system developed by Optical Systems, Inc., if the subscriber does not want to see something, he takes his plastic card out of the box. So if he doesn't want his children to watch a certain program, he can make sure they don't. There is no need for self-appointed censors in this system as there is in regular film and television.

Public access to programs will be another important issue. Should cable owners, programmers, or advertisers be allowed to decide which of their subscribers will be allowed to see any given presentation? Let's take a problem that currently exists. Doctors in Louisiana can watch on cable films related to various medical topics. The films are produced and paid for by drug companies. The general public is not offered the option of deciding whether it wants to watch or not. With public access prohibited, the doctors can maintain their closed professional group. The hard-sell (or soft-sell) techniques of the drug companies are beyond the scrutiny of the general public. And the public is denied the benefits of watching and perhaps gaining some useful information.

As with any new industry, policies and price structures have yet to be decided. In the MATV field, the industry is charging what the market will bear rather than cost plus a reasonable profit. What the effect of these pricing policies will be on future costs of entertainment is yet to be determined, but it is worth some consideration.

VALUES UNDER SCRUTINY

Cable technology will provide custom-made services and conveniences. But, like its predecessors, the car, the airplane, the telephone, the TV set, and the satellite, it will also be a redefiner of social, political, and economic boundaries. Thus, for the next decade it is sure to be a prime area of controversy in the courts, the FCC, and Congress. Cable could give us services written of only in science fiction; it could also produce the Orwellian nightmare long before 1984.

Notes

1. 20 F.R. 988.

2. 23 FCC 531.

3. 16 Pike and Fischer, R.R. 1539.

4. 26 FCC 265.

5. Connecticut Committee Against Pay TV v. FCC, 301 F.2d 835 (D.C. Cir., 1962) cert. denied, 371 U.S. 816.

6. Weaver v. Jordan 411 P.2d 289, (Calif. 1966).

7. 15 FCC 2nd 417 (1968).

8. FCC figures released March 3, 1972, report that as of January 1, 1972, there were 5,008,580 cable subscribers. FCC News Release, March 3, 1972.

9. Interview with Tom Wilson, Public Information Director, National Cable Television Assoc., Washington, D.C., January 17, 1972.

10. Address by Geoffrey M. Nathanson, President, Optical System Corp., to National Assoc. of Theatre Owners, Mid-Continent Convention, Pfister Hotel, Milwaukee, Wis., August 18, 1971.

11. Hearings before the Subcommittee on Communications and Power of the Committee on Interstate and Foreign Commerce 427. House of Representatives, 91st Congress, first session on H.R. 420. December 12, 1969.

12. Letter from FCC to Pierson, Ball and Dervo, Washington, D.C., September, 1971.

9. PRESENT AND PROBABLE CATV/BROADBAND-COMMUNICATION TECHNOLOGY

John E. Ward

I. INTRODUCTION

The impetus for this report came from the Sloan Commission on Cable Communications, which felt that as part of its deliberations on the desirable future role of broadband cable networks, and how best to achieve that role, it needed a comprehensive review and evaluation of the state-of-the-art in CATV technology and expected future capabilities.

Of particular interest to the Commission in connection with such a review and evaluation were questions such as the following:

Should future broadband systems for home communications services be organized in frequency multiplex tree networks, as in present CATV systems, or in switched hub networks, as in the telephone system?

What are the prospects for obtaining 40, 60, or 80 program channels, and what are the respective cost expectations?

Is nationwide standardization of CATV systems in the near future a desirable or necessary condition if a nationwide broadband network is the goal?

These questions cannot be answered definitively in a short six-month study, particularly when one is trying to peer ten or twenty years into the future in an environment of fast-paced technological evolution and when some of the alternatives may represent a fair share of the gross national product in terms of required investment. However, the study has attempted to provide the best information possible to the commission by thoroughly examining the underlying framework of video/digital transmission and switching techniques applicable to combined CATV/communications services, and making estimates of future technical/cost trends.

This paper investigates present techniques and probable future extensions for "high-channel-count" systems, that is, those capable of providing 20 or more program distribution channels. Two-way video and digital data services are also examined.

It soon became apparent that no valid comparison of the alternate techniques and the motivations behind them could be made without constant reference to the exceedingly complex interchannel interference effects in frequency multiplexed TV channels. It was also found that a vigorous industry/government dialogue is currently underway on the question of possible CATV-system standards, which could influence the relative balance between techniques so far as these interference effects are concerned. However, no comprehensive source reference on these questions could be found. Thus to provide the necessary base for system comparisons, considerable time was devoted to the compilation, analysis, and organization of available data from a variety of sources, as reported in the next section.

Switched CATV systems are quite new and little understood, and because of their importance in connection with the questions posed by the commission, the two switched systems extant were examined perhaps more thoroughly than the more conventional technology. Finally, some attention was paid to the fact that there are two other sets of wires leading into every CATV home — the telephone line and the power line — and that some efforts are underway in the direction of meter reading and possibly other data services over one or the other of these lines. Data gathered on these efforts are given in Part VI.

Note that in the following analyses all cost figures for cable distribution plant (including subscriber drops but not head-end equipment) are on a per-subscriber basis and on the assumption of 100 percent penetration in medium-density residential areas (aerial plant).

Traditional single-cable CATV systems with TV-set tuning have offered a maximum of 12 program channels (VHF Channels 2-13) but often cannot deliver more than six or seven usable channels in the major cities because of interference caused by local VHF-TV transmitters. Three different approaches to increased capacity are being pushed by various manufacturers and/or system operators: dual-cable systems with a different group of programs on each cable, subscriber set-top converters to permit use of additional cable channels, and switched systems in which channel selection is performed remotely and only the selected channel appears on the subscriber cable (drop). Of major interest to the commission was the comparison of these systems in such matters as channel capacity, relative cost, and advantages or disadvantages for various two-way services. The results obtained (in summary) are as follows.

A program capacity of 20-26 channels is now available on a single cable with converters at a per-subscriber cost of $80, of which $30 is the converter cost. Dual cable VHF-only systems with 24-channel nominal capacity (16-20 actual) have a comparable cost ($70). With the addition of converters ($30), dual-cable capacity can be extended to 40-52 channels at a total per-subscriber cost of $100. The two present hub-network, switched systems have capacities of 20 and 36 channels, with per-subscriber costs of $113 and $186, respectively.

Single-cable converter-system technology appears to limit at about 35 channels ($50 per converter, or $100 total per subscriber). This capacity may be reached in the next few years, but extension beyond 35 channels per cable is clouded by many technical factors. Use of such improved converters in dual-cable plants would yield 70 channels ($120). Costs of doubling the two switched systems just cited to provide 70-80 channels are not as well defined but would be about $200 and $280, respectively. The greater costs of switched systems should be balanced against other capabilities.

These cable plant costs for tree-network systems are for downstream capability only and would be about 30 percent higher with two-way amplifiers for a two-way capability (not including terminal cost). The percentage increment in adding two-way capability to a switched cable plant is lower but cannot yet be accurately defined.

II. INTERCHANNEL INTERFERENCE AND THE CABLE-FREQUENCY ALLOCATION PROBLEM

1. Introduction

Community antenna (CATV) systems started out as just that — carrying TV and FM-radio programs on the same frequency bands that are used in over-the-air broadcasting, and permitting subscribers to receive these programs on their standard TV and FM receivers without additional equipment. Since the practical maximum frequency that can be carried on a CATV cable (currently 300

MHz) is well below the UHF television band (470-900 MHz), a traditional single-cable system of this type can offer only the 12 VHF television channels, which occupy two frequency bands from 54 to 88 and 174 to 216 MHz. Actually, 12 channels is an upper limit, which is seldom if ever obtained, particularly as CATV systems have departed from their original role of extending TV coverage into remote, marginal-reception areas and become "urbanized," that is, they have moved into areas covered by strong VHF-TV stations. In such areas, direct pickup of broadcast signals in the cable and/or subscriber TV sets creates an interference with cable signals that can make the channels occupied by strong local stations unusable for cable transmission and reduce system capacity below 12 usable channels — to as little as five or six in some top-market TV areas.

The CATV industry, in its effort to break out of the traditional pattern of 12 channels (or less) per cable, has only recently begun to make use of channel frequencies that are different from the standard VHF-TV channels for which all TV receivers are designed. The initial move in this direction was to take advantage of 66 MHz of unused bandwidth that was already being carried in all cable systems — that between the upper edge of the FM band (108 MHz) and the lower edge of Channel 7 (174 MHz). By general industry acceptance (but not specific standardization), a "mid-band" has been defined from 120-174 MHz, divided into 9 channels designated A through H (the 108-120 MHz region is apparently avoided because of possible interference effects of cable leakage on aeronautical radio and navigation services, although these services actually extend to 136 MHz). Use of these mid-band frequencies, which requires a special channel converter device at each subscriber TV location, permits 21-channel capacity (12 VHF plus 9 midband), and a few cable systems now offer such service.[1]

Once started on this tack, it was natural to define a "superband" starting at 216-MHz (the upper edge of Channel 13), with the upper limit of this essentially open ended. Many of the newer cable systems are engineered to carry frequencies up to 240 MHz, permitting four super-band channels (I through L) and 25-channel total capacity (with appropriate converters). With 300-MHz cable technology now becoming available, the super-band can be extended to provide nine additional channels M through U, or a total of 35 channels per cable (with mid-band as presently defined). Similarly, a "sub-band" has been defined in the region below 50 MHz, with the general consensus that this band will be used for upstream channels (signals moving from the subscribers toward the head end). Some proposals are for four upstream channels between 6 and 30 MHz, but no general agreement has been reached as yet on channel assignments or usage, and the first upstream experiments are barely underway.

Unfortunately, adding these extra channels on a cable can and does raise problems of interchannel interference that do not arise with use of only VHF Channels 2 through 13 (either broadcast or on the cable), largely because the frequencies for these VHF channels were chosen by the FCC partly on the basis of avoiding such problems. The intent of this discourse is not to say that these interchannel interference problems are insurmountable in augmented-channel cable systems but to clearly define them, explain how they come about, and point out the precautions that must be taken in the design of subscriber equip-

ment (converters or special cable receivers) and in the final agreement on
cable-channel frequencies.

Since most of the interference questions revolve around the basic U.S. 6-
MHz television channel standard and the carrier frequencies assigned within
it, this subject is taken up first in Section 2, followed by a brief background
in Section 3 on the historical development of the VHF allocations. The interchannel interference effects are then discussed in detail in Section 4, including examples of their impact on the rules adopted by the FCC for geographic
separations of channel assignments in VHF and UHF television broadcasting.

Cable systems are also subject to two other interference problems not
found in TV broadcasting: the leakage of strong, local TV station signals into
the cable or receivers to interfere with cable signals (a real problem in present cable systems) and the leakage (radiation) of signals out of the cable to interfere with other radio frequency services (a potential problem in the use of
non-VHF-TV channels). These problems are discussed in Section 5.

Finally, Section 6 discusses the intensive industry/government debate just
getting under way on frequency allocations for broadband cable systems.

2. The 6-MHz Television Channel

The origins of 6-MHz channel bandwidth standard for U.S. TV are rather interesting. In the early 1930s, the first experimental television systems used
the same double-sideband amplitude modulation technique that is still used in
AM radio. The highest picture frequency component then transmitted was 2.5
MHz, and the picture signal thus required a 5-MHz channel, with the picture
carrier in the center. To provide space for the sound signal, an extra 1 MHz
was tacked onto the high side of this picture channel to create a total 6-MHz
channel, with the FM sound carrier 0.25 MHz below the upper band edge (as
it is now).

In the late 1930s, vestigial-sideband modulation techniques were perfected,
which permitted an increase in picture component frequencies (and thus in
picture resolution) without increasing the 6-MHz channel bandwidth. With the
change to vestigial-sideband modulation, all but 0.75 MHz of the lower picture
sideband was eliminated, and it became possible to move the picture carrier
down to 1.25 MHz above the lower band edge (as it is now) and use modulation
frequencies up to 4 MHz in the upper sideband. This of course made all existing receivers obsolete, but only a handful had yet been produced (television
broadcasting was still largely experimental). Although these revised standards in 1939-1940 were for monochrome transmission at only 441 lines per
frame,[2] the 6-MHz channel and the sound and picture carrier assignments
within it have survived both the later increase to a 525-line standard in the
1940s (again making prior receivers obsolete) and the addition of compatible
color transmission in 1953.

In regard to color, the majority technical opinion up to the late 1940s and
early 1950s was that either a wider channel (perhaps 12 MHz) or a slower picture frame rate would have to be adopted in order to transmit each picture
frame as three separate, field-sequential red, blue, and green images, as
was then thought to be required for reconstruction of a color picture at the
television receiver. Indeed, the quite controversial color television standard
initially adopted by the FCC in September 1950 was a field-sequential "color-

wheel" system with scanning standards quite different from those in use for
monochrome receivers then in the hands of the public. The subsequent intensive industry/government effort, which found that it was possible to develop
and standardize by December 1953 a new method of color transmission that
fitted the 6-MHz bandwidth and was compatible with the existing monochrome
standards, was a major engineering feat and finally got color TV off the
ground. This dot-sequential system retains the standard monochrome picture
modulation as a brightness signal (for either monochrome or color sets) and
inserts a third carrier at 3.5795 MHz above the picture carrier to carry all
color-related information needed by color receivers, but which can be ignored
by monochrome receivers. The only change made in the previous monochrome
standards was a 30 to 50 percent reduction in amplitude of the sound signal
relative to the picture signals so as to reduce the possibility of any interference of the sound signal with the color carrier (the sound carrier and the new
color carrier are separated by only 0.93 MHz in frequency). The final transmission standard, which is in use today, is shown in Figure 1.

The gist of this brief history is that in the period 1937-1953, three major
modifications were made in U.S. television transmission techniques to raise
picture quality and add color, and by improved know-how and clever engineering, each of these modifications was accomplished without changing the basic
6-MHz channel bandwidth that had been chosen in the early 1930s for quite different transmission techniques. That is not to say, however, that all aspects
of the present U.S. television standards are necessarily what one would now
choose if it were ever possible to make a change. Three-quarters of the countries in the world have opted for better picture resolution by adopting higher
line counts (625 or 819 lines), higher picture frequencies (5 or 6 MHz) and
wider channels (7 or 8 MHz), although the one-quarter of the countries using
U.S. standards account for over half the world's TV sets (132 million out of
the total 254 million).[3]

Figure 2 shows the characteristics of the television systems in use throughout the world. System M, the U.S. 525-line, 6-MHz standard, is used in most
of the Western Hemisphere and in Japan, the Philippines, Okinawa, Korea,
Taiwan, Cambodia, and Vietnam. Three countries (Bolivia, Jamaica, and
Barbados) use 625 lines in a 6-MHz channel bandwidth by using 25 instead of
30 pictures frames per second (System N), which increases vertical but not
horizontal resolution (maximum picture frequency is still 4 MHz). The 405-
line, 5-MHz standard (System A) is used only in Hong Kong (a closed-circuit
system) and in parts of Ireland and the United Kingdom. The 14-MHz System
E is quite wasteful of spectrum space and is used in only three countries
(France, Algeria, and Monaco), but France and Algeria also use Systems L
and B, respectively, for at least half their stations. Most other countries in
Europe, Asia, and Africa use Systems B, D, or G.

3. The VHF Channel Allocations

In the original allocation of transmission frequencies for television in 1937,
the FCC adopted the standard 6-MHz channel width and set aside 19 television channels as shown in Table 1, although only the eight lowest were considered technically feasible within the radio art of the day.[4] Three years later,
an FCC revision reallocated the 44-50 MHz channel to FM broadcasting and

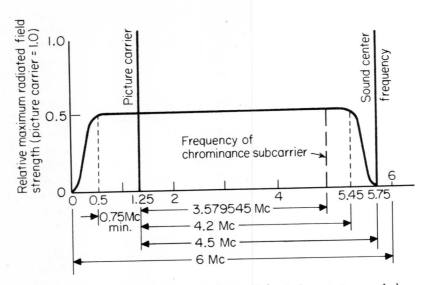

Figure 1. The NTSC color television channel (not drawn to scale).

added 60-66 MHz as a television channel (FM was to be later moved to 88-108 MHz). Although these 1940 vintage channels were allocated in pairs, separated mostly by 6- or 12-MHz frequency gaps assigned to other services (government, amateur, police, etc.), the basic pattern of the present VHF channel allocations was beginning to emerge, and six of the channel allocations (the present Channels 3, 4, 8, 9, 12, and 13) are still the same. It is also interesting to note that most of these 1937 frequency allocations that did not become part of the final postwar VHF-TV allocations are identical with presently accepted mid- and super-band (letter) channels for CATV (156-168 MHz and those above 234 MHz).

The FCC again revised the frequency allocation charts in the mid-1940s to create the present lineup of VHF television Channels 2-13 (see Table 2), at the same time moving the FM broadcast band from 44-50 to 88-108 MHz. One of the factors in these final frequency shufflings was a desire to choose the television channel frequencies so as to avoid as much as possible four types of potential interference in home receivers (other than adjacent-channel) that can be caused by particular relationships among channel transmission frequencies: interference from local oscillator radiation (leakage) by neighboring TV receivers, image interference, intermodulation interference, and IF beats. These interference effects, which are described in the next section, were almost completely eliminated in the VHF allocations because of the way the channels are grouped into two isolated blocks of 6 and 7 channels each, separated by a large frequency band (86 MHz, equivalent to over 14 channels). Note that except for Channels 5 and 6, the VHF channel boundaries are all integer multiples of 6 MHz, the channel bandwidth (Channel 5 is offset from Channel 4 by an intervening 4-MHz band assigned to radio services). As will be explained in Section 6, a uniform integer-multiple pattern is highly desirable in multi-

System	No. of Lines	Channel width MHz	Vision bandwidth MHz	Vision/sound separation MHz	Vestigial sideband MHz	Vision modulation	Sound modulation
A	405	5	3	-3.5	0.75	Pos.	AM
B	625	7	5	+5.5	0.75	Neg.	FM
C	625	7	5	+5.5	0.75	Pos.	AM
D	625	8	6	+6.5	0.75	Neg.	FM
E	819	14	10	±11.15	2	Pos.	AM
F	819	7	5	+5.5	0.75	Pos.	AM
G	625	8	5	+5.5	0.75	Neg.	FM
H	625	8	5	+5.5	1.25	Neg.	FM
I	625	8	5.5	+6	1.25	Neg.	FM
K	625	8	6	+6.5	0.75	Neg.	FM
L	625	8	6	+6.5	1.25	Pos.	AM
M	525	6	4.2	+4.5	0.75	Neg.	FM
N	625	6	4.2	+4.5	0.75	Neg.	FM

Notes:

Field frequency: System M, 60 cycles per second; all other systems, 50 cycles per second.

Picture frequency: System M, 30 cycles per second; all other systems, 25 cycles per second.

Line frequency: System A, 10.125 kHz; systems E and F, 20.475 kHz; system M, 15.75 kHz; all other systems, 15.625 kHz.

U. S. standard is system M.

Source:

C. C. I. R. Report 308, Xth Plenary Assembly, Geneva, 1963.

Figure 2. Characteristics of television systems in use throughout the world.

channel systems, and the present FCC frequency assignments for Channels 5 and 6 represent a problem in broadband cable TV.

4. Interchannel Interference Problems

This section discusses four types of interchannel interference that can be directly created within the circuits of a television receiver as a result of its operation in an environment of TV signals on channels other than the one to which it is tuned—adjacent channel interference, image interference, intermodulation interference, and IF beats—and one that can be indirectly created by local oscillator radiation from neighboring TV receivers. All of these are related both to the particular frequency relationships among the TV channels and the receiver intermediate (IF) frequency and to the quality of receiver design (shielding, filtering, etc.). Except for the adjacent-channel problem, these have not been of much concern in VHF television reception because of the particular choice of channel frequencies, but they are all of concern in augmented-channel cable systems using mid- and super-band channels, as will

John E. Ward

Table 1. 1937-1940 Television Channel Allocations

MHz	Present VHF or Cable Channel	Notes
44-50		Reassigned to FM
50-56		June 1940
(60-66)	3	Added in June 1940
66-72	4	
78-84		
84-90		88-108 is present
94-102		FM broadcast band
102-108		
156-162	(G)	Present mid-band
162-168	(H)	cable usage
180-186	8	
186-192	9	
204-210	12	
210-216	13	
234-240	(M)	Present super-band
240-246	(N)	cable usage
258-264	(P)	
264-270	(Q)	
282-288	(T)	
288-294	(U)	

Table 2. Present VHF Television Channel Allocations

VHF Channel Number	Frequency Band	Picture Carrier	Receiver Local Oscillator	Receiver Image Band	Channel in Image Band (Beat Frequency)
2	54-60	55.25	101	148-142	
3	60-66	61.25	107	154-148	
4	66-72	67.25	113	160-154	
5	76-82	77.25	123	170-164	
6	82-88	83.25	129	176-170	7(-.5)
7	174-180	175.25	221	268-262	
8	180-186	181.25	227	274-268	
9	186-192	187.25	233	280-274	
10	192-198	193.25	239	286-280	
11	198-204	199.25	245	292-286	
12	204-210	205.25	251	298-292	
13	210-216	211.25	257	304-298	

Note: All frequencies in MHz

be discussed in the following sections. They also have all been taken into account by the FCC in its rather stringent rules for geographic isolation of UHF-TV channel assignments and, as will be illustrated in an interesting example, have led to rather inefficient use of the UHF spectrum.

a. Adjacent-Channel Interference As has been discussed in Section 2, the transmitted television signal contains three separate pieces of information (picture, color, and sound) modulated onto three different carriers within the 6-MHz channel (see Figure 1), and the receiver must sort these out as cleanly as possible by means of appropriate filters, as well as reject all spurious out-of-band signals.

Filter technology is such that it is difficult to achieve sharp, distortion-free boundaries between an accepted and a rejected band of frequencies, particularly if one is trying to hold circuitry costs to a minimum as in the majority of home TV receivers. Thus in order to satisfactorily pass the desired band of picture frequencies, most TV receivers accept some transmission of frequencies beyond the edges of this band. Note, however, that for any particular channel within a contiguous group of channels, the sound carrier frequency for the next lower channel is only 1.5 MHz below the picture carrier of desired channel, and the picture carrier of the next higher channel is only 1.5 MHz above the sound carrier of the desired channel, both rather close so far as band-pass filtering technology goes. However, since these adjacent carriers are discrete interference sources and their exact relative frequencies are known, additional, narrowband rejection filters called "traps" are used in most (but not all) TV sets to improve adjacent-channel interference rejection.

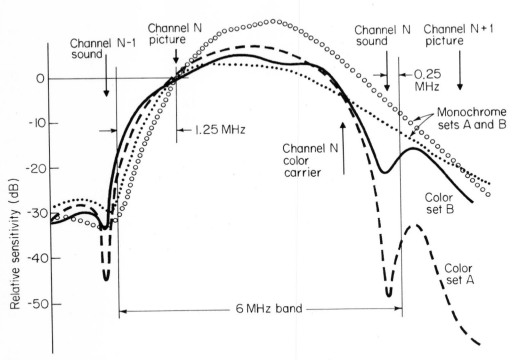

Figure 3. Picture-IF selectivity characteristics of typical television receivers (from NCTA Supplementary Comments to FCC, January 6, 1971).

Figure 3 shows typical picture IF response curves measured for several monochrome and color receivers.[5] Note that color set A includes strong frequency traps for the lower-adjacent and the in-channel sound carriers and for the upper-adjacent picture carrier. Color set B also has traps but has substantially (28 dB) less protection against the in-channel sound and the upper-adjacent picture carriers, and 10 dB less against the lower adjacent sound carrier (note also that the monochrome sets, which don't need to protect the 3.5795-MHz color carrier, don't bother with traps for the in-channel sound carrier). The point here is that adjacent-channel rejection characteristics vary widely from set to set, not only as originally manufactured but perhaps even more so as circuitry ages and gets out of adjustment (for example, trap filters may shift in frequency). Rejection characteristics also vary as a function of signal levels.

Recognizing the difficulty of designing receivers to properly cope with strong adjacent-channel signals over their useful life, the FCC has followed a strict policy of never assigning adjacent channels to TV transmitters separated by less than 60 miles at VHF, or 55 miles at UHF. Of course, many receivers are located such that they are able to pick up stations on adjacent channels, but because of this separation rule, one of the signals will usually be weaker than the other and not interfere with the stronger one. When objectionable interference is encountered (perhaps in trying to view a distant station), many set owners are able to use a directional antenna to strengthen the desired signal relative to a stronger undesired one.

In CATV, adjacent-channel transmission is the norm and cable systems must thus be carefully engineered and maintained both to keep all signal levels balanced with respect to one another and at an optimum overall level at each subscriber's tap. The sound carriers have given particular trouble, and most cable systems have had to keep them at lower levels relative to the picture carriers than prescribed by the FCC for broadcast TV, leading to marginal sound on some receivers. Where individual receivers give trouble, they may require special adjustment or even modification or replacement.

b. Oscillator Radiation   The oscillator interference problem is as follows. Each TV receiver acts like a little transmitter, radiating a small amount of power from its local oscillator at a frequency that depends on the channel to which the receiver is tuned. This radiation can escape either back through its antenna (or into the CATV cable), into the ac power line, or directly from the receiver chassis. If the oscillator frequency for one channel falls within the frequency limits of another higher channel, and two neighboring TV receivers are respectively tuned to these particular channels, the oscillator radiation from the receiver tuned to the lower channel might be strong enough to be picked up as a spurious signal at the neighboring receiver and interfere with its reception of the higher channel. The first four columns of Table 2 show the 12 VHF channels, their transmitted picture carriers, and the frequencies to which a receiver local oscillator must be tuned to receive each channel.[6] Note that none of the local oscillator frequencies for Channels 2-13 fall within the transmission band of any other channel, and thus no particular safeguards were necessary either in geographic assignments of VHF stations or in stringent limits on oscillator leakage from VHF receivers. The only possible oscillator problem in the VHF allocations is that the second harmonics of the oscillator frequencies for Channels 2 and 3 fall within the bands of Channels 11 and 13, respectively, but second- and third-harmonic radiation is generally 10-20 dB weaker than the direct oscillator radiation in any given receiver.

Although oscillator radiation interference is not a problem within the 12 VHF channels because of the way in which the channels are allocated, it is a problem in both UHF-TV and in cable systems using mid- and super-band channels. As will be discussed later, many oscillator/channel overlaps occur in both these cases, and means must be taken to keep oscillator interference below troublesome levels.

c. Image Interference   Image interference in any superheterodyne radio or TV receiver is caused by the fact that the mixing of the local oscillator signal with incoming radio-frequency signals to convert the desired signals to the IF frequency band for amplification can also create a number of other sum and difference frequencies. As noted previously, the television standard is that desired frequencies are lower than the receiver oscillator frequency, and the IF amplifier accepts the frequency band equal to the oscillator frequency minus the frequency limits of the desired channel. However, the mixer also creates the same band of difference frequencies if any incoming signals present at its input are the same amount <u>higher</u> than the oscillator frequency, thus the term <u>image</u>. Good receiver design dictates that the input radio frequency circuits (part of the channel tuner) discriminate against image reception by passing only the desired channel frequencies and rejecting all others, thus preventing any significant input signal energy in the image frequency band from ever reaching the heterodyne mixer. The extent to which this goal is achieved in a particular receiver (or CATV converter) design is called its "image rejection ratio."

The fifth column in Table 2 shows the receiver image band for each VHF channel. Note that the channel frequency relationships are inverted in image reception, and that to receive a correct picture on an image frequency the picture carrier would have to be 1.25 MHz below the upper image band edge, with the picture modulation band extending downward. Thus the image frequency range for objectionable interference beats for any channel is from perhaps 0.75 to 2.75 MHz below the upper edge of its image band. Only one potential image problem exists in the VHF-TV allocations — Channel 7 partially overlaps the image band for Channel 6. In this case, however, the Channel 7 picture carrier is 0.5 MHz <u>above</u> the image frequency of the Channel 6 picture carrier; thus any signal energy resulting from image reception of Channel 7 when viewing Channel 6 will appear in the lower sideband of the true Channel 6 signal and be attenuated at least 10 dB in the IF amplifier (see Figure 3). This is called a "negative" beat and should cause no trouble, except in receivers with very poor image rejection characteristics, but even so, the FCC has apparently tended to avoid allocation of both Channels 6 and 7 in the same market (there are currently only four cases of 6-7 broadcast coallocation: Denver, Miami, Omaha, and Spokane). Channels 6 and 7 have been routinely carried side-by-side in traditional 12-channel cable systems, however, apparently with little or no trouble. The inference here is that most TV receivers have image rejection characteristics that are satisfactory, at least in regard to this particular "negative" image beat between Channels 6 and 7. However, "positive" in-band image beats can occur when mid- and super-band channels are added, and the channel converters for such augmented cable systems must be designed to cope with them.

The preceding discussion has considered only image interference from picture carriers. Each TV signal also carries significant energy in its sound carrier located 4.5 MHz above the picture carrier. With an oscillator offset of 45.75 MHz, image interference from a sound carrier affects the next higher channel than the one affected by its associated picture carrier. For example, the picture and sound carriers for mid-band cable Channel E (144-150 MHz) are at 145.25 and 149.75 MHz, respectively. Table 3 shows the spurious IF signals that these can create in image reception. Since the receiver IF passband is from 41.25 to 46.5 MHz, the Channel E picture carrier image can affect Channel 2, and the sound carrier image, Channel 3. The other two beats are outside the IF pass-band.

Finally, it should be noted that any given channel can appear as an image to <u>lower</u> channels, at the same time that <u>higher</u> channels appear as an image to it. Thus image relationships to be considered in contiguous channel blocks with 6-MHz spacing are ±15 channels (picture carrier image) and ±14 channels (sound carrier image). These will be discussed further in Section 6.

Table 3. Image Interference from a Sound Carrier

Receiver Tuned to Channel	Oscillator Frequency	IF Signal from Channel E Picture Carrier	IF Signal from Channel E Sound Carrier
2	101	<u>44.25</u>	48.75
3	107	38.25	<u>42.75</u>

d. Intermodulation Intermodulation is a term describing the interaction of signals on different frequencies to create various sum and difference signals on other frequencies. There is no mechanism for such interaction as signals propagate through the air, but it can and does occur when multiple signals pass through any circuitry containing active elements (amplifiers, mixers, detectors, etc.). Thus in broadcast TV the presence at the receiver antenna of multiple station signals on various carrier frequencies results in some unavoidable interaction between them in the receiver circuitry, which can create visible picture interference depending on the frequencies of the intermodulation products and their strength (which depends on the quality of the receiver design).

The most troublesome (but not the only) intermodulation effect is the third-order beat involving the second harmonic of one interfering signal minus the frequency of another interfering signal. Take, for example, VHF Channels 9 and 11 (picture carriers 187.25 and 199.25 MHz). The two possible third-order beats between these two carriers fall precisely on the picture carriers of other channels:

$$2(187.25) - 199.25 = 175.25 \text{ MHz (Channel 7)}$$

$$2(199.25) - 187.25 = 211.25 \text{ MHz (Channel 13)}$$

A Channel 7 beat is also produced by the combinations 8-9 and 10-13, and a Channel 13 beat by 7-10 and 11-12; but 8-9 and 11-12 problems are unlikely in broadcasting because of the non-adjacent-channel rule.

Another important but apparently less troublesome intermodulation effect is the second-order beat involving one frequency plus or minus another one. For example, 140 MHz can be produced by 60 plus 80, or 220 minus 80.

The intermodulation performance of receivers at VHF frequencies is generally good enough so that no special precautions against it were necessary in VHF station assignments (7-9-11-13 is a common comarket assignment pattern). In cable systems, however, these intermodulation products can be created in the cable system itself since all VHF channels are carried side by side through various head-end and cable amplifiers. Here again, careful design and maintenance of the cable system is necessary, particularly as additional mid- and super-band channels are added and the total bandwidth of the amplified frequencies is increased.

e. IF Beats A particular form of second-order intermodulation called an "IF beat" results if two channels separated approximately by the receiver IF frequency (45.75 MHz) should somehow interact in the circuitry of a receiver to create difference frequencies at the input to the IF amplifier that fall within its pass-band and will be amplified along with the desired IF signal. The most probable such situations are the beats between the desired channel and the ones located 7 and 8 channels above and below it in frequency. This cannot occur in the frequency assignment pattern of Channels 2 through 13, but can in UHF-TV and in cable systems using mid- and super-band channels in addition to the VHF channels. UHF-TV avoids the problem by geographic isolation of IF-beat channels. Cable systems will either have to live with it or, as some people now propose, design cable converters or special cable receivers with a much higher IF frequency. This would also help the oscillator leakage and image problems.

f. Application of Anti-Interference Rules in UHF-TV

Although UHF transmission and reception are not of concern in cable systems, the cumulative effect of the above mutual interference problems in large blocks of contiguous television channels are perhaps best illustrated by the UHF "taboos" (non-co-assignment rules for UHF channels based on the possibility of mutual interference). In a recent paper proposing a new look at the question of geographic/frequency allocations,[7] Norman Parker (Motorola) traces the history of the UHF "taboos" and gives the following example. If a particular UHF channel, say, 29, is assigned in a given area, FCC rules state that a total of 18 other UHF channels cannot be assigned to transmitters operating within distances ranging from 40-120 miles from the Channel 29 transmitter. These channels, their potential interference effect, and the minimum mileage separations are shown in Table 4. This process is, of course, repeated for each channel. Thus in any 31-channel span of frequencies, almost 60 percent of the channels can cause interference to (or receive interference from) the one in the center of the span, and their use is also restricted for UHF broadcast purposes. Although the "taboos" were adopted by the FCC at a time (1952) when the design of UHF-TV receivers was in its infancy and were deliberately ultraconservative, they are still in effect. Part of Mr. Parker's thesis is that in the light of modern UHF-receiver performance, some of these restrictions could perhaps now be relaxed. They still are a good blueprint, however, for potential interchannel interference problems in 20- 40-channel cable systems, particularly since the cable system itself represents an additional mechanism for the generation of intermodulation products.

5. Interference Problems Peculiar to Cable

The interchannel interference problems described in Section 4 occur in both broadcast TV and cable TV and are all generated in or by television receivers as a result of the interaction of receiver design parameters and particular channel-frequency relationships. As has been discussed, their deleterious effect on received pictures can be eliminated or reduced (at least below levels discernible to the viewer) by some combination of proper receiver/converter design, proper choice of channel frequencies and transmission standards, and, in broadcasting, by geographic isolation of troublesome station frequency assignments. There are two additional interference problems that are peculiar

Table 4. Channel 29 Interference Example

Relative Channel Spacing	Actual Channels Idled If Channel 29 Is assigned	Type of Interference Potential	Minimum Separation Radius (in miles)
±1	28, 30	Adjacent channel	55
±2, 3, 4, 5	24, 25, 26, 27 31, 32, 33, 34	Intermodulation	40
±7	22, 36	Oscillator	120
±8	21, 37	IF beat	40
±14	15, 43	Sound image	120
±15	14, 44	Picture image	75

to cable systems and have a somewhat different impact on cable and receiver engineering and the choice of cable frequencies: cochannel interference with cable signals by broadcast TV signals and possible interference of cable signals, should they escape from the cable, with other radio services.

a. Cochannel Interference by Broadcast Signals. In the early days of cable TV, systems were usually installed where there were no strong television stations, and programs were generally carried on the cable on the same channels on which they were broadcast. As cable expanded into areas nearer the stations, it was found that with such on-channel carriage, many subscribers were troubled by interference caused by dual reception of the same program: a direct, unwanted signal leaking into the cable or their receiver from the broadcast station, and the same signal as picked up by the cable system antenna, processed by the cable head end, and transmitted through the cable. Because the cable signal travels a greater distance, at least part of which is through a slower medium (1.2 μsec per 1,000 feet in the cable as opposed to 1 μsec per 1,000 feet through the air), the interfering direct-pickup signal arrives first and creates various effects depending on its strength relative to the cable signal and the relative signal delays. Particular effects in received pictures are "left-hand ghosts," vertical black synch bars, or random stripe patterns.

As a result of the on-channel carriage problem, it has been common cable practice to carry strong local stations on different "quiet" channels on the cable. This represents both an inconvenience for the subscribers (these stations are not found at their normal positions on the dial), and a waste of channel capacity. In some metropolitan areas with many local stations, as many as five or six cable channels are unusable because of on-channel interference, leaving only six or seven channels in what should be a 12-channel cable. It was this problem that has prompted the installation of many dual-cable systems (with subscriber A/B selector switch) to increase capacity, but note that the same interference rules apply to both cables, and some dual-cable systems can still offer only 12 to 14 channels.

Broadcast signals can leak into any part of the cable system (up to and including the final connection to the antenna terminals of a subscriber receiver)
or directly into the receiver itself via pickup in its internal wiring or circuits
(note that on-channel leakage is of little concern in broadcast reception). On-
channel leakage pickup into the cable system is susceptible to correction in a
large number of cases by careful engineering and installation; the methods for
this have been described by Archer Taylor.[8] However, in this same article,
Mr. Taylor also states, "Where cable subscribers are located less than about
10 miles from a local TV transmitting antenna, it is probable that nothing
short of modification of the TV receiver will cure some cases of direct pickup. ...In the final analysis, the best answer is the redesigning of TV receivers so that they are isolated (shielded) from ambient fields. Work on this is
in progress. We stand a better chance of success when we have 20 million
subscribers than we do with 3 1/2 million (1969)." These comments of course
apply to cable systems in which receivers are directly connected to the cable,
that is, without use of subscriber (set-top) converters. It would probably take
at least ten years to turn over the present inventory of sets in use, even if
only shielded sets were to be produced from now on.

When certain (but not all) types of converters are used, the receiver is left
tuned to a "quiet" VHF channel, eliminating any cochannel problem in the receiver. However, the converter must then be designed with sufficient shielding to prevent cochannel interference effects within it, but this is relatively
easy to do, and the requirement is clear-cut, which it hasn't been for TV receivers. At the same time, converters can be designed to avoid all of the interchannel interference effects described in Section 4.[9] In this regard, it is
interesting to note that the first CATV converter appeared less than six years
ago, and that the number in service is still quite small (in the tens of thousands). Most of these have been assembled largely from standard TV tuner
components, modified only for the new channel frequencies, and it is only recently that more than cursory attention has been paid to the possibilities af-
forded by new designs optimized for broadband cable purposes.

b. Cable Interference with Other Services. The signals carried in a cable system are at radio frequency and thus are quite capable of propagating in all directions through the air if not properly confined within the coaxial cable and
shielded equipment boxes that form the system. Although the cable signal levels
are quite low compared to those normally applied to the antenna of a transmitter, they can still create significant field intensities (and thus interference)
up to a few thousand feet from the cable in the case of the worst possible type
of shielding break—one that somehow acts as a perfectly efficient isotropic
antenna and is located immediately following a cable amplifier where the signal levels are highest.[10] Even without a major cable fault of this sort, cable
systems can never be perfectly shielded, and there will always be some radiation, even if it is only detectable within a few feet of the cable.

Radiation from a cable system carrying only the 12 standard VHF channels
should, of course, cause interference only with television reception and only
in receivers located nearby. However, when additional mid- and super-band
frequencies are carried, cable radiation could possibly affect other radio
services as follows:

Mid-Band

88-108 MHz — FM broadcast (may actually be carried on the cable)
108-136 MHz — Aeronautical
136-144 MHz — Government, space research, meteorological
144-148 MHz — Amateur
148-151 MHz — Radionavigation
151-174 MHz — Land-mobile, maritime, government

Super-Band

216-225 MHz — Government, amateur
225-329 MHz — Government

The FCC has long recognized the problem of cable radiation and specifies in subpart D, Section 15.161, on "Radiation from a Community Antenna Television" of its Rules and Regulations:

Radiation from a community antenna television system shall be limited as follows:

Frequencies (MHz)	Distance (feet)	Radiation Limits (micron v/m)	
		General requirement	Sparsely inhabited areas[1]
Up to and including 54	100	12	15
Over 54 up to and including 132	10	20	400
Over 132 up to and including 216	10	50	1000
Over 216	100	15	15

[1]For the purpose of this section, a sparsely inhabited area is that area within 1000 feet of a community antenna television system where television broadcast signals are, in fact, not being received directly from a television broadcast station.

Although not immediately apparent because of the fact that the specified distances are not all equal, the protection in the band between 54 and 216 MHz is considerably higher than for frequencies above and below it. Note also that permissible levels are much higher in the TV bands (and mid-band) in areas remote from TV stations where every TV receiver is expected to be on the cable and not using an antenna.

Now that substantial use is beginning to be made of mid- and super-band transmission on cable systems, the FCC and the FAA, among others, have become concerned about potential interference dangers with the radio services just listed, particularly the aeronautical (ATC) services. One situation

postulated is that an aircraft in taking off or making a landing approach may
fly quite low over a CATV cable at the same time that it is a considerable distance from the ATC transmitter. If the cable were faulty and radiating at the
ATC frequency, the resulting interference at the aircraft receiver might have
critical consequences, even though the time exposure to it would be brief (on
the order of 40 seconds maximum). One analysis to date concludes that this
should not be a serious problem, but that certain safeguards should perhaps
be established:[11]
1. Adoption of upper signal level limits for CATV cables
2. Installation of break-detection devices in all cable systems
3. Placement of mid-band CATV carrier frequencies between ATC channels.
   This subject is just coming under intensive study and discussion by the
FCC, the FAA, the Offices of Telecommunications and Telecommunications
Policy, the IEEE Cable Television Task Force Committee, and others, as
part of the current debate on cable standards. The intent here is not to shed
any light on the solution but only to point out that this question adds another
dimension to the cable frequency allocation problem, and that the ultimate decisions, particularly in regard to item (3), may have a deleterious effect on
the channel capacity of future systems.

6. The Cable Channel Allocation Question

So long as cable systems simply carried television signals on the same channel frequencies assigned by the FCC for VHF broadcasting, the only two of
the six interference effects discussed in Sections 4 and 5 that represented any
real problems for cable operators were adjacent-channel interference and on-
channel pickup; the former was handled by careful signal balancing and level-
control, and the latter by idling the troublesome channels.[12] However, as new
cable channels began to be added in a few systems starting about 1968, all of
the effects applied, initiating a debate that has just reached full intensity in
the past few months and in which this study has had a small measure of involvement. The following are indicative of the questions now under consideration.

a. A Noncontiguous Channel Scheme

In a paper presented at the 1970 NCTA
National Convention, M. E. Jeffers (Jerrold Electronics Corp.) set forth the
oscillator and image beat effects in the generally accepted mid- and super-
band assignments (shown by the boxes in the last two columns of Table 5) and
proposed a realignment of the mid- and super-band cable-channel assignments
to eliminate these problems.[13] His proposal for realignment was to alter the
band limits for the mid- and super-bands and insert frequency offsets between
these channels at appropriate points, with the objective of placing all oscillator and image beats at band-edge (between channels) where their interference
effect would be minimal.
   Since the new channel assignment charts in Mr. Jeffers' paper did not cover
the entire 54-300 MHz cable spectrum, an overall chart based on his mid- and
super-band assignments was prepared as shown in Table 6. This gave the total
channel count (35) but also brought to light the fact that because of the complexity of the interchannel relationships, it is almost impossible in such a fre-
quency-juggling exercise to avoid creating new problems while trying to cor-

Table 5. Present VHF, Mid-Band, and Super-Band Assignments

Channel	MHz	Video Carrier	Oscillator Frequency	Image Frequency	Channel Affected by Oscillator	Video Carrier on Image Frequency
2	54-60	55.25	101	146.75		E (1.5)
3	60-66	61.25	107	152.75		F (1.5)
4	66-72	67.25	113	158.75		G (1.5)
	----- 10					
5	76-82	77.25	123	168.75	A (1.75)	—
6	82-88	83.25	129	174.75	B (1.75)	7 (-.5)
	----- 32					
A	120-126	121.25	167	212.75	H (3.75)	13 (1.5)
B	126-132	127.25	173	218.75	I (3.75)	J (1.5)
C	132-138	133.25	179	224.75	7 (3.75)	K (1.5)
D	138-144	139.25	185	230.75	8 (3.75)	L (1.5)
E	144-150	145.25	191	236.75	9 (3.75)	M (1.5)
F	150-156	151.25	197	242.75	10 (3.75)	
G	156-162	157.25	203	248.75	11 (3.75)	
H	162-168	163.25	209	254.75	12 (3.75)	
I	168-174	169.25	215	260.75	13 (3.75)	
7	174-180	175.25	221	266.75	J (3.75)	
8	180-186	181.25	227	272.75	K (3.75)	
9	186-192	187.25	233	278.75	L (3.75)	
10	192-198	193.25	239	284.75	M (3.75)	
11	198-204	199.25	245	290.75		
12	204-210	205.25	251	296.75	3.75 MHz	
13	210-216	211.25	257	302.75	is a color	
J	216-222	217.25	263	308.75	beat	
K	222-228	223.25	269	314.75		
L	228-234	229.25	275	320.75		
M	234-240	235.25	281	326.75		

Mid-band

Super-band

Table 6. Suggested New Mid- and Super-Band Allocations by M. F. Jeffers

Channel	MHz	Video Carrier (6-MHz separations except as noted)	Oscillator Frequency	Image Frequency	Channel Affected by Oscillator (Beat Frequency)	Video Carrier on Image Frequency (Beat Frequency)
2	54-60	55.25	101	146.75	—	—
3	60-66	61.25	107	152.75	—	—
4	66-73	67.25	113	158.75	—	—
		------ 10				
5	76-82	77.25	123	168.25	B (-1.25)	—
6	82-88	83.25	129	174.75	C (-1.25)	7 (-.5)
		------ 35				
A	117-123	118.25	164	209.75	H (- .75)	—
B	123-129	124.25	170	215.75	—	I (-.5)
C	129-135	130.25	176	221.75	7 (+ .75)	I (-.5)
D	135-141	136.25	182	227.75	8 (+ .75)	J (-.5)
		------ 6.5				
E	141.5-147.5	142.75	188.5	234.25	9 (+1.25)	K (0)
F	147.5-153.5	148.75	194.5	240.25	10 (+1.25)	L (0)
G	153.5-159.5	154.75	200.5	246.25	11 (+1.25)	M (0)
		------ 10				
H	163.5-169.5	164.75	210.5	256.25	13 (- .75)	O (-2)
		------ 10.5				
7	174-180	175.25	221	266.75	I (-1.25)	P (-2)
8	180-186	181.25	227	272.75	J (-1.25)	Q (-2)
9	186-192	187.25	233	278.75	K (-1.25)	R (-2)
10	192-198	193.25	239	284.75	L (-1.25)	S (-2)
11	198-204	199.25	245	290.75	M (-1.25)	T (-2)
12	204-210	205.25	251	296.75	N (-1.25)	U (-2)
13	210-216	211.25	257	302.75	O (-1.25)	
		------ 11				

Jeffers channels (A–H)

Jeffers channels

I	221–227	222.25	268	313.75	
J	227–233	228.25	274	319.75	
K	233–239	234.25	280	325.75	
L	239–245	240.25	286	331.75	
M	245–251	246.25	292	337.75	
N	251–257	252.25	298	343.75	
O	257–263	258.25	304	349.75	
	---------- 10.5				
P	267.5–273.5	268.75	314.5	360.25	P (-.75)
Q	273.5–279.5	274.75	320.5	366.25	Q (-.75)
R	279.5–285.5	280.75	326.5	372.25	R (-.75)
S	285.5–291.5	286.75	332.5	378.25	S (-.75)
T	291.5–297.5	292.75	338.5	384.25	T (-.75)
U	297.5–303.5	298.75	344.5	390.25	U (-.75)

35 channels

rect other ones. For example, Mr. Jeffers was successful in placing all mid-band/low-VHF-band and super-band/high-VHF-band image and oscillator beats at band edge, but in the process he created three dead-beat images of super-band channels on mid-band channels and five undesirable ("positive") oscillator beats on high-VHF band channels by mid-band channels, as shown by the boxes in the last two columns of Table 6.

These findings were brought to Mr. Jeffers' attention, and he has agreed with them. However, he feels that these particular problems can be eliminated by proper converter design — either by holding oscillator radiation and image rejection to very tight limits or by using a different, very high IF frequency in converters to completely eliminate oscillator and image beats. These measures would not help reduce intermodulation interference effects, which become exceedingly complex with such noncontiguous channel spacing.

b. A Contiguous Channel Scheme   The interchannel interference effects discussed in Section 4 have long been a factor in the analog frequency-multiplex systems (L_1, L_2, etc.) utilized by the telephone company in its intercity trunk circuits, and methods have been developed to cope with them. One of these is the careful control of channel frequency ratios, including derivation of all carriers by suitable frequency multiplications from a common master oscillator.

Mr. I. Switzer (Maclean-Hunter Cable TV Limited) has presented a paper advocating use of such techniques in CATV systems.[14] Mr. Switzer has also corresponded with the author on the mid- and super-band frequency allocation question, advocating a contiguous set of channels at constant spacing to help overcome intermodulation effects. If such a plan were carried through completely, Channels 5 and 6 would have to be moved 4 MHz lower in frequency to fit into the plan. This would not be of any consequence in a system where all subscribers have converters or special cable receivers but would be significant if "dual-class" service is offered, as will be discussed next.

A contiguous-channel allocation scheme would yield a total of 41 CATV channels if the spectrum between 54 and 300 MHz were completely filled. As has been discussed, some frequencies in this range may be unusable for CATV if it is determined that a serious interference hazard exists with other services, particularly with aeronautical radio-navigation aids in the 108-136 MHz band.

c. Dual-Class Service Considerations   One factor that further complicates the whole frequency allocation issue is that many operators may desire or need to offer two classes of compatible service on the same cable: VHF-only, at one price for subscribers with standard receivers and no converters, and VHF plus cable channels, at a higher price for subscribers using converters or special cable receivers. Thus a channel-utilization scheme that depends primarily upon the ability of converters or special cable receivers to eliminate interference effects might be unacceptable for such dual-class service — those standard TV receivers that are directly connected to the cable might not be able to cope with interference beats resulting from the presence of non-VHF cable channels. Here an allocation scheme that provides additional protection to the VHF-TV channels may be required, such as the proposal by M. F. Jeffers. On the other hand, the optimum solution for broadband cable service may turn out to involve relocation of the present VHF Channels 5 and 6 downward by 4 MHz to fit into a contiguous channel allocation scheme as discussed previously, and a nonconverter subscriber to such a system would be unable to receive Channels 5 and 6 and be "channel poor." Current discussions include the possibility that there may have to be several alternate cable-frequency allocation standards to fit different situations,[15] but this would be unfortunate from the point of view of cost-effectiveness in equipment manufacture and interchangeability or portability of equipment from system to system. Eventual large-scale production of augmented-channel TV sets for cable use would seem to depend upon having a single nationwide standard, such as now exists for broadcast TV.

III. AUGMENTED-CHANNEL SYSTEMS

1. Use of Converters in Multichannel systems

Multicable systems operating directly into standard TV sets should provide 12 channels per cable but lose the same direct-pickup channels on both cables. Thus a more typical capacity in upper-market areas is eight or nine channels per cable, down to as little as five or six in major markets such as New York City and San Francisco. This hardly provides the channel capacity desired for "wired nation" services.

Although converter systems will be discussed more fully in the next section, it should be noted here that the current state-of-the-art for converter systems is 21-25 nominal channels per cable; thus adding them to a dual-cable system immediately jumps usable capacity to the 42-50 channel region (depending on the type of converter, some of these may be lost due to direct pickup, harmonic problems, etc.). In fact, this combination seems much more promising for obtaining such capacity than foreseeable extensions of single-cable converter technology. If converter systems improve to 30 or more channels per cable in the future, combination dual-cable/converter systems can then provide 60 channels or more. The other alternative for very large capacity is a switched system.

The addition of converters to a dual-cable system adds about $30 (current price range for 20- to 25-channel converter units) to the other per-subscriber cable-plant costs. Using the figure of $70 per subscriber for a modern two-way, dual-cable plant, the total distribution cost of a dual-cable/converter system with upstream channel capability is about $100 per subscriber on a 100-percent saturation basis. This is the figure that is used as a basis for comparison with the single-cable and switched systems.

Use of subscriber converters to solve the direct-pickup problem in cable reception dates from late 1965, when shielded VHF-VHF pretuners for the 12 standard VHF channels were first developed by the International Telemeter Corporation as a solution to the direct-pickup problem.[16] The function of these units was to perform the channel tuning in a shielded environment (instead of in the set tuner) and convert to a "quiet" channel (usually 12 or 13) to which the TV set is left tuned. Channels 12 and 13 are the best converter-output channels for a variety of reasons described in Mr. Court's paper, providing alternate output channel choices that are adjacent to ensure that one or the other of them will be "quiet" in all areas. Later, converters were constructed with the capability for tuning additional non-VHF frequencies, leading to greater channel capacity per cable.

2. Characteristics of Present Augmented-Channel Converters

The cable industry has generally adopted a set of nine "mid-band" cable channels between 120 and 174 MHz (see Table 5); thus a converter designed to handle these mid-band channels in addition to the standard 12 VHF channels has a maximum capacity of 21 channels. Capacity beyond 21 channels is obtained by also using the "super-band" channels, which start at 216 MHz (above Channel 13) and run as high as cable bandwidth will allow. Since few presently installed cables will handle frequencies higher than 240-246 MHz, the current state of the art in tuner-converters is to include the first four or five super-band channels (216-246 MHz), yielding 25- or 26-channel capacity (these numbers were also convenient in converter design).

Not all converter designs on the market are of tuner type. One "tunerless" type block-converts seven mid-band channels to high-VHF band (Channels 7-13), with a switch to determine whether the TV set tunes from this group or the standard, unconverted Channels 7-13. Another type is similar in operation but block-converts seven super-band channels to Channels 7-13. Both of the above yield a 19-channel capacity. Still another unit block-converts nine mid-band channels to some part of the UHF band for tuning by the TV-set UHF tuner. This latter unit yields 21 channels, but does not provide detent-tuning for the mid-band channels (tuning is the same as for UHF broadcast stations). Since the converted channels are all adjacent, which they never are in off-the-air UHF, one might expect some difficulty in tuning, especially since the selectivity of UHF tuners is usually not as good as that of VHF tuners (actual performance of this unit in practice has not been investigated). All of the tunerless converters that convert to the high VHF band are still subject to direct-pickup problems on the VHF channels, since all tuning is done by the TV receiver.

3. Converter Costs

Present 25-channel tuner-converter costs are quoted in the range from $30 to $50 depending on the type and manufacturer. Actually, converter technology is in a period of transition. Only a few tens of thousands of converters of all types are yet in use, and although many of these have been cleverly devised from available VHF-TV tuner components, which are standard, highly tooled production items and thus reliable and low in cost,[17] they do not represent an optimum solution to the interchannel interference problems in augmented-channel systems or to further expansion of channel capacity. The industry is now moving toward new all-channel designs that can take full advantage of interference rejection possibilities and the new extensions in cable bandwidth (to 300 MHz). Until large-scale production builds up over the next few years, such new designs will undoubtedly cost more than present 25-channel designs, perhaps in the $50-60 region.

Assuming that per-subscriber installation cost of a single-cable trunk and feeder system, including drops, is about $50 on a 100-percent saturation basis, the addition of converters to provide 25-channel capability (or more) brings the total per-subscriber cost of an augmented-channel, single-cable system to $80-$100, depending on actual converter cost.

4. Ultimate Single-Cable Capacity

There are obvious limits on the maximum number of downstream channels that can be carried on a single cable, including the following:

Total cable system bandwidth (largely determined by cable amplifiers)

Channels that may be unusable for one reason or another (interference effects)

Channel needs for upstream transmission.

Within the past six months, cable amplifiers giving good performance up to 300 MHz and higher have become available. One series of cable-mounted trunk and distribution amplifiers, offered by Anaconda Electronics, incorporates a postage-stamp-sized hybrid amplifier chip made by the Hewlett-Packard Company. This amplifier chip sells for only $50 and provides excellent performance from 40 to 330 MHz (±1 dB from linear slope over this range, ±0.3 dB from 40-270 MHz), with very low cross modulation and second-order intermodulation products (-89 and -80 dB, respectively).[18] Thus 300-MHz bandwidth is now a reality.

The question of how many usable channels can be fitted into the available downstream bandwidth between 40 MHz and 300 MHz on a single cable remains open. At face value the maximum would be (300-42)/6 = 43 channels, but there are a number of factors that appear to militate against obtaining this exact number. These are listed here:

The FCC/FAA may proscribe some frequencies in the 108-136 MHz region by FCC/FAA as a possible hazard to aircraft navigation/control.

The present VHF Channels 5 and 6, if retained on the cable, don't fit into a contiguous-channel allocation scheme.

Converters to handle such a large number of channels and cope with all the interference effects possible in a bandwidth of almost three octaves remain to be demonstrated. Some particular channels may present unsolvable problems and have to be left idle.

It is certain that converter technology will progress beyond the present plateau of 25-26 channels and may reach 30-35 channels in the near future. How close one can come to completely filling the available bandwidth with usable channels remains to be seen.

IV. THE REDIFFUSION SWITCHED TV DISTRIBUTION SYSTEM

The previous sections have discussed progress and prospects in the extension of traditional CATV technology, which amounts to broadcasting on a cable network rather than through the air. At least two CATV equipment companies have recently decided that the many complex interference problems encountered in the frequency multiplexing of large numbers of TV channels on an imperfect medium (cable) and the selection among them by subscriber-end "de-multi-plexers" (converters and/or TV-set tuners) are better solved, at least in the urban multi-VHF-station environment, by adoption of a completely different approach—remote channel selection in "centralized" switching equipment and transmission of only one (or a few) channels per cable at sub-VHF-band frequencies (less than 50 MHz) selected for minimum interference problems. This, of course, requires that the subscriber "drops" must be extended to a common point where the selection switching is performed, instead of simply being tapped into the nearest point on a multichannel, frequency-division-multiplex trunk/feeder system that carries the same signal menu past all subscribers.

The radiating structure of the drops in switched systems characterizes them as "hub networks," as opposed to the conventional CATV "tree network." Since the telephone system as a point-to-point communication service is also a hub network, considerable speculation has developed about this similarity and the possibility of enlarging and/or merging functions up to and including two-way, point-to-point "videophone" services on the same broadband subscriber cables used for CATV distribution.

1. Introduction

This section presents a description and cost analysis of the "Dial-a-Program" cable TV distribution system being actively promoted in the U.S. market by Rediffusion International, Ltd., a London-based company. The system was first announced in several papers published in 1968 and was demonstrated in the United States and Great Britain in mid-1969 and early 1970. Subsequently, arrangements were made with Mr. Richard Leghorn, the owner of Cape Cod Cablevision in Hyannis, Mass., to make an experimental installation in Dennis Port, Mass., for which he held an unexercised franchise. Installation got under way in April 1970, and service began in August, providing 12 program channels taken off the Hyannis cable system, plus additional experimental channels. There are presently 122 subscribers connected (January 1971).

2. Background

Rediffusion International, Ltd., has had many years of experience in Great Britain in radio and TV distribution over twisted-pair cables, using "HF" cable frequencies in the range from 0-10 MHz. In their British installations, the trunks, feeders, and subscriber drops carry each TV and radio program (up to six of each maximum) on a separate wire pair, and each subscriber has a manual selector switch. The TV picture carrier used is 5.9 MHz (or 8.9 MHz), and both TV and radio sound signals are carried at baseband at a level sufficient to drive a loudspeaker without amplification.

For those subscribers with a standard TV set, a fixed-frequency inverter translates the selected TV channel from the HF transmission frequency to a convenient VHF channel frequency (the TV receiver is left tuned to this frequency), and the video signal is recovered in the TV set in the normal way. The base and audio signal directly drives the speaker through a volume control (requiring a special connection into the receiver). There is also apparently substantial use in Great Britain of simplified TV receivers without VHF

tuners. In these "wired" receivers, the HF picture carrier is simply amplified and detected to recover the video signal.

In the United States, with its many TV channels, such a distribution system using a wire-pair per channel with home selection would be uneconomic.
There is also interest in Great Britain in expanding the number of available
channels. Rediffusion has thus come up with the "Dial-a-Program" concept,
in which a single HF distribution channel is brought to each subscriber and
program selection of up to 36 channels is performed in a remote switching
exchange under subscriber control via a telephone-type dial. This of course
implies a hub-type network between the switching exchange and the subscribers, as opposed to the tree structure used by Rediffusion in Great Britain and
by the present VHF-CATV distribution systems in the Unites States.

This hub-type switched network has a number of potential advantages over
the tree-structured VHF distribution system when considered in the context
of the many proposals for advanced communications services in the "wired
city" of the future. Among these are the ability to expand the number of subscriber program channels without limit by simply enlarging or cascading the
selector switches in the exchange, the ability to set up the switching so that
privileged information (for example, medical news) can be seen only by selected subscribers, the demonstrated capability (in the Rediffusion system)
for simultaneous "upstream" TV transmission using a 15-MHz vision carrier
on the signal pair of any subscriber cable, and the many possibilities for two-
way, narrow- or medium-band communications services using either or both
of the unique wire pairs for each subscriber (a subscriber cable actually consists of four wires—a signal pair, and a control pair for program selection).

There are also a number of disadvantages to the hub system as compared
to present VHF systems. Among these are the more complex cabling of a hub-
type network, the fact that each TV set to be used independently requires a
separate cable and exchange switch (not just each household), the initial cost
and maintenance problems of the switchgear, and the need for dispersal of
switching locations roughly every one-third mile throughout the served area
because of the 1800-foot length limitation on the unamplified subscriber cables,
or every half-mile if small exchange amplifiers ($18.00 per line) are used,
which permit a 2500-foot maximum subscriber distance. Actually, the use of
remote switching locations aids the hub-type cabling problem by limiting the
number of subscriber cables brought to any one hub, but real estate must be
found and purchased or leased for each of these many remote switching installations.

The relative balance of the above advantages and disadvantages vis-à-vis
those of VHF tree networks is of course of great interest, particularly in any
attempt to predict the most desirable type of system to provide some of the
communications services that have been postulated for a decade or two hence.
The present paper does not attempt to strike such a balance but provides an
input to such an evaluation by documenting the technical features and cost data
for this particular system.

3. The Dial-a-Program System

The following describes the novel technical features of the Rediffusion system,

including the special distribution cable, the program switching exchange, and
the present and possible two-way communications services.

a. Distribution Cable   As previously described, the heart of the Rediffusion
approach is the transmission of TV signals in the high-frequency (HF) region
of the frequency spectrum — 2-15 MHz, rather than at the VHF frequencies
(50-240 MHz) at which they are carried on conventional cable systems. Rediffusion also calls this "Super Video" because of the small frequency trans-
lation (7.94 MHz) of a baseband 6-MHz video signal to produce their HF signal.
The advantage of using such low transmission frequencies is that it permits
use of twisted-pair wires in place of the coaxial cable needed at VHF frequencies. It is well known that standard telephone cable pairs have a usable frequency range (limited by attenuation) up to about one MHz, which is the bandwidth of Picturephone signals. Rediffusion has designed special twisted-pair
cables, which have an attenuation in the 2-15 MHz region comparable to that
of coaxial cable at VHF frequencies — about 2 dB per 100 feet.

The Rediffusion distribution circuit actually consists of four wires twisted
together — what they call a Qwist^{TM} — with two of the wires forming a signal
pair and the other two, interdigitated with the signal pair, serving as a control pair. The 25-gauge signal pair has a usable frequency range of 0-15 MHz,
and the 26-gauge control pair is usable from 0-1 MHz. Special attention must
be paid in manufacturing to produce this performance, but present cable costs
are not unreasonable.

The Qwist cable is produced in two versions: a one-way cable (one Qwist),
used for house drops, and a six-way cable (six Qwists in one sheath), used for
feeders from the switching exchange. These cables are produced by several
British manufacturers in both screened and unscreened types, with the former
having a copper tape shield. Both types are jacketed in a conducting polythene
sheath. The unscreened type, which is less expensive, is used in underground
runs or in aerial runs where interference pickup is not a problem. Where pickup is a problem, the shielded type is used. Adjacent-channel coupling was, of
course, a problem in the six-Qwist cable since the Qwists are in physical contact, but the cable has been carefully engineered (for example, use of different twist pitches on adjacent Qwists) to keep cross-coupling below -46 dB in
the longest cable run. The outer diameter of the six-way cable is about 3/8
inch, and in practice seven of these are laid together to form a 42-circuit
bundle 1-1/4 inches in diameter. Discussions are under way with several
United States manufacturers for production of these cables.

The present costs of Qwist cables (without U.S. duty) are as follows:

Six-way unscreened	$ 87.00	per 1000 feet
Six-way screened	117.00 " " "
One-way unscreened	21.00 " " "
One-way screened	42.00 " " "

There is thus a cost advantage in running as much as possible of the distribution in the six-way cable, since the per-subscriber cost is then $15.00 and
$20.00 per 1000 feet for unscreened and screened cable, respectively.

The six-way cable attenuation is 1.4 dB per 100 feet at 10 MHz, and maximum run of this cable from the exchange to a subscriber is 1500 feet without an amplifier, set largely by the maximum allowable attenuation (21 dB) and partly by the maximum permitted "crossview" between adjacent cable pairs (-46 dB). Individual drops on one-way Qwist may extend 300 feet beyond this (or up to five miles with single-channel repeaters at 3/4-mile intervals, but the repeaters cost $61 each). Except for occasional isolated houses, an attempt would be made to site exchanges so that all runs would be less than 1800 feet and therefore not require amplifiers.

Interexchange trunks in the Rediffusion system use a separate coaxial cable for the HF composite video (7.94 MHz picture, 3.44 MHz FM sound) of each program channel and telephone-type wire pairs for the associated baseband sound signals. The composite video and baseband sound signals are separately amplified in each exchange (thus avoiding any possibility of cross-modulation) and mixed for application to the exchange busbars. At HF, the 0.3-inch coaxial cables used for the trunks have an attenuation of only 40 dB per mile, so no line amplifiers are needed between exchanges, which are generally located about one-third of a mile apart.

Rediffusion quotes a cost of 2.5 cents a foot for the coaxial cable used in the trunks, and one cent per foot for the audio twisted pair (in 12-pair cables). Cable costs for a 36-channel trunk are thus $6,500.00 per mile, plus installation.

b. The Program Exchange   Rediffusion has apparently settled on 336 subscribers as the best size for a program-switching exchange, although no justifications for this choice are given. To some extent, this size seems to have been arrived at both from practical considerations in switch design and from the fact that the service area of an exchange is limited to one-fifth of a square mile or less by the 1800-foot maximum length of subscriber cables. Some of the Rediffusion documents discuss ten exchanges (3360 TV sets) per square mile as a typical installation plan. For very high-density areas, either the 336-size exchanges could be installed in multiples or larger exchanges may be planned (in his IEEE paper,[19] R. P. Gabriel states that an exchange for 5000 subscribers would require a room 27 by 15 feet, but he does not discuss its implementation). The following describes the present 336-subscriber switch as installed at Dennis Port.

The exchange is designed around a rotary, 36-pole switch of novel design. Thirty-six reed switches are arranged in radial configuration and are activated one at a time by a magnet on a rotatable selector arm. The arm is moved by a solenoid-driven ratchet that responds to subscriber dial pulses. Two of these switches are mounted on a printed-wiring card 9 × 15 inches in size, which plugs into the exchange distribution bus as shown in Figure 4. The printed wiring is laid out with alternate wires at ground potential for shielding; thus three 24-pin connectors are used on each card for the 36 channels and associated grounds. Even with this shielding, there can still be some coupling between lines on the cards and in the buses, and Rediffusion has provided for use of vision-carrier offsets of 5244 Hz (one-third the horizontal scan frequency) between alternate lines in the event of any problem with visible beats between programs (they have not had to use this at Dennis Port). This would also aid in

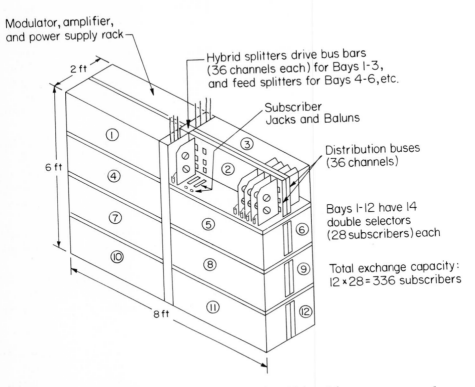

Modulator, amplifier,
and power supply rack

2 ft

6 ft

8 ft

Hybrid splitters drive bus bars
(36 channels each) for Bays 1-3,
and feed splitters for Bays 4-6, etc.

Subscriber
Jacks and Baluns

Distribution buses
(36 channels)

Bays 1-12 have 14
double selectors
(28 subscribers) each

Total exchange capacity:
12 × 28 = 336 subscribers

① ④ ⑦ ⑩ ③ ② ⑤ ⑧ ⑪ ⑥ ⑨ ⑫

Figure 4. Mechanical configuration of a 336 × 36 program exchange. This illustration represents an eventual packaged configuration. At Dennis Port the
switch frame is as shown, but there are several additional racks of "head-
end" signal processing equipment needed to convert Cape Cod Cablevision
VHF channels to the rediffusion HF channel frequency.

reducing visibility of any beats with the second harmonic of the sound carrier,
which appears at 6.88 MHz.

The distribution buses also use stripline techniques, and each bus can accommodate 14 switch cards. As shown in Figure 4, the exchange is arranged
in two sets of back-to-back racks, with the bus feeds in the center. Three of
the racks hold four bays of 28 switches each, and the fourth bay contains the
amplifiers and other electronic equipment.

Figure 5 shows the signal distribution scheme in the exchange. One of the
problems was to avoid interactions between subscriber lines, that is, signal
level changes as a function of subscriber load on a particular program. This
seems to have been successfully accomplished, since it is stated that signal
levels are maintained within 1 dB for 0 to 100 percent of the subscribers
choosing the same program. Also a short on any subscriber line causes only
a 1/2 dB reduction in bus signal level, so other subscribers are not affected.

Note that the signal level decreases 24 dB between the first and last banks
of switches, and that a subscriber must therefore be assigned to a switch bank

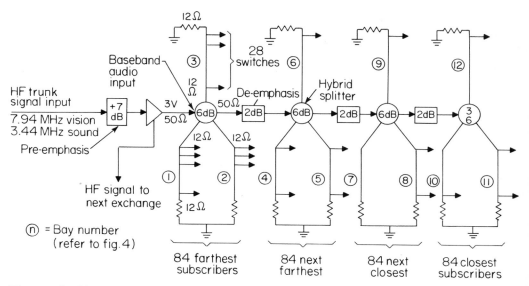

Figure 5. Signal distributions in 336 subscriber exchange (typical of each 36 channels)

on the basis of his cable length from the exchange. It is apparent that careful planning and balancing of the cable network is required for each particular area served. Also, if an exchange were to serve, say, just one large apartment building (which could have 336 TV sets), there would not be a 24-dB variation in cable loss between the nearest and farthest subscribers, and attenuators would be required in many of the subscriber lines.

The buses and selector cards of the present exchange are designed for a maximum of 36 channels. Rediffusion states that 72- or 108-channel systems are easily obtained, but it is not easy to see how they would do this except by paralleling exchanges, each with 36 different channels, and assigning a selector in each exchange to each subscriber. Present costs for 336 by 36 exchange equipment (see Section 4) work out at $75 per subscriber, or roughly $2.00 per channel per subscriber. Assuming parallel exchanges, 72- and 108-channel systems would have exchange costs of about $150 and $225 per subscriber, respectively.

c. Operational Features   Each subscriber is provided with a combination HF/VHF inverter and control unit the size and shape of a 500-series dial telephone without the handset. This unit is connected to both the incoming Qwist cable and to the antenna terminals of the TV set and is usually installed with wires long enough so that it can be conveniently carried around the room for armchair program selection.

The dial uses one of the two control wires to operate the subscriber's selector switch in the exchange on an incremental basis; that is, each dial pulse advances the ratchet one position. A subscriber can thus quickly step through all the channels by dialing successive "ones." For dialing a desired line directly, the subscriber first pushes a reset button, which uses the second control wire to operate a homing solenoid that returns the selector to channel 0, where a channel directory card is shown (by means of a camera at the exchange). He then dials the desired number. Because of the incremental nature of the dial system, channel 10 is reached by dialing 0, 11 by 01, 33 by 0003, etc. It is also possible with a little mental arithmetic to dial only the difference between the present and any greater channel number without going through the reset cycle, for example, dialing 7 will advance the selector from channel 6 to channel 03 (13). Note, however, that there is no indication of what channel number is currently selected.

The present dialing at Dennis Port is rather convenient since there are only 14 channels in use — one program directory, 12 Cape Cod cablevision, and one local origination. This latter is a static camera in a delicatessen, which shows their special-of-the-day, primarily to demonstrate the two-way capability of the system. With these few channels, only two digits are required in direct dialing, but even so, many subscribers apparently prefer to step through one at a time. Dialing would be less convenient for a full 36-channel system, since it takes about 10 seconds to push the reset button and dial 0005, or 30 seconds to step through 36 channels one at a time. It would seem that Rediffusion may eventually want to go to a digit-decoding dial system, particularly if channel capacity is increased to 72 or 108 channels.

I was also shown the channel lock-out feature, which is accomplished quite simply by clipping small magnetic shields over the selector reeds that are to be disabled for a particular subscriber. Alternately, the reeds can be removed (or not installed). This could be used, for example, to permit access to a medical news channel only by doctors, or to set up easily changed private, two-way TV networks for the police, and so forth.

d. Two-Way Capabilities and Potentialities   As previously mentioned, the frequency band between 9.2 and 15.2 MHz on any subscriber cable can be used for simultaneous upstream TV transmission, what Rediffusion calls "program injection." To use this capability, hybrids and filters must be inserted in the cable at both the subscriber and exchange ends to prevent signal interactions. A camera and modulator may then be plugged in at the subscriber end, and its signal picked off at the exchange end. One use that was demonstrated is to demodulate the received signal, remodulate it at the distribution frequency, and put it on an unused bus for viewing throughout the network. Thus local origination from any point in the network is quite convenient. An interesting maintenance and troubleshooting aid results from this capability — a portable camera can be pointed at a subscriber's receiver, and the receiver pictures monitored at the exchange as the selector is stepped through manually.[20]

No provisions exist at the moment for connection of an injected signal to any other subscriber except through a distribution bus of the exchange. Also, the necessary hybrids, filters, and modulators are installed on only two lines for demonstration purposes. Although the transmission capability for private two-way TV links could be obtained by fitting all lines, the switching capability for subscriber-to-subscriber connections on a general basis does not exist, and its implementation with the present types of switch cards would require expansion and a completely different organization of the exchange.

Rediffusion also stresses the availability of one or both wire pairs for other two-way uses, such as meter reading, voice links, subscriber response to programs, and so forth. One possibility they mentioned is to shift the present dc control functions to the signal pair, leaving the 1-MHz control pair completely free for other uses. Here again, additional switching would have to be added to the exchange to make use of these lines in any way. The interesting question to be answered is the relative difficulty and cost of adding such capability in a tree network (VHF cable) by time-division multiplex techniques.

4. Cost Data

Rediffusion has published a schedule of provisional prices, dated September 15, 1970, which gives detailed costs for items of head-end equipment, the equipment items needed in a remote exchange, the per-foot cost of cables for trunks, feeders, and drops, and the cost of each subscriber control unit. From this it is apparent that head-end equipment (except for antennas and low-noise preamps) works out at about $1,000 per channel, roughly comparable to that of VHF head-end equipment.

What is of more importance in any comparison with VHF systems is the per-subscriber cost of a subunit of the distribution system: a local exchange serving 336 TV sets, its trunk connection, and the distribution cabling, junction boxes, and control units. The Rediffusion literature has example equipment costing for 12-channel systems, but it was of more interest to examine a 36-channel system.

a. Exchange Equipment

The following lists equipment needed for a 336 × 36 exchange (note that the prices shown do not include U.S. customs duty, which might add 6 to 12 percent):

Switchframe and busbars				$3,390
Selector power unit				158
TV repeaters	36	at	$122.50	4,392
Power supplies	6	at	72.00	432
Sound amplifiers	36	at	55.00	1,980
Power supplies	3	at	72.00	216
Hybrids (sets of 4)	36	at	33.12	1,192
Control panels	168	at	9.36	1,572
Double selectors	168	at	72.00	12,096
Total exchange cost (with audio)				25,428

The per-subscriber cost of the exchange itself is thus $25,428 : 336 = $75, without provisions for housing it and controlling its environment. If mounted outdoors (probable in suburban areas), a weatherproof kiosk would be used (estimate $1000), and 100-200 square feet of land would have to be purchased or leased. If mounted indoors (which it might be in urban apartment areas),

space would probably be available without cost, as it is for telephone company
equipment, on a user's premises. The exchange requires one kilowatt of electrical power (about $300 per year).

b. Cabling   For purposes of estimating cable costs, it is necessary to consider the type of area served and a typical cable layout. It is also necessary
to remember that subscribers with multiple TV sets that they wish to use independently will have to have a separate selector and cable for each set, that
is, they will have to be multiple subscribers. Rediffusion has estimated that
40 percent of the subscribers will be multiple users; thus it is assumed that
a 336-channel exchange can serve 240 households, 96 of which have two TV
sets.

   Figure 6 diagrams a typical block and street layout for a medium-density,
single-dwelling area. It is assumed that lots are 50 by 100 feet, 20 to the
block. Each exchange can thus serve an area 2 × 6 blocks (240 house lots)
with an allowance for 96 multiple TV sets. The average length of the 56 6-
Qwist feeders is a + b/2 = 125 + 1650/2 = 125 + 825 = 950 feet, and total 6-
Qwist cable required is 56 × 950 = 53,200 feet.

   It is assumed that single-Qwist housedrops will average 100 feet; thus
single-Qwist requirements will be 336 × 100 = 33,600 feet. One junction box
($5) is required to connect a 6-Qwist cable to six house drops.

   The total distribution cable and junction box costs for the layout of Figure 6
would thus be (assuming shielded cable for aerial installation):

6-Qwist	53,200 feet	at	$0.117/foot	$6,210
1-Qwist	33,600 feet	at	$0.042/foot	1,412
Junction Boxes	56	at	$5.00 each	280
Total distribution cable cost				$7,902

   This probably represents close to a maximum cost for distribution cable
per exchange. Exchanges in more densely populated areas (apartments or
two-family housing) would require less 6-Qwist cable because they would
serve a smaller area. In suburban areas with larger house lots, the same
length of 6-Qwist cables could fan out over a larger area. More single-Qwist
(and perhaps line repeaters) might be needed in the latter case, however.

   We must also figure in the cost of the trunk cable from the previous exchange as part of the subunit cost. In the layout of Figure 6 the air-line distance between exchanges is 500 feet. Using a 36-channel trunk cost of $13,306/
mile and an estimated terminal-to-terminal cable length of 700 feet, trunk
cable cost per exchange would be $1,764. Note that there is a premium, as
Rediffusion points out, on keeping trunk distances short because of the relatively high cost of trunk cables — $2.52 per foot for 36 channels.

   These figures pertain to cable purchase costs only. Good cost figures for
the installation of Rediffusion cables have been difficult to arrive at. In one
Rediffusion paper,[21] a labor cost of $1.60 per foot for trunk and feeder instal-
lation is given, without making it clear whether this applies to route-footage
or to the total footage of individual cables, which is about ten times greater

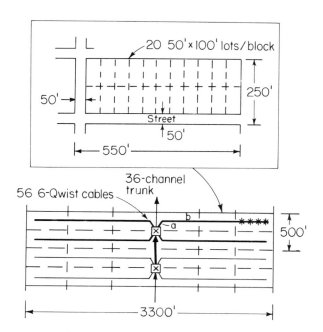

Figure 6. Possible exchange layout in medium-density single-dwelling area. The following assumptions are made: a = 125'; b = 1650'; average length of 6-Qwist = a + b/2.

than the route footage. Thus in the original version of this present paper, the geometric mean of these two interpretations was tentatively assumed for want of a better figure. This resulted in a labor-cost estimate of $70,000, which was felt to be high, but was somewhat supported by the indication in Mr. Gargini's paper that two-thirds of total remote selection costs are in labor and overhead for trunk and feeder installation.

Subsequently, the installation cost question was discussed with Mr. H. F. Goodwin of Rediffusion by telephone. He said that he also did not understand the basis for the $1.60/foot figure in Mr. Gargini's paper and kindly offered to prepare an approximate labor cost estimate for a specific layout. The exchange layout of Figure 6 was described to him, and he called back a day later with an estimate of $30,000 for installation labor costs. He also furnished figures for cable and equipment purchase costs, which agreed to within a few dollars with those derived here.

Subsequent to submission of this report, the $30,000 estimate for cable installation labor was reduced to $15,500 by Mr. R. W. Lawson. For a 336-subscriber exchange, this factor alone lops off $43 per subscriber, more than half the net reduction in the latest Rediffusion per-subscriber cost.

c. Per-Subscriber Distribution Costs   From the figures which have been established, we now can sum up the total cost of installing a 36-channel Rediffusion exchange for 336 subscribers under the assumptions inherent in Figure 6 and establish the per-subscriber cost:

Cable and equipment purchase (catalog prices)

Exchange equipment	$ 23,232
Qwist cables and junction boxes	7,902
Trunk cable (36 coax plus 36 audio pairs)	864
Subscriber control units (336 at $ 34)	10,416
	$ 42,414
Installation labor (Rediffusion estimate)	15,500
Land and kiosk (estimate)	5,500
Total distribution cost per 336 subscribers	$ 62,414

The per-subscriber installation cost for the Rediffusion switched distribution system is thus $ 186. Note that this does not include any share of head-end equipment costs (which should be prorated over all subscribers in all exchanges served by it), or any special installation costs such as make-ready charges on poles, etc. These omitted costs depend on total system size and local conditions and tend to be about the same for any type of system.

As discussed previously, the program buses and rotary selector switches of the present exchange are designed for a maximum of 36 channels. Although some economies might be possible in a redesign for more channels, the incremental cost of doubling or tripling the number of channels can be fairly accurately estimated on the basis of adding an exchange and trunk for each 36 channels, with distribution cables and subscriber control units remaining the same. (One easy way to expand the control function without altering the present control unit would be to have the reset button cycle the subscriber's line through the channel 0's of the exchanges available to him. He could then read the channel listings for each exchange and dial the desired program in that exchange or shift exchanges by pushing the reset button again, etc.) On this assumption, the total distribution costs for 72- and 108-channel systems might be:

	36 Channels	72 Channels	108 Channels
Exchange preselector	-	$ 3,360	$ 5,000
Exchange equipment	$ 23,232	46,464	69,696
Qwist cable and junction boxes	7,902	7,902	7,902
Trunk cable	864	1,728	2,592
Subscriber control units	10,416	10,416	10,416
Installation labor (estimate)	15,500	17,500	19,500
Land and kiosk (estimate)	5,000	7,500	10,000
Total for 336 subscribers	$ 62,914	$ 94,870	$ 125,106
Per-subscriber cost	$ 187	$ 283	$ 374

V. THE AMECO DISCADE SWITCHED TV DISTRIBUTION SYSTEM

1. Introduction

Ameco, Inc., Phoenix, Arizona, is making installations of a switched, sub-channel TV distribution system of its own design in Daly City, California, and at Disneyworld, Florida. Although the basic principle of this system is quite similar to that of the Rediffusion "Dial-a-Program" system, there are significant differences in system implementation that affect both the installation requirements and the per-subscriber costs. It was therefore felt that a comparable analysis should be made of the DISCADE system in order to have the best possible basis for comparison of switched versus nonswitched systems.

Preliminary information on the technical features of the DISCADE^{TM} system (<u>DIS</u>crete <u>C</u>able <u>A</u>rea <u>D</u>istribution <u>E</u>quipment) was provided by an informal writeup (dated February 1, 1971) obtained from Ameco in mid-February. In order to clarify a number of details that were not evident in the writeup and to obtain cost data, I visited Ameco on March 3, 1971, where I met with Mr. Bruce Merrill, President, and Mr. Earl Hickman, Chief Engineer. They were most cooperative in answering all my questions and in discussing and/or demonstrating all the hardware used in their system. The revisions to the original memorandum are based on comments received from Mr. Merrill in a letter dated April 29, 1971, and primarily concern minor corrections to frequency capabilities, channel assignments, and cost estimates for subtrunks and drops.

2. Technical Features

The DISCADE system has been developed as a means of providing 20-40 usable channels while avoiding: (1) the on-channel interference problems of VHF-distribution and (2) the harmonic, oscillator-beat, and image problems that can arise in the use of mid- and super-band channels. It was obviously influenced by the earlier Rediffusion system but also has its roots in previous Ameco experience in subchannel transmission techniques to avoid ambient signal interference in links between CATV antenna sites and their associated head ends. DISCADE represents an alternate engineering solution to the design of a switched distribution system that appears to have a number of advantages over the Rediffusion system. On the other hand, it has the same primary disadvantage —that a household with multiple TV sets to be used independently must be a multiple subscriber. DISCADE utilizes coaxial cables throughout.

<u>a. Trunks</u> Trunks in the present design are made up of eleven sets of cables and amplifiers designed for 5-50 MHz transmission, of which ten carry 2-4 TV channels each, and one carries the FM band, block-converted to 20-40 MHz from its normal 88-108 MHz region in the spectrum. Because frequencies no higher than 50 MHz appear on a trunk cable, smaller cable may be used than in usual practice, and amplifiers may be more widely spaced. For example, with 0.412-inch cable, amplifier spacing is 4,000 feet, and trunk lengths up to 25 miles are feasible. These figures double if 0.750-inch cable is used; in this case, <u>50-mile</u> trunks are feasible. Trunk amplifier assemblies are 10 × 7 × 22 inches in size, and the 11 amplifiers are modular, plug-in assemblies, designed so that any cable may be changed to <u>upstream</u> transmission by simply inverting its amplifiers when plugging them in.

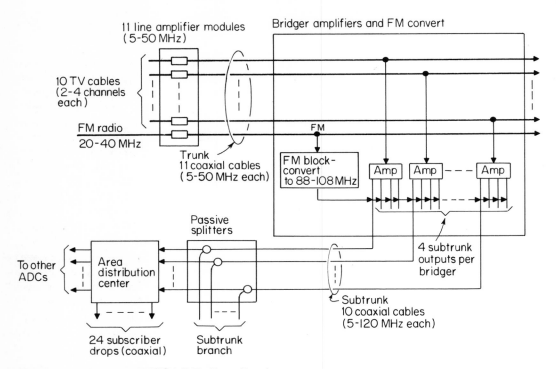

Figure 7. Ameco DISCADE distribution system.

For a 20-channel system, Ameco has used two channels per cable, choosing ones that are nonadjacent and not harmonically related, but both of these conditions can not be met in a 40-channel system. For example, at Daly City the two initial channels (22-28 MHz and 34-40 MHz) are neither adjacent nor harmonically related. If there is a requirement in the future to expand this particular system to 40 channels, Ameco has already verified that it is possible to add 5-11 MHz (note that use of frequencies between 11 and 22 MHz could create harmonic problems) and 28-34 MHz (adjacent to both the original channels). Such additions require no changes to the cable distribution system — only expanded head-end and subscriber selector equipment. For a new system with 40-channel capability initially, Ameco feels that they might use 24-48 MHz. These channels would all be adjacent, but not harmonically related. Another growth possibility mentioned is to keep two channels per cable but install more trunk and subtrunk cables and larger switches, but this would increase costs linearly with added channels.

Ameco gives an approximate installed cost for 11-cable trunk of $11,000 per mile, of which about 35 percent is labor.

b. Subtrunks   Subtrunks are connected to the trunk by sets of bridger amplifiers, each of which will drive the corresponding cables in four subtrunks. An unusual feature of DISCADE is that the subtrunk cables carry different signals than the trunk cables. As shown in Figure 7, one of the functions of the bridger

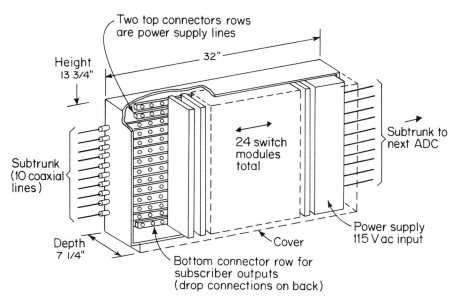

Figure 8. Ameco DISCADE area distribution center (ADC)
Size of 24-subscriber ADC = 7 1/4" × 13 3/4" × 32"
Weight of 24-subscriber ADC = 90 lb.
Subtrunk loss per ADC = 0.5 dB
ADC provides power for subscriber selectors
ADC designed for cable mounting
Cost = $192/ADC housing plus $60/switch module and $15/subscriber selector
Each cable carries 2-4 TV channels plus FM radio (20-40 TV channels total).

amplifier assembly is to block-convert the 20-40 MHz FM radio signals on
the FM trunk cable to their correct frequency band (88-108 MHz) and add them
to each of the ten subtrunk cables. The subtrunk cables thus carry both TV
and FM signals, and all parts of the subtrunk and subscriber distribution systems are designed for 5-120 MHz. Subtrunks typically use 0.240-inch cable
and are not amplified but may be divided by means of passive splitters. Ameco
gives an approximate cost figure for installed subtrunk of $7,500 per mile, in-
cluding the Area Distribution Centers (without switch modules) described next.[22]

c. Subscriber Switching   The main component of the DISCADE system is the
Area Distribution Center (ADC), which is a cable-mounted switching unit,
presently designed in sizes for 8, 16, or 24 subscribers. As many as ten ADCs
may be spliced into each subtrunk either at initial installation or later, as
needed to meet subscriber hookup requirements. Standard CATV-type drop
cable is used to connect a subscriber to an ADC, and drops may be up to 2,000
feet in length.
    The physical configuration of a cable-mounted, 24-subscriber ADC is shown
in Figure 8. The switch modules are of solid-state design and connect a sub-

scriber drop to any one of the ten subtrunk cables as controlled by direct-
current pulses from the subscriber selector unit fed back to the ADC on the
subscriber drop. An incremental form of control is used, with the switch advancing one position for each pulse (actually every other pulse, see discussion next). When cable number 10 is reached, the next pulse starts the sequence again at cable number 1, etc. The major cost item of the entire DIS-
CADE system is the switch module, which costs $60 per subscriber drop but
need only be installed as subscribers are actually connected.

Note that since the FM radio band is on all subtrunk cables, it will appear
on the subscriber drop no matter what position the switch is in and will be un-
affected by channel selection.

d. Subscriber Selector The subscriber selector unit, which costs $15.00, is
slightly smaller than a desk telephone and has a click-stop rotary channel selector knob on top, which sends a pulse to the ADC for each "click." A film-
strip coupled to the knob provides a very legible display of the selected channel number in large (3/4-inch) illuminated numerals projected in a window on
the front of the unit (this is a plus over the Rediffusion telephone-dial selection system, which provides no indication of what channel is currently being
viewed). Because of the incremental-type switch control, the knob has a mechanical ratchet so that it can only be turned in the direction of increasing
channel numbers, and there is no reset function like that in the Rediffusion
system, except an automatic one on the transition from Channel 20 to Channel 1 that provides knob and switch resynchronization in case they should ever
get out of step. Thus to go back one channel, the selector knob has to be
turned several complete revolutions to cycle through all 20 (or 40) channels —
a minor inconvenience.

The selector unit also contains two (or four) conversion oscillators, and as
the selector knob is advanced, these are turned on alternately to convert one
of the two (or four) channels available on the drop cable at each ADC switch
setting to the clear VHF channel chosen for the TV receiver input. This combination of frequency and space switching sounds somewhat complicated, but
it is completely hidden from the user — all he sees is one dial labeled from 1-
20 or 1-40, and the proper coordination of the cable switching and frequency
selection takes place automatically as the selector knob is advanced.

As previously mentioned, the selector sends a pulse to the ADC for each
knob "click," but the ADC switches cables only every second (or fourth) pulse.
The reason for having a pulse sent to the ADC for each knob "click" is so that
a channel-monitoring system could be implemented in the ADC if desired.
Note, however, that since the selector unit is cable powered, it works whether
or not the TV set is on, and some means of also monitoring TV set power
would be required for meaningful monitoring of channel viewing.

DISCADE Installations
Two different DISCADE installations are currently in progress, one using an
interesting variation in the basic technique just described.

a. Daly City, California Vista Grande Cablevision is installing a 20-channel
DISCADE system exactly as described here in Daly City, with a potential sys-

tem size of 16,000 subscribers. The main impetus for choice of this system by the operator was a firm requirement for 20 channels, plus an unusually severe local-signal problem — seven of the 12 standard VHF channels would be subject to ghosting from direct pickup. A field trial during 1970 with proto-type hardware was entirely successful but led to some equipment redesign to reduce costs. Installation of final hardware is now underway, and equipment sufficient for about 500 subscribers has been shipped to date.

b. Disneyworld, Florida   The Disneyworld installation will involve about 2,000 TV receivers and will be a 10-channel system with only one channel per cable. In this case, the set manufacturer (RCA) is incorporating a simplified Ameco selector unit designed to fit into the sets in place of the normal VHF-UHF tuner, and no frequency conversion is necessary in the selector because the signal on each subtrunk cable is at the 45.75 MHz IF frequency of the receiver. The standard RCA motor tuning feature is retained, with its remote control unit.

4. Two-Way Considerations

The DISCADE system as presently implemented does not incorporate any two-way features. It is not known what plans Ameco has for two-way use, but there are a number of possibilities.

As previously mentioned, one or more trunk cables can be converted to up-stream transmission toward the head end by simply reversing their line am-plifier modules, providing 2-4 upstream channels per cable. If reverse bridger modules were developed to permit upstream transmission on one or more sub-trunk cables to feed into these upstream trunk cables, and any passive split-ters were suitably modified, then these channels would be available at all switching units (ADCs).

At this point, the way to proceed isn't as clear, but one would have several options for video origination. One option would be simply to make one or more switch positions be for upstream use only, essentially reversing the subscrib-er's drop and permitting him to insert a video signal on one of the channels for transmission to the head end. However, he then wouldn't be able to see any-thing at the same time. A second, and perhaps more attractive, possibility would be to install a second, upstream-only drop for each subscriber that is not switched but added to an upstream subtrunk cable at the ADC along with all other upstream drops. This would permit simultaneous video origination plus viewing of any downstream channel for as many subscribers as there are up-stream channels. A third possibility, which would require new types of am-plifiers and perhaps switches, would be to use frequency splitting techniques to permit bidirectional transmission on one or more cables. There are per-haps other possibilities. Either the second or the third schemes could permit one or two "videophone" conversations between any two points in the system, but these would be nonprivate.

Downstream and upstream digital transmission channels for a variety of uses (monitoring, meter reading, data, channel control, etc.) can also be es-tablished by various techniques such as those just mentioned. In this case the presence of the switch is somewhat of a hindrance because the polling equip-ment at the head end would not know what cable to address a particular sub-

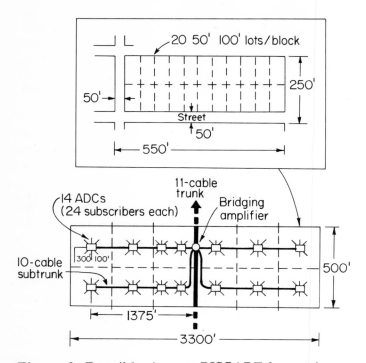

Figure 9. Possible Ameco DISCADE layout in same unit area used for Rediffusion analysis

Average drop length = $\dfrac{100 + 300}{2}$ = 200'

12 × 20 = 240 houselots
14 × 24 = 336 subscriber capability (240 + 40%)

scriber on unless it continuously monitored his switch. One possibility to a-
void this complication would be to handle the downstream digital channel the
same way as the FM band is handled at present, perhaps adding it to the FM
trunk cable in the 16-20 MHz band so that it would always appear on all subscriber drops. Another would be to provide a separate, tree-structured bidirectional digital data cable that runs throughout the system. Here again,
these suggestions do not exhaust the possibilities, and Ameco's particular
plans are not known.

In summary, DISCADE can be augmented to provide the same two-way capabilities as the nonswitched systems at about the same incremental cost, but
direct provisions for such augmentation are not readily apparent in the present
hardware, except for the reversible line amplifiers and the switch-control design for eventual channel monitoring capability.

5. Cost Data

For purposes of relative cost comparison, the DISCADE system has been
costed out for the same 12-block, 240-houselot unit area used in the analysis
of the Rediffusion system. A possible DISCADE layout to provide 100-percent

service to this area, with an allowance for 40-percent multiple subscribers, is shown in Figure 9. As shown, it is assumed that one trunk-bridging amplifier would drive four subtrunks, each with either three or four Area Distribution Centers. The 40-percent multiple-set allowance would require 28 drops per block; thus the ADCs must be spaced not quite one block apart, and 14 are required in all to provide 336 total drop capability, the same as the Rediffusion exchange. Subscriber drops would range from 100 to 300 feet, or 200 feet average, and are figured at an installed cost of $0.05 per foot (Ameco estimate). Trunk and subtrunk installed costs are figured at the Ameco per/mile estimates given earlier. The distribution costs per unit area (as defined) are thus:

Trunk	500 feet	at	$11,000/mile	=	$ 1,050	
Subtrunk	6,000 feet	at	7,500/mile	=	8,520	
Drops	67,200 feet	at	0.05/foot	=	3,360	
Total cable costs				=		$12,930
Switch modules	336	at	$60	=	20,160	
Subscriber selectors	336	at	15	=	5,040	
Total equipment costs				=		25,200
Total distribution cost per 336 TV sets (not including head end)				=		$38,130

This works out at $113.50 per subscriber for a 20-channel system and would vary little (if at all) for a 40-channel system, since only a minor change in the subscriber selector unit would be required.

VI. TWO-WAY CONSIDERATIONS AND SYSTEMS

1. Introduction

Except for a few experimental installations and trials during the past year or so, existing cable plants are only equipped with one-way downstream amplifiers in the trunk lines for the normal TV distribution band from 54 to 216+ MHz. By use of suitable frequency splitters at each downstream amplifier location, it has been found possible to add amplifiers to selectively transmit frequencies below this band (that is, from roughly 5-40 MHz) in the upstream direction.[23] A number of CATV equipment manufacturers are now beginning to offer such devices, either for new systems or the upgrading of older ones, with the exact frequency range provided varying with the manufacturer.

The availability of, say, 30 MHz of upstream bandwidth permits up to five 6-MHz upstream channels, some of which can be used for remote TV originations from any point in the system and some for a variety of digital data purposes. Whether five upstream channels is sufficient depends upon what services one wants to provide. To go beyond this, some proposals are to install a separate cable with full upstream bandwidth, use a greater share of one or both cables in dual-cable installations, or, as in the Rediffusion Dial-a-Program switched system, provide a separate upstream channel from each sub-

scriber to his program exchange. Whatever the needs, it is clear that substantial upstream cable capacity can be provided within the state of the art
and within a factor of two (or less) of the cost of downstream-only configurations.

2. Upstream Television Channels

The first paragraph of the next section discusses the current trends in the industry, that is, what is now possible or will be shortly. The second paragraph
treats the question of expanding CATV systems to include full two-way, private
video transmission on a subscriber-to-subscriber basis and concludes that
this would be very expensive and would not fit within tree-structured CATV
systems.

a. Present Capabilities   The most obvious use for an upstream TV capability
is to permit cablecast originations from any point in the system, transmitting
the camera signal back to the head end or the cable casting studio for taping
and/or live retransmission on a regular downstream channel for general viewing. Needs for such service can probably be satisfied by a few upstream channels on an occasional-usage basis.

   The next level of service is to provide certain restricted-access, sub-
scriber-origination video transmission services not connected with general
CATV program distribution, such as the interconnection of TV-visual services between schools, municipal or police visual nets, etc. This would add
to the needs for upstream channels, both in the number of simultaneous channels required and in average usage. Note also that in order to provide such
point-to-point or "one-to-few" services passing through the head end, a con-
trolled-access downstream channel is required for each upstream channel so
used. An interesting example of this class of service is a community conference hookup permitting a controlled group of subscribers (the conferees) to
view and participate verbally with their chairman (another subscriber who has
a camera and means for controlling viewing access).

b. The Cloudy Future   Beyond the types of video services that can now or very
shortly be offered, there has been speculation about personal point-to-point
video services, such as remote medical diagnosis, general "video-phone"
service, etc. These of course imply permanent installation of cameras and
cable modulators at subscriber locations, which would represent a major cost
escalation—by at least $500 per subscriber even when such devices are in
large-scale production and perhaps more. More important from the viewpoint
of the cable plant, the number of independent two-way channels needed would
be far in excess of the foreseeable extensions of present cable technology, except perhaps for the Rediffusion Dial-a-Program system. Note, however, that
in the Rediffusion system as presently implemented, the individual two-way
subscriber lines have a "reach" of only 2,000 feet, and the largest hub is 336
lines, which is a rather small base for a generalized point-to-point switching
network (the balance of interexchange lines to subscriber lines would be very
poor). Whether or not the subscriber lines were extended (with two-way amplifiers and/or a change to coaxial cable) to permit larger hubs, total switching gear at least comparable in complexity to the usual 10,000-subscriber tele-

phone exchange would be needed within a typical head end, since the switching requirements would be the same.

Large-scale point-to-point switching apparatus for 6-MHz channels is probably feasible, particularly if the signals can be handled at baseband-video or at very low carrier frequencies such as used by Rediffusion and Ameco. The Bell System already can switch 1-MHz Picturephone signal on modified No. 5 crossbar and ESS equipment, and solid-state video crossbar switches (116 by 211 lines) have been constructed for NASA. Large-scale broadband switchgear would have to be developed, however, and would certainly be more costly than present telephone switchgear. Also, interconnections between exchanges would require broadband trunk circuits (in the telephone system sense), which would become extremely costly as the size of the interconnected system expands (one 6-MHz channel occupies the same bandwidth as 1,000 voice-grade telephone channels). Minimum investment costs for the most modern long-haul terrestrial telecommunications systems are $1,750 per channel-mile for 6-MHz television channels (TD-2 Microwave Relay).[24] Present proposals for a domestic satelite (with a backup satelite) are based on providing eight 6-MHz TV channels (or 10,560 voice channels) at an investment cost of $47 million. Annual lease of a two-way TV circuit (two channels) is expected to be $1.8 million, or $5,000 per day (from FCC filings by GT and E/Hughes, 1970).

In this connection, one study by Complan Associates, Inc.[25] has estimated the added capital cost of a complete nationwide 1-MHz Picturephone service serving 100 million subscribers to be $3,000 per subscriber, about five times greater than the investment in the existing voice-grade telephone system, and that a 6-MHz service on the same basis would cost about 1.2 trillion dollars ($12,000 per subscriber). In both estimates, "out-of-plant" and local exchange costs (subscriber lines and terminals and first-level switching) account for one-third to one-half of the total; thus two-way 6-MHz "videophone" service just within a typical CATV head end (10,000 subscribers) would cost at least $4,000 per subscriber, perhaps 20 times more than the most probable types of CATV configurations over the next few years.

No analysis comparable to that of the Complan study has been made for generalized point-to-point services during the course of this study. Whether the estimates presented here are correct or not, however, it is clear that the cost multiplier for expansion of CATV systems to include generalized point-to-point switched "videophone" service is quite large, and that a hub-type network would be required.

3. Digital Channels for Control, Monitoring, and Data Services

There is a wide range of services one can imagine for digital communication via the cable plant. Some of the simpler ones include providing the head end with information concerning the operation of the cable plant via equipment sensors or the monitoring of the tuner of each subscriber to gather viewing statistics. Several test installations of this sort are now in progress, and new equipment for these purposes is now coming on the market. Moreover, if current computer communications techniques are employed, it is possible to provide (in order of ascending cost) such services as push-button opinion samples or voting, meter reading, data entry and retrieval from local or remote data banks, electronic mail, and so forth.

Utility companies have recently become interested in automating the process of reading utility meters. Two alternate systems under development have been investigated in an effort to determine what role the CATV cable could play in providing these or similar services.

The first system is being developed by Shintron Company[26] to read electric meters using the power lines as the communications medium. The information is coded as a several millivolt signal carried directly on the power line. Since bit rates are very low (they are transmitting 0.05 bits per second), the small signals can be recovered from the much larger power signals by appropriate processing techniques. They use a single transmitter at the power station to service 2,000 installations and control 25 transmitter units with a small computer that tabulates return data. The equipment required in the home can be largely fabricated in MOS integrated circuit form and should fit inside an electric meter. Projected costs for the unit at the home are in the $20 to $30 range.

The second system is being developed by Bell Laboratories at Holmdel, New Jersey, in cooperation with a number of manufacturers who have developed encoders for transmitting meter readings via the switched telephone network. A simple alerting circuit answers the meter reading request without ringing the customer's telephone. The coded readings are then converted into tone signals and routed through a telephone company central office to a data center serving one or more utility companies. This is done using a meter reading access circuit[27] in the telephone company office. The total time required to read the meter(s) in a customer's home is roughly 10 seconds. Thus 300-400 customers can be polled per hour.

The two systems presented are not designed to be expanded to provide any type of general-purpose digital communication service to the customer beyond the simple reading of utility meters but are no less and no more costly than performing just this same function over a CATV cable. However, the meter reading task can be easily and cheaply integrated into an existing digital communications service, since the incremental cost of reading meters would probably be smaller than that for implementing a parallel system only for reading meters. Note, however, that the inclusion of meter reading via the cable plant would require some sort of standardization of digital cable systems either at the meter interface or of the digital system itself. Note also that there would probably be little interest in reading meters via a CATV cable unless cable penetration (including two-way data capabilities) was virtually 100 percent in a given area. For some time to come, the power and Telco lines will reach more meter locations.

Notes

1. A variety of conversion schemes have been developed for mid-band. Some are simple block-converters that transpose the mid-band to UHF or to high-band VHF for selection by the respective tuners of the TV receiver. Others are complete 21-channel tuners (more if super-band is included), which perform the channel selection and convert to an unused VHF channel, to which the TV receiver is left tuned.

2. D. G. Fink, Principles of Television Engineering (McGraw-Hill, 1940).

3. Derived from data in 1970-1971 Television Factbook No. 40.

4. D. G. Fink.

5. From "Supplemental Comments of National Cable Television Association (NCTA)" on FCC Docket No. 18894, January 6, 1971.

6. The current standard intermediate frequency (IF) amplifier pass-band for picture signals is from 41.5 to 46.5 MHz, and the oscillator is always tuned 45.75 MHz above the desired picture carrier. The prewar IF pass-band was 8.5 to 12.75 MHz, and the oscillator was tuned 12.75 MHz above the desired picture carrier. After the war, an IF frequency pass-band of 21.5-26.5 MHz was standard for a period, with an oscillator offset of 25.75 MHz, and some of these receivers are still in service.

7. Norman Parker, "A Proposal for the Modernization of the UHF Television Taboos," paper presented at the IEEE-PGVT Conference, Washington, D.C., December 2, 1970.

8. Archer S. Taylor, "On-Channel Carriage of Local TV Stations on CATV," IEEE Transactions on Broadcasting, Vol. BC-5, No. 4 (Dec. 1969), pp. 102-104.

9. Patrick R. J. Court, "Design and Use of CATV Converters," Information Display, March/April 1971.

10. J. E. Adams, "Possible Radiation from Broken CATV Cables in the 118-136 MHz ATC Band," informal working paper prepared for the Office of Telecommunications Policy and circulated to the IEEE Spectrum Allocation Subcommittee, January 1971.

11. Ibid.

12. Hubert J. Schlafy, "The Real World of Technological Evolution in Broadband Communications," report prepared for the Sloan Commission on Cable Communications, September 1970.

13. Michael F. Jeffers, "Best Frequency Assignments for Mid- and Super-Band Channels," included as Exhibit C in Supplemental Comments of NCTA on FCC Docket No. 18894, January 6, 1971.

14. I. Switzer, "Phase Lock Applications in CATV Systems," paper presented at NCTA Annual Convention, Chicago, Illinois, June 1970.

15. "Minutes of Cable Spectrum Allocation Subcommittee," Cable TV Task Force, IEEE, January 2, 1971.

16. Patrick R. J. Court, "Design and Use of CATV Converters," Information Display, March/April, 1971.

17. For example, at least one of the 25-channel tuner-converters is constructed from two modified 13-position VHF tuner mechanisms in cascade, the first of which tunes the 12 VHF channels and the second, selected by the 13th (UHF) position, tunes 13 mid- and super-band channels (ibid.).

18. HP Specifications "Exhibit C" dated March 15, 1971 for H01-35602A and H02-35602A amplifiers.

19. R. P. Gabriel, "Dial-a-Program — an HF Remote Selection Cable Television System," Proceedings of the IEEE, July 1970, pp. 1016-1023.

20. It should be noted at this point that a similar subscriber injection capability can exist in any VHF tree-structured system that is equipped for upstream channels. One important difference is that in a VHF cable, there are a limited number of upstream channels shared among all subscribers (four per cable if only sub-band is used, perhaps 30 if a separate upstream cable is used). Also, it would be difficult to prevent clandestine monitoring of upstream transmissions in a tree-structured system, since the cable would be readily accessible to all subscribers between the injection point and the head end. Note that there is an additive noise problem in transmitting upstream in tree-structured systems, but it is assumed that this will be overcome.

21. E. J. Gargini, "Dial-a-Program Communication Television," paper delivered to the Royal Television Society, February 12, 1970, Figure 20.

22. This estimate is for "reasonable density" and would vary somewhat depending on circumstances.

23. See, for example, Schlafy.

24. R. D. Swensen, "Investment Cost of Terrestrial Long-Haul Telecommunications Facilities," IEEE Transactions on Aerospace and Electronics Systems, Vol. AES-7, No. 1 (January 1971), pp. 115-121.

25. President's Task Force on Communications Policy, Staff Paper I, Part 2, Appendix I, Clearinghouse No. PB 184413, June 1969.

26. The information presented here was obtained from Mr. Larry Baxter of Shintron Company, Cambridge, Mass.; details of the system other than those given here are considered proprietary.

27. R. E. Cordwell and P. J. McCarthy, "Communications Facilities for Automatic Meter Reading," paper presented to the Power Distribution Conference, Austin, Texas, October 20, 1970.

10. WHAT BELONGS ON THE CABLE

John E. Ward

Many papers and articles over the past two or three years have suggested that the 300-MHz coaxial CATV cable carries sufficient bandwidth into (and out of) the home that it can serve almost every conceivable home communications need — providing as many television viewing channels as anyone can think of a use for; two-way data, audio, and data services, and perhaps eventually taking over the functions of the present voice telephone system and augmenting it to provide full nationwide videophone service with TV bandwidth. At the same time, the cable is touted as the way to also provide data and facsimile services among businesses, video and audio interconnections among schools (and school systems), and a wide variety of municipal communications services, such as police and fire networks, traffic surveillance and control, and so forth. The purpose of this paper is to examine some of the realities of this communications utopia from the standpoints of performance and cost, to suggest those areas where the cable seems best suited to providing new or improved communication functions, and just as important, to point out those areas where the drive to "cable-ize" may be inappropriate.

First of all, let's talk about cable capacity. It is true that the 300-MHz bandwidth of a single coaxial cable, or the 600-MHz bandwidth of two cables, can theoretically carry about as many entertainment, educational, or citizen-information viewing channels into the home as anyone would probably ever want. As is well known, there are a number of complicated technical problems in trying to use every scrap of the available cable bandwidth, but these do not seem to represent a serious limitation at present and should gradually be solved as cable technology improves. The present cost increment for just increasing downstream capacity per cable to 25-30 channels is not great, but cable system costs (including frequency multiplex/demultiplex equipment) can rise quite steeply if many additional signals of various types are to be carried; in other words, the solution to the technical problems of adding these signals without mutual interference may be quite expensive in terms of system cost.

It is also true that through use of presently available time-division-multiplex (TDM) techniques, just a few channels used for two-way digital data services, plus possibly addressed-frame video (sometimes called "frame grabbing"), can provide individualized communications services of enormous capacity for all subscribers in a system. A single one-megabit, two-way data channel (which requires no more than 4 MHz downstream and 4 MHz upstream) can, for example, drive about 4,000 printers simultaneously at 10 characters per second, at the same time accepting the same simultaneous input rate from 4,000 keyboards or other data input devices. Since every home obviously won't be receiving or sending all the time at this rate, such a data channel should easily serve twenty to thirty thousand homes (or more, depending on how much access delay is permissible), each of which can also be polled at least every few seconds to see if they wish to initiate communications. If in addition, just one downstream TV channel is devoted to addressed-frame video, as many as 216,000 different text or picture frames per hour can also be transmitted, surely sufficient for the on-demand information access needs of our twenty to thirty thousand homes.[1] These individualized services, however, do carry a high incremental capital cost for the necessary home terminal equipment. Added head-end equipment costs are nominal on a per-subscriber basis, except as the services become more sophisticated. More will be said about this later.

The seemingly bountiful cable capacity begins to look less bountiful when one adds requirements for any significant number of non-TDM, dedicated, upstream or downstream TV-bandwidth channels that are for "private" use of some sort, and not for general viewing. For example, 4-15 TV channels per trunk seems a reasonable range of upstream capacity, depending upon whether one has a one- or two-cable system (this assumes that in a two-cable system about half of one cable would be used for upstream signals). Allowing a few of these upstream channels for remote program origination use, a few each for school and municipal TV interconnections (note that these interconnections would probably be two-way and thus also need a <u>downstream</u> channel per upstream channel), there isn't room left for very many traffic or security surveillance camera channels (for example), or for much private TV channel usage by individual subscribers.

Carrying this last point to the extreme, the total upstream and downstream channel capacity of a tree-structured cable system is clearly insufficient to support large numbers of simultaneous, individual, two-way video conversations (that is, videophone service). For that a switched hub-structured cable plant with individual subscriber lines would be needed, organized like the telephone system. This is the cable configuration of the Rediffusion, Inc., "Dial-a-Program" CATV system,[2] and may certainly be cited as a potentially great advantage of that system. However, their present one-of-86 program-distribution switchgear was not designed for line-to-line switching and would have to be very greatly reorganized and augmented in order to provide the switching functions needed for many simultaneous subscriber-to-subscriber two-way connections on a dial-up basis. The cost of such switchgear would be substantial.

Returning to the tree-structured cable, it is interesting to speculate what would have happened if cable systems of this type had been invented one hundred years ago and installed before any other type of communication system. I am sure that the bottleneck represented by the finite simultaneous cable channel capacity per trunk would eventually have forced many full-time users (or potential users) off the cable and onto circuits dedicated for their own use and many other users desiring part-time, private circuits onto a switched network of some sort. Thus, one must certainly examine closely potential "nonbroadcast" cable communications functions or services that cannot be time-division multiplexed on just a few channels (in the manner discussed earlier) to see if it is technically practical (or economic) to devote the necessary cable bandwidth continuously to their needs, as opposed to services available to all subscribers. In many cases, dedicated facilities may represent a better solution for users with heavy communication needs among just a few locations —for example, if closed-circuit, interclassroom usage among all schools in a city requires more than one full-time two-way channel, a private interschool cable may be advisable, in addition to CATV cable tie-ins for <u>its</u> program and general communication services. Of course, if a given cable system does have unused operable channels, adding such functions or services up to capacity makes sense, at least for the short term. The intent of this discourse is, not to say that such services don't belong on the cable, but to point out that there <u>are</u> limitations in total capacity and that the various "closed-circuit" services that one hears discussed can't <u>all</u> be handled without going to special cable configurations for this purpose. The question then arises as to whether these special, dedicated cable facilities should be handled by a CATV operator or, in some cases, even be interconnected with a CATV system at all.

On the other hand, it is worth re-emphasizing that the cable does have a unique, very practical capability for providing, at little cost in bandwidth, a class of communications services that can be handled only awkwardly (if at all) by existing communication systems — the simultaneous, rapid, two-way interaction among a very large number of individual subscribers and a central information processor/source. The key concept here is that the network configuration of a cable system permits the placing of all subscribers on one or more gigantic, broadband party lines on which addressed, time-division-multiplex messages can be transmitted at very high rates. This clearly represents a new dimension in communications capability, and one that has great potential as an addition to other cable services. The main problem is: What sorts of services can be made to pay for themselves, given the high costs of the necessary terminal equipment for home and possibly business uses (and for certain services, of the associated head-end equipment and/or interconnections with other data bank systems)?

Unfortunately I have no ready answer to this question, but it may be of value to cite the factors that must be considered. First of all, I see this as a "chicken or the egg" situation. The economics of two-way communication has a serious critical mass problem, both in the number of subscribers equipped and connected for a given polling-type service (say, meter reading) and in the number of access services (banking, shopping, etc.) available to a given subscriber. (As an AT&T official was recently quoted in relation to the decision to temporarily shelved Picturephone, "It's no good if there is no one else to talk to."[3] Thus two-way hook-ups and services may have to be subsidized in some way to ever get off the ground, and their eventual self-supporting viability would seem to depend on the totality of a large number of different services each with modest fees, rather than a few services with high fees (a possible exception is pay TV).

Second, there is still a large human factor question in regard to the sorts of services under discussion, which are generally services never before available. Given that a broad spectrum of possible services is set up and offered to subscribers in their "electronic fortresses," will they use them to the extent necessary for economic viability of the services? For example, even if a service such as home shopping is immensely popular as a concept, would enough subscribers pay a monthly terminal fee for the service, plus possibly a price premium on purchases due to the added sales costs for goods presentation on camera, automated order taking, and home delivery after purchase? Large-scale experiments, conducted for a substantial period of time (sufficient for novelty transients to die out) seem the only way to really shed some light on these difficult questions, and these will require extensive cooperation and commitment on the part of many organizations for a given experiment.

A brief word about head-end costs for new types of individualized services. Simple polling and data store-and-forward interconnections with other computerized data services (banks, stores, reservation systems, etc.) can be handled by a small computer with a capital cost of only a few tens of dollars per subscriber. Similarly, the head end could forward subscriber requests for addressed-frame video signals originating elsewhere (say, in a library) and put the signals on the cable at little cost in head-end equipment. However, if the cable operator wishes to provide digital data bank, data processing, or video frame services himself, his head-end costs could go up by an order of magnitude or more, depending on the services provided (rapid-access memory devices of all types tend to be very expensive). Only time and experience will show how some of these services can best be provided (assuming that there is a real market for them).

This discussion has been deliberately couched in rather general terms and has not attempted to predict costs for specific services to the penny for some future period. Many such predictions are available, and the author has made some of his own in another chapter in this book. What seemed more valuable for the present purposes was an exposition, from a communications systems viewpoint, of what types of services the cable seems best at and of what other services it could provide but with performance difficulties or at high cost.

Notes

1. See, for example, John Volk, "The Reston, Virginia, Test of the Mitre Corporation's Interactive Television System," Report MTP-352, Mitre Corp., May 1971. This system digitally addresses each interlace field separately and thus transmits 60 262-line images per second, each containing up to 800 alphanumeric characters.

2. R. P. Gabriel, "Dial-a-Program—an HF Remote-Selection Cable Television System," Proceedings of the IEEE, July 1970, pp. 1016-1023.

3. "Picturephones Shelved Due to Lack of Demand," Washington Post, April 13, 1972.

11. SCREEN FEEDBACK FROM HOME TERMINALS

Noam Lemelshtrich

INTRODUCTION

A survey of the literature on home terminals for two-way communications shows that the scientists who design these devices have only a vague idea of uses of these terminals. All list some possibilities for the usage of such terminals, yet it is clear that the guiding principles are mainly economic.

The home terminal is a vital part of two-way communications and will put constraints on the communications between the home and the broadcasting center. Establishing guidelines for the design of a home terminal through examination of the broad picture of two-way communications can help the designer answer such questions as: How important is it to enable the listener to state specific requests, and would the optimal terminal be the one that is most versatile, or yet, the cheapest?

The major objectives for two-way communications can be considered in four major areas: politics, education, social relations, and services.

In politics, the major objective is to make government more responsive and improve its decision making by improving the communications channels for feedback from the citizenry. Technological advance in this vital area will for the first time allow new channels of communications, which may challenge long-held theories, such as the popular conception and misleading actuality of the democratic idea. These channels will allow citizens to participate directly in the political decision-making processes, which they have considered since childhood the goal of democracy. Social scientists may be aware of dangers in this participation. But once such channels are installed, will the public listen to these social scientists, telling them that democracy works better if it limits participation?

A second major objective of two-way communications is education. New developments in man-computer interaction and computer memories and the potential for interaction with the TV make this objective very attractive. Computations and books from libraries (stored in computer memories and computer graphics, making actual designs on the TV screen) can be transmitted to the home TV. Individual self-teaching of skills and trades can be made available.

A third major objective of two-way communications is to improve social relations within a community.

The fourth major objective of two-way communications and the one that will probably be the main reason for its introduction is services. The successful services will include those that will yield most profit (entertainment, shopping, games) and others of social significance, which are presently being developed by research (medical services, computer interaction).

High versatility in home terminals is needed in the areas of politics and education and for services such as computer interaction (computer graphics) and medicine. A versatile home terminal should have the following features:

Independence of oral communications

Frame grabber

Printer

Signal for confirmation of input

Key to allow the TV owner to control the usage of the terminal in his absence.

Key to set a dollar limit for purchases (such as a computer signal when this limit is reached).

In the following paragraphs I will examine six different techniques that can be employed for using the entire television screen for two-way communications. Transmitting a signal from the home to the studio by touching the TV screen directly (by user's finger) or indirectly (by holding some device like a pen) is defined as <u>screen feedback</u>. Using the entire screen for feedback will allow projection of keyboards of different types and sizes on the screen and will save the cost of constructing an actual keyboard.

None of the techniques were originally designed for application to cable TV two-way communications. The use of the techniques for this purpose introduces new design dimensions in the human factors and the technical and economic fields. Furthermore, most of the developers of these techniques are still not aware of the potentials their techniques have for cable TV two-way communications and consequently are not developing these techniques toward that end. Mass production will allow for considerable improvements in the present hardware, which makes it difficult to evaluate the techniques in their present design. Two of the devices, although technically feasible, have not yet been built, and some of the others are in their initial development stages.

Using a standard home TV as a functional part of the home terminal has the following advantages:[1] (1) economy — it makes use of the huge capital society has already invested in TV sets; TV receivers are much cheaper than conventional computer CRT terminals (due to relative volumes of production) and consume less power (random scanning, which is used in computer CRT terminals, requires more power); (2) maintenance — component parts are cheap and well distributed in neighborhoods and trained servicemen are available; (3) universality — TV receivers can be used with inputs from TV cameras (color signals can be easily synchronized) and unlimited number of feedback displays can be projected (like keyboards, adding machines, or engineering as well as artistic design); (4) human factors — users can construct their own "keyboards" or display for feedback; TV is a familiar device to people and can be expected to be more accepted and understood than complex terminals for communications.

In the following sections I will first discuss the basic hardware system needed for screen feedback. This will be followed by a description of the technologies that could be used for screen feedback.

Two comparisons will be made: first, among the different technologies available for screen feedback and second, between the proposed screen-feedback method and the push-button feedback method, which is presently used in developing two-way home terminals.

In the conclusion, a terminal will be proposed based on relative advantages of the different screen-feedback techniques.

Using the Entire Screen for Feedback
Figure 1 illustrates the main technological features of the communications system needed for using the entire TV screen for eliciting feedback. Identi-

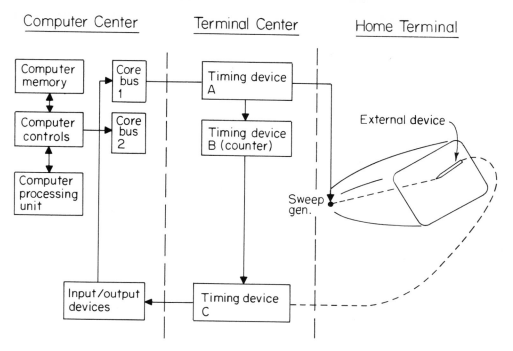

Figure 1. Screen feedback system

fying the choices made by the users (at their homes) involves the following
coding and synchronization: (1) the image generation must be synchronized by
pulses derived from a timing track of the video disc recorder, which also
controls the TV sweep generator; this keeps the computer and the TV in step.
(2) A special digital counter should keep track of the location of the sweep generator (the electronic beam) on the screen (the line number for vertical identification and line-region number for horizontal identification). (3) The signal
generated by the user should be synchronized with the above information, (1)
and (2), to properly identify the choice made. The choice of where to perform
the timings (whether at the home or the head end) depends on the specific technologies used for feedback and the number of subscribers. This choice will be
discussed in a separate section.

The timing devices used to locate the electron beam on the screen are defined as internal addressing devices. The devices used for eliciting the response at the home end (by the listener) will be defined as external addressing
devices. It is the synchronization of the internal and external devices that
identifies the specific request made by the user.

In all of the screen-feedback technologies, except the light pen, the synchronization done at timing devices A and B can be combined by computer
codification of the image and synchronization of it with a screen-coded scale
kept in its memory.

Since internal addressing is common to all of the screen-feedback technol-

ogies, it will be discussed in the next section. The external technologies will
each be discussed separately.

Internal Addressing

The internal addressing system consists of two timing devices: timing device
A, which codes the image (records which information is stored at a given line
and segment of line), and timing device B, a counter, which identifies which
line is being formed at a given instant by the sweep generator.

Timing device A is the more complex of the two and requires a large memory and video disc recorder. Since such a memory and disc can be time-
shared by many users (which reduces its costs per user), it should be located
at the computer center.

The structure of timing device B and its location depend on the specific
technology used and the number of users. In most screen-feedback technologies, the TV screen scale can be precoded and kept in the computer memory.
The x, y coordinates of the user's response can be synchronized with this information and the image code to identify the user request. For example, in
the light-pen technology, it is important to know the location of the sweep generator at the position of the pen. A simple digital counter can be built for this
purpose (at an estimated cost in mass production of $20). If the counter has
to be installed at the home end, the cost of the skilled labor needed for this installation should be added. Figure 2 describes the main features of this counter. The physical location of the sweep generator on the TV screen is determined by counting horizontal synchronizing pulses to yield the vertical dimension, then counting pulses from a high-frequency oscillator to yield the horizontal dimension. One counter can be used, its input being alternated at each
field interval from either the horizontal synchronizing pulse line or the output
of the high-frequency oscillator. The count accumulated during a given field
interval is transferred to a shift register at the end of the field. The vertical
synchronizing pulses are the signal for the pulses. The count can then be converted into a serial pulse train of low frequency about 500 Hz which can be
transferred by ordinary telephone lines. Whether to locate these counters at
the home end or at the head end depends on the number of counters to be installed. This decision is discussed in the following section.

Cost Curves for Counter B as a Function of Its Location

Figure 3 describes the cost curve. These curves are based on the following
assumptions:

1. Beyond a certain volume (point x) the cost of the counter per user will be
lower if the counter consists of a computer unit located at the head end and if
it is shared by a large number of users.
2. At a certain level (point y) the number of users may exceed the capacity of
the computer used, which may require an additional computer, a memory unit,
or a processing unit. This could be expected to increase sharply the cost per
user.
3. Connecting a home TV to a central counter will require considerably less
skilled labor at the home end.
4. The cost of hardware needed to interpret the information received from the
home counters is lower than the cost of a central computer counter.

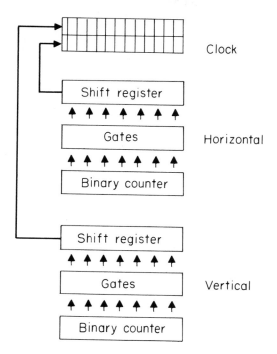

Clock

Horizontal

Vertical

Figure 2. Horizontal-vertical digital counter

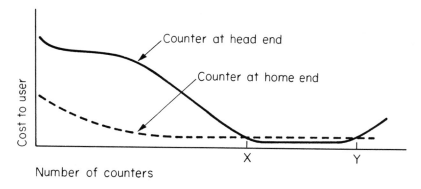

Figure 3. Cost of counters

EVALUATION OF ALTERNATIVE SCREEN-FEEDBACK DEVICES

Light Matrix Technique

Matrices of intersecting light beams can be used for screen feedback.[2] A frame containing light emitters and receivers is mounted on a standard TV screen (see Figure 4). The receivers (or emitters) are calibrated such that each intersection of the light beams will have xy coordinates [point 0 is therefore (3, 2) or (110, 010) in binary coding]. By touching the screen with his finger, the viewer causes the light receivers to detect a change in the light intensity, which causes them (both vertical and horizontal) to send a signal to timing device A (Figure 5). This technique was developed at MIT's Draper Laboratory for the space shuttle. It is to be used by the astronauts for computer interaction.

The electronic circuitry needed for the frame is simple and inexpensive. The frames can be mass produced from plastic material with proper sockets for the electronic circuitry and light emitters and receivers. No special technical skill is needed to mount these frames on the TV receivers.

One of the main advantages of this technique is that no intermediate devices (like light pens) are needed to elicit a response. This advantage is made clear when a keyboard is being projected on the screen for interaction with a computer. A response can also be elicited from a distance by using a sticklike device to intersect the light matrix.

The frame can also be used to mount transparencies on the screen. These transparencies could be used to define input zones or to add to the versatility of the terminal. For example, prior to a discussion on a certain issue a transparency that shows standarized responses can be sent to the viewers by mail and mounted on the screen during the discussion. The transparencies do not have to be flat. A tactile cue to define the input zones can be added. This could greatly increase the precision of the input signals by decreasing finger positioning errors. Figure 6 shows a transparency that can be used for a keyboard projection. The frame can be easily removed from the TV set and be mounted on a flat board for remote control operation.

The main disadvantage of this technology is low resolution, which limits the terminal versatility for computer graphics. The technique is useful for any service that consists of a selection among alternatives, be it in politics or entertainment. These services can range from selecting among products in a supermarket, to conducting a library search, to selecting keys from a projected keyboard. However, no continuous digitizing, which is necessary for computer graphics, is possible. The resolution can be increased by adding more light emitters and receivers, but the number of light emitters needed for approximating continuity will be economically prohibitive.

Another disadvantage of this technique is that any erroneous disturbance of the net of light beams will send false signals to the head end; the user must be careful to point his finger at the right location.

A third disadvantage is cost. This technique is economically feasible only in mass-production quantities. An estimated cost for a mass-produced frame would be $75-85. It will be useful to establish a curve of cost as a function of the number of emitters and receivers in order to determine the optimum size

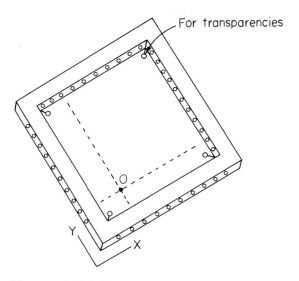

Figure 4. Light emitters mounted on standard TV screen

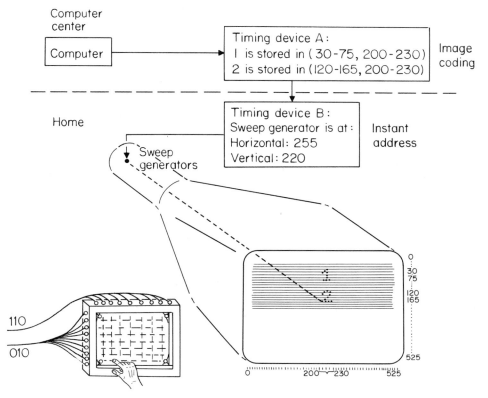

Figure 5. Internal addressing

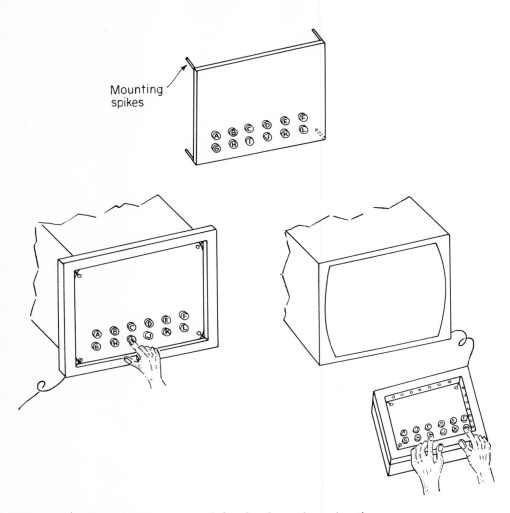

Figure 6. Transparency used for keyboard projection

of the light matrix. A minimum size light matrix should allow the projection of an entire keyboard on one-third of the TV screen.

A full-size prototype input array of 80 input zones has been constructed and mounted on the face of a TV display at MIT's Draper Laboratory. Tentative specifications for the large array are as follows:

Emitters: Light-emitting diodes

Mode: AC or pulse mode

Receivers: Silicon photo transistor receivers

Power consumption: 5 watts or less

Frame Weight: Maximum 20 ounces.

Sound Technique

This screen feedback technology[3] employs a "pen." This pen transmits ultrasonic sound waves, which are detected by sensor strips (microphonic strips) mounted on the edges of the TV screen. The position of the pen can be calculated by measuring the time it takes the sound waves to reach the sensor strips (see Figure 7).

The main advantage of this technique is its high resolution and accuracy. The entire space enclosed within the sensor strips can be used for feedback purposes since the sound detectors are continuous (that is, each point within this space can be located accurately). The different modes of operation also allow a discrete or continuous feedback, which makes this form of communication very versatile.

Some of the applications of the pen mode that have been tried are as follows:

Computer graphics —without keyboard, mouse, light pen or curser, rough sketches or menu-selected symbols are transferred to paper as finished engineering drawings by a controlled computer-aided drafting system.

System analysis —a flow diagram or electronic circuit can be sketched on the TV screen (or a tablet) and the component parameters entered into the computer; performance is read out; parameters or dimensions can be altered and new values reported.

Data reduction —digitized visual records such as EKGs, EEGs, response curves, statistical graphs and Polaroid photos of transient phenomena can be transmitted for computer analysis.

Hand printed character recognition —the interpretation of the dynamics of alpha numeric characters can be formed by the pen stylus and converted to machine-usable form (or typewriter form).

The frame can be designed so that it will be possible to remove it from the TV screen and mount it on a wooden tablet that will allow the user to sit at a distance from the TV screen (see Figure 8). The use of the tablet for input has the advantage of producing a hard copy during input.

The main disadvantage of this technique is its sensitivity to disturbances. Strong noise in the location of the terminal could cause errors in the feedback. This is especially significant in computer graphics. In discrete selections among alternatives (keyboard, etc.) errors due to noise will be insignificant.

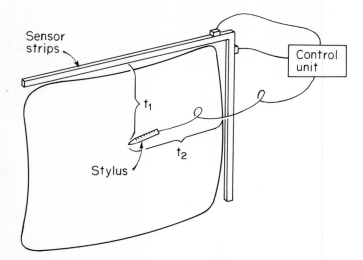

Figure 7. Diagram of sound technique

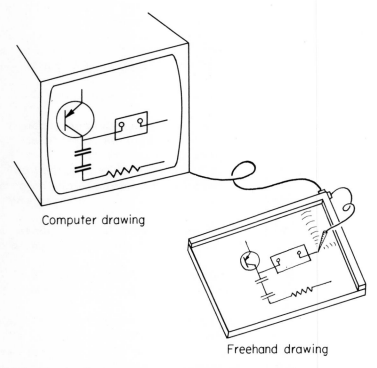

Figure 8. Computer graphics used for drawing

Another disadvantage is that any object placed between the pen and the sensor strips (such as the hand of the user) would cause a discontinuity in the feedback.

A third disadvantage is the noise the sound-emitter makes in continuous digitizing. This noise could be bothersome.

The sound technique is also sensitive to room temperature and humidity. However, if it is operated indoors and at normal temperature the disturbance will be minimal.

An intermediate device, the pen, must be used by the operator in order to elicit a response, slowing keyboard operation considerably.

Only flat transparencies can be used, which further limits the versatility of the technique.

The technical characteristics of the sound technique are as follows:

Resolution: 2000 × 2000 line pairs

Data rate: Variable, one to 200 word pairs per second

Power: on-off

Modes: free run, single shot, pen control, remote

Dimmer: Continuously variable adjusting brightness for comfort in dim environment

Left hand/right hand: interchanges X and Y axes for more convenient use by left-handed operators

Produced by: SAC — Science Accessories Corporation, Southport, Connecticut.

Potentiometer Technique

A position on the screen can be determined by using two potentiometers based on polar coordinates. A potentiometer is a device producing variable voltages, which are functions of a controlled electric resistance. The voltage is directly proportional to the resistance. In the potentiometer technique, one such potentiometer controls angle (the angle between the lines xx and yy) and another controls the distance R (the distance between box A and the location on the screen), as shown in Figure 9.

The electric wire is controlled by a spring in box A, and pin B is located in socket ZZ in zero position (see Figure 10).

To operate, the user takes the pin out of its socket, pulls the string out of box A, and locates it on the screen. When the pin is positioned at the desired location, the user presses a button (located at the side or top of the pin), which signals the computer to read the voltages accumulated in the potentiometers. These voltages are then transferred (possibly by telephone lines) to digital counters for final identification of the request. Once the signal is transmitted, the wire will be pulled back automatically by the spring into box A unless the user wishes to move it to another location.

Figure 11 describes the simple technical construction of the potentiometer technique. By moving the pin on the screen, the users move points A and B along the resistors changing the output voltages. These output voltages are

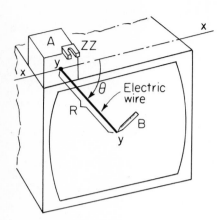

Figure 9. Potentiometer technique

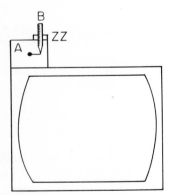

Figure 10. Potentiometer pin in socket

then transferred to a voltage oscillator, which converts the voltage levels into sine waves for the purpose of transmission over telephone lines. Before leaving the box, the voltages are amplified for better transmission.

This technique for screen feedback is cheap and yet highly versatile: The home hardware consists of cheap and simple electronic circuitry; yet both discrete and continuous position digitizing is possible.

Both two- and three-dimensional transparencies can be used. An additional advantage is the possibility of attaching different heads at the end of the wire so that a finger or a pen can be used in order to elicit feedback (see Figure 12). Using a pen will allow the user to keep an ink copy of his input. This technique has a high resolution and low power consumption.

The main disadvantages of the potentiometer technique lie in the human considerations. A projected keyboard cannot be operated with ease since the finger has to constantly pull the string from box A and press a button for each

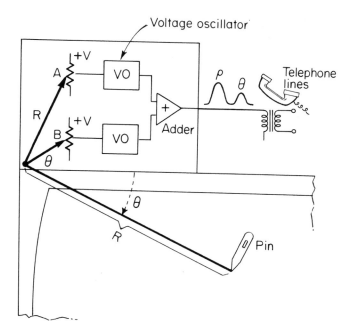

Figure 11. Technical construction of potentiometer technique

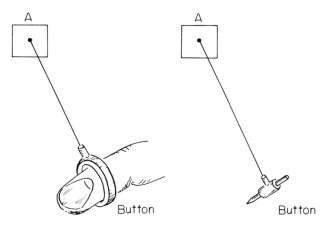

Figure 12. Using a pen and a finger to indicate a choice

letter. The user must sit close to the TV screen for eliciting feedback. The
wire that stretches across the screen may disturb the view.

In addition to the human considerations, the technique has the disadvantage
of moving mechanical parts (the potentiometers), which decreases the reliability of the technique and may increase the cost of maintenance.

Potentiometer Technique B — The "SRI Mouse"

A second potentiometer technique for screen-feedback, developed at the Stanford Research Institute, is known as the "SRI mouse."[4] The mouse is a roughly
box-shaped object about four inches on its longest side. Mounted on wheels, it
rolls on any flat surface and is moved by hand. The wheels drive two potentiometers, which are read by an analog to digital converter at the computer end;
this causes a tracking spot ("bug") to move on the screen in correspondence to
the motion of the mouse (see Figure 13).

To move the arrow from point A to point B on the screen, the user has to
move the mouse from point A to B on the tablet. The tablet can be located anywhere in the room.

The three buttons located at the top of the mouse (C, D, E) are the control
buttons. Button A is the command that signals the computer that a selection is
being made. These buttons provide eight binary combinations for control instructions (for example, A = 001). On the underside of the SRI mouse, two
wheels rotate in proportion to the x or y movement over a flat area (Figure 14).
A device similar to the mouse is the joy stick, which is operated in a similar
manner. Experiments have shown that the mouse is easier to operate.

The main advantage of the mouse is that the user does not have to sit close
to the screen. He can control the display from anywhere within the room. Another advantage is its low price compared with other screen feedback techniques. In mass production its price would be $15-20. The mouse can be used
for both discrete and continuous digitation. Most effectively it can be used for
discrete "menu" selection.

The main disadvantage of the mouse is that a continuous downstream computer display is required. The spot on the screen is controlled by the computer
and should move simultaneously with the mouse for effective usage. If a frame
grabber is used, a delay could result between the hand movements and the
screen reaction. This could make the selection process a tedious one, since
the user can't know if he has placed the mouse in the right position until the
TV picture changes. (One solution for this problem will be proposed in a later
section.)

Although possible, it would be difficult to operate a projected keyboard from
a distance by using the mouse. Also, it would be difficult to draw as accurately
with the mouse as we are accustomed to draw with a pen. Due to its cheap price
and limitation, the mouse should augment other screen-feedback techniques,
since it can be used effectively for remote "menu" selection.

Light Pen Technique

Another device that can be used by the home viewer is the "light-pen" (or raster
stylus). The light pen consists of a ballpoint pen containing a photodiode. To
elicit feedback the user has to hold the light pen against the face of the TV. The
photodiode collects light from a small area of the TV screen and transfers a

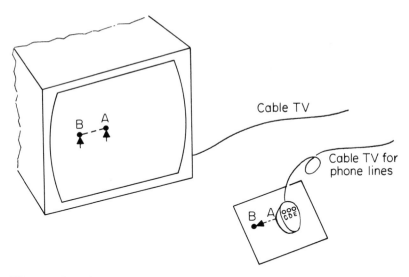

Figure 13. SRI mouse

Figure 14. Underside of SRI mouse

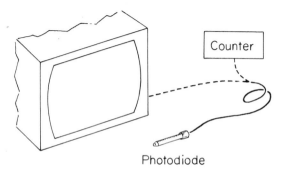

Figure 15. The light pen

signal to the computer or to a timing device. This signal is synchronized with the counter and the image code in order to identify the choice of the viewer (see Figure 15).

Any standard TV can be converted into a graphics terminal by the addition of the vertical and horizontal counters and the light pen. The physical location of the pen on the TV screen is determined by counting the horizontal synchronizing pulses to yield the vertical dimension and counting pulses from a high-frequency oscillator to yield the horizontal dimension.

The light pen must perform the following functions:
1. Collect light from a small area of the TV screen
2. Notify the computer (by user operation) that the light pen is positioned at the desired location and is to be tracked
3. Provide the user an indication that the pen is in the proper location.

The advantage of the light pen technique is its high resolution and its computer-graphics versatility. Its resolution is bounded by the number of vertical lines (525 in American TV) and the number of horizontal synchronizing pulses. This technique can be used for both discrete and continuous digitizing (that is, both for discrete selection among alternatives and computer graphics). Furthermore, the electronic circuitry needed for the light-pen technique is simple and cheap.

The light pen sensitivity[5] is a function of the instantaneous brightness of the display, the luminous intensity of the light at the photosensor, and the photosensor response. The pen is operated by light change rather than steady-state light, which makes it dependent on the fluorescent rise time. The user may be required to adjust the level of light since if the light is too low, the spot will either disappear completely or "dance" on the screen.

There are a few other disadvantages. The user has to sit close to the TV for eliciting feedback. A projected keyboard cannot be easily operated since the light pen must be used for sending signals. Only two-dimensional thin transparencies can be mounted on the screen. No ink copy can be kept of the feedback input.

Graphic Tablet Technique[6]

A transparent tablet containing thin invisible wires (see Figure 16) can be mounted on the TV screen and by using a "pen," the operator can "write" on it. The control unit, which is attached to the tablet, converts the position of the stylus on the screen into Cartesian coordinates in digital or analog form and transmits the information to a computer.

The determination of the position is achieved as follows: during the X measurement, a zero degree 1 KHz signal with 20V amplitude is connected to the left edge (G) and a 90° 1 KHz signal to the right edge (H). These signals are derived from a 2 MHz crystal clock with a 12-bit scaler. This information modulates a 64 KHz carrier to permit capacitive pickup and effective filtering of power noise. An electric field is produced as a result of the conducting wires, and each point along the X axis will have the same voltage but different phase (which results from the sum of the two square waves). The signal varies linearly in phase between 0 to 90° as the X position is increased. The phase is detected by the pen, which contains a capacitor. The higher the frequency of the waves, the thicker could be the paper mounted on the tablet (64 KHz → pa-

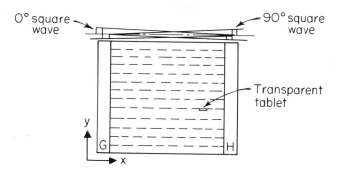

Figure 16. Graphic tablet technique

per pad of up to $\frac{1}{2}$" thickness — not applicable when the tablet is mounted on the TV, as the picture will be blocked). The Y measurement is similarily accomplished. The mass production cost of the transparent tablet plus the additional required electron circuitry is $175-200.

This is a very useful and versatile technique for computer interactive graphics with a computer-refreshed CRT display. It has high resolution and accuracy. It enables long-distance graphics communications with a digital modem over telephone lines. Both discrete and continuous digitizing are possible. The tablet can be easily removed from the TV screen and be mounted on a flat surface for remote operation. When it is mounted on a horizontal plane, the observer watching the display on the vertical TV surface can put a thick paper on top of it, which can serve as input ink copy (the thickness is a function of the field frequency). Horizontal operation causes less operator fatigue.

The ability of the tablet to trace feedback directly from a hand copy allows the operator simply to handwrite his input, a mode of communication that is "second nature" to him.

The main disadvantages of this technique are its relatively high mass-production cost and the necessity of an intermediate device to send a signal. The intermediate device makes it difficult to operate a projected keyboard. Software for this technique is just beginning to be developed.

The technical details are as follows:

Resolution: digital — 10 bits both x and y

Resolution: graphic — 70 lines per inch

Accuracy: better than 2% peak to peak

Graphic position sampling rate: 800 samples per second

The pen signal is transmitted to the computer either when the pen touches the tablet or when it is within 1/2" of it. This distance is adjustable.

Analog signal format: 1 V peak to peak both x and y

Analog bandwidth: 20 Hz

Power: 110/220V, 50/60 Hz, 25 VA

Weight: 15 pounds.

Discrete versus Continuous Digitizing

The versatility of a screen-feedback technology could be constrained by the type of cable system used to carry the signals between the home and the computer center (or TV studio).

There are two main alternatives:[7] rediffusion and a trunk-type system. Rediffusion is a switch-network that resembles the phone system. Each subscriber is provided with a <u>private wire</u> connecting his house to a central program exchange. A trunk-type system — wide band coaxial cable (carrying at present up to 24 channels but able to carry up to 43 channels) is laid in a community, and subscribers who want CATV service are provided with a branch that carries all the available channels. Using a refreshing device, a "frame grabber," at the home end will permit a few hundred subscribers to receive individualized TV pictures on a TV channel that would normally carry only a single program.

In the rediffusion system, a continuous user-computer interaction is possible. In the trunk type, this interaction is more limited. The upstream digital signals (which the user sends to the information center) can be continuous (on a time-share basis), but the downstream signals, which display to the user the response to his input by means of a frame grabber, will be periodic depending on the number of users per channel (see Figure 17).

The significance of this difference becomes clear if we consider computer graphics or any service that requires continuous digitizing (immediate computer display response). Suppose the user types the letter A on the screen. Using a trunk system, the user can type on the screen (on a projected keyboard) continuously (as the home to computer communications is continuous), but he will see what he types only after a fixed time interval. This limitation becomes more severe when graphic communication is desired. A possible solution to this problem could be reserving special downstream channels for computer graphics and other services that require an almost instant computer display response. If a user requires a new picture 3 times per second, the number of users per channel is reduced to 20. However, this reserved capacity can be time-shared by many users, since the actual time of the man-computer interaction is negligible compared to the time the user spends on thinking about his project.

This reserved capacity will increase the costs of the trunk-type system and should be considered when comparing it to the switch systems where the problem does not exist.

Comparison among the Different Screen-Feedback Methods

In this section, a comparison among the different screen-feedback technologies will be done in a matrix form (see Tables 1-4). Four aspects of design will be considered: human factors, economics, versatility, and technology. Human factors include (1) how people react to operating the different techniques and (2) what kinds of operational difficulties different services create. Versatility means how feasible and practical it is to develop software for the

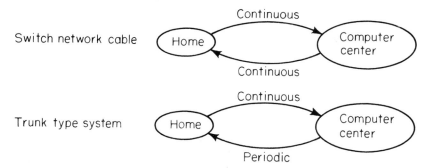

Figure 17. User-computer interaction in the switch and trunk systems

different techniques. Economic considerations involve (1) the extent of appeal to public tastes, (2) what kind of services are the most attractive to the public, (3) the costs of the head-end hardware needed for each of the techniques, (4) the marginal cost of software programming as a function of the complexity of the different techniques, (5) the added cost of TV repairs due to the additional electronics, (6) the cost of labor required to install the feedback mechanism in the home, (7) further examination of the home hardware costs in light of the new developments in hardware and construction of quantity price curves, and (8) estimates for life expectancy and cost of maintenance for the terminals employing the different techniques. Technological evaluation includes (1) determining what is the minimum resolution needed as function of the service required and (2) developing a low-price frame grabber that will allow control of the screen during the "stationary" stage (without a computer link).

The comparison should be carried out at two levels: First, within a certain design aspect (for example, human factors) and second, across the design aspects (for example, human factors versus economics). However, a comparison across design aspects would require a comprehensive market survey and is beyond the scope of this paper. Such an analysis should answer the following questions:

TECH-ECON: What would be the economical advantage of high versus low resolution, or what would be the electricity bill for the user as a function of the power consumption of the different techniques?

VERSATILITY-ECON: What would be the economic value of having a computer graphics capability versus not having it, or what is the relative market value of the different services?

HUMAN-ECON: What is the market value of allowing a remote operation of the home terminal (one in which the user does not have to sit close to the TV), or what would be the expected demand for the different technologies as a function of people's subjective preferences among them?

The main problem in comparing the technologies within a certain design aspect is to assign relative weights to the different features, thus showing which feature is more desirable than another. In this section I will attempt to describe the technologies in objective terms. A final choice among them should be based on market studies.

Table 1. Human Factors
1 = yes;   0 = no;   A = advantage;   D = disadvantage

	Can a projected keyboard be operated with ease?*	Any intermediate device needed to elicit feedback?	Can the terminal be operated from a distance?	Can tactile feelings be simulated?	Is there any active field the user has to be aware of?	Can the technology be modified for remote operation?	Can user keep an ink copy of his input?	Is there any noise associated with eliciting a response?	Does the user have to adjust TV controls?	Are there any wires across the screen that could disturb user?	Is the user required to exert some pressure whenever holding the input device?	Can the terminal be operated in horizontal position?	Human subjective preference for the specific technique**	Relative ease of operation**
Light matrix	1A	0A	1A	1A	1D	1A	0D	0A	0A	0A	0A	1A		
Sound pen	0D	1D	1A	1A	1D	1A	1A	1D	0A	0A	0A	1A		
Light pen	0D	1D	0D	0D	0A	0D	0D	0A	1D	0A	0A	0D		
Potentiometer	0D	1D	0D	0D	0A	0D	1A	0A	0A	1D	1D	0D		
SRI mouse	0D	1D	1A	0D	0A	1A	0D	0A	0A	0A	0A	1A		
Graphic tablet	0D	1D	1A	0D	0A	1A	1A	0A	0A	0A	0A	1A		

*In all screen-feedback techniques a projected keyboard can be operated. Easy operation is defined as one where the user can simulate real keyboard operation (using ten fingers).

**People can be expected to prefer one technique over the other for subjective reasons, which would not necessarily be rational from an efficiency standpoint. The relative ease of operation is also a subjective judgment. It is important to research these attitudes.

Table 2. Economic Aspects

	Cost of hardware at home end*	Electricity bill for user	Cost of installation at home end	Added cost of TV repair	Cost of training for user	Cost of software and hardware per user at head end	Projected demand for the specific technique
Light matrix	$75-85						
Sound pen	$100						
Light pen	$60-70						
Potentiometer	$30-40						
SRI mouse	$20-30						
Graphic tablet	$175-200						

*These costs are estimated based on only mass-production volume.
**If yes, the technology cannot be used for computer graphics.

Table 3. Versatility
1 = yes;   0 = no;   A = advantage;   D = disadvantage

	Can be used for discrete feedback only*	Both two- and three-dimensional transparencies can be used	Replaceable heads can be mounted cheaply**	Ink copy can be easily kept
Light matrix	1D	1A	—	0D
Sound pen	0A	0D	0D	1A
Light pen	0A	0D	0D	0D
Potentiometer	0A	1A	1A	0D
SRI mouse	0A	0D	1A	0D
Graphic tablet	0A	0D	0D	1A

*If yes, the technology cannot be used for computer graphics.
**For less than $10.

Table 4. Technological Features
H = High;   L = Low;   1 = Yes;   0 = No;   A = Advantage;   D = Disadvantage

	Resolution*	Dependent on environmental conditions?	Sample rate	Digital signal format?	Power consumption	Is mechanism for position interpretation complex?	Any moving mechanical parts?**	Are there many software programs for applications?
Light matrix	100×100	0A			3 watts	0A	0A	0D
Sound pen	2000×2000	1D	200 pair/sec			1D	0A	0D
Light pen	525×525	0A				1D	0A	1A
Potentiometer	H	0A			L	0A	1D	0D
SRI mouse	H	0A			L	0A	1D	1A
Graphic tablet	H	0A	500 pair/sec	10 bits both x & y	L	1D	1D	0D

*High resolution is defined as one that is sufficient for computer graphics.
**Moving mechanical parts at the home end could increase the cost of the maintenance. Solid-state devices are more durable.

Comparison between Presently Developing Home Terminals and the Screen-
Feedback Method

The main difference between home terminals using screen feedback and presently developing home terminals is in the method used for eliciting nonverbal
responses from the user. Other peripheral devices, such as microphones, re-
freshers (frame grabbers), home cameras, and printers, are features that
could be added to both the screen and push-button feedback techniques.

There are two techniques for eliciting nonverbal feedback response: (1)
screen-feedback (Figure 18) and (2) push-button feedback (Figure 19) — the
only technique used in presently developing terminals. Technically, the main
difference between the two is that in screen feedback, the TV screen is used
both as a display and as a feedback device. With push-button feedback an external device is used for eliciting the response, and the screen serves merely
as a display.

The comparison between the two techniques will be done only in general
terms of the generic characteristics of each technique. A specific comparison
is possible only after a specific screen-feedback technique is chosen since all
have somewhat different characteristics. The comparison will be made in the
following areas: (1) human factors, (2) versatility, (3) economics, and (4) technology.

HUMAN FACTOR CONSIDERATIONS

Screen Feedback

1. User can design his own feedback
display to fit his needs and understanding.

2. User does not have to switch his
attention from the screen for
eliciting feedback.

3. Display can be designed such that
no extra coding will be needed (for
example, the user can buy a hat by
just touching a displayed hat).

4. Since screen feedback is a new
form of man-machine interaction, it
can be expected that longer period

Button Feedback

1. User is limited to whatever button-
box he acquires.

2. User has to switch his attention between the screen and the button-box.

3. User must remember codes in his
head before eliciting a response. This
can become difficult if codes change
often. The smaller the number of buttons available to the user, the more
complex is the coding.

4. Since "button pushing" is a man-
machine interaction that is familiar
to people, the period of learning to

of adjustment and learning will be required by the public before it becomes "second nature."

VERSATILITY CONSIDERATIONS

Screen Feedback

1. Since the screen is both the display and the feedback device, unlimited feedback displays and codings are possible.

2. Keyboards of all sizes can be projected on the screen for unlimited modes of interaction.

3. Both continuous and discrete feedback are possible.

4. An ink copy of the input can be made simultaneously with the input.

Button Feedback

use the home terminal should be relatively shorter.

1. Having a button feedback requires that all displays must be coded accordingly. The number of buttons limits the amount of information that can be displayed at one time.

2. Any terminal that has less than a full typewriter keyboard capability severely limits the use of the TV for man-computer interaction.

3. Only discrete feedback is possible, which does not allow the use of the terminal for computer graphics. Since it is reasonable to expect that computer graphics will become the dominant mode of man-computer interaction, this is a severe limitation.

4. No ink copy can simultaneously be kept with the input.

5. While listening to a discussion the listener can respond at his own choice of time and ideas.

ECONOMIC CONSIDERATIONS

Screen Feedback

The cost of the home terminal using the screen-feedback technique ranges from $20 to $175 in mass-production quantities. The addition of three control buttons could add up to $20 to a terminal.

2. Since the display is the feedback device, complex feedback displays are allowed without the need to simplify them for the purpose of de-

Button Feedback

1. The lowest price available for home terminal hardware is $85. This terminal includes only four buttons (A. D. Little). Other terminal prices, which include the costs of other peripheral devices and have up to 14 buttons, range from $200 to $415. The John Ward[8] report quotes a price of $50-$100 for a simple yes-no button, meter reading, and alarm system. According to the report, a gen-

coding, which can be expected to lower the cost of programming.

3. Timing synchronization hardware is more expensive due to the need to code the image.

eral-purpose data capability, which will include access to other computer systems for information retrieval, banking, shopping, and electronic mail, will cost between \$200-\$1000, depending on the desired terminal display and/or hard-copy capabilities.

2. The constraint of having a fixed number of buttons makes display planning more complex and expensive.

3. Hardware coding as opposed to human coding is simpler than in screen feedback and, thus, less expensive.

4. The versatility advantage of the screen feedback over the button feedback will make the first more commercially attractive, which can be expected to yield a much higher return on the initial investment.

5. Since the screen feedback is an unfamiliar man-machine interaction to most people, the cost of the public learning to use the screen-feedback terminal can be expected to be higher than the cost of learning to use the "push button" terminal.

TECHNICAL CONSIDERATIONS

1. Hardware devices for both techniques are similar except that an additional timing device will be needed for the screen feedback. The screen-feedback techniques require image coding in the computer memory and the synchronization of this coding with the x, y coordinates of the input. In the push-button feedback, only the codes of the displayed alternatives have to be kept in memory and synchronized with the corresponding buttons.

2. Additional peripheral devices could be added to both techniques at equal costs.

3. Analog to digital converters can be used in both methods.

CONCLUSION — A SUGGESTED HOME TERMINAL
The advantages of screen feedback over push-button feedback are in versatility and human factors considerations. The difficult question to answer is which of the screen-feedback technologies should be used. In the light of the presently known facts about the different screen-feedback techniques, I recommend the following home terminal:

Using a Switched Distribution System
Using a switched distribution system (Figure 20) allows the use of the home terminal for computer graphics. The light matrix technique has the following

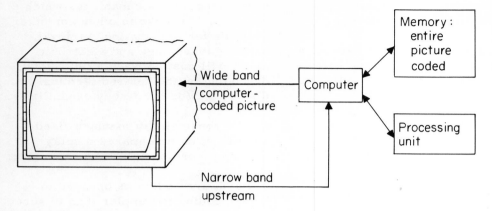

Figure 18. Screen-feedback network

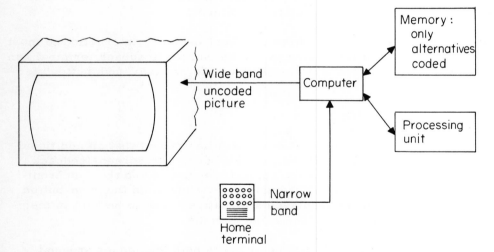

Figure 19. Push-button feedback network

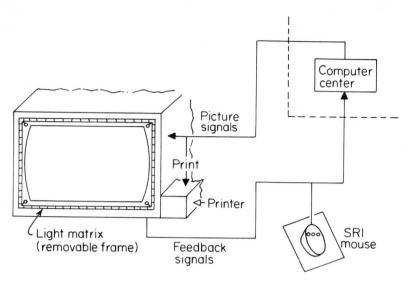

Figure 20. Switched distribution system

important advantages over the other methods in the areas of human factors
and versatility:

It is easy to use.

It is invisible to user.

It allows the simulation of real typing with projected keyboards.

No intermediate device is needed to elicit feedback.

It can be modified for remote feedback.

It is not sensitive to environmental disturbances.

It is a simple mechanism for software design.

It allows the use of two- and three-dimensional transparencies.

The main disadvantages of the light matrix are the following:

It is relatively expensive.

It has low resolution.

No simultaneous ink copy can be kept from input.

The user must be careful not to disturb the light field in areas other than the
intended position.

Adding the SRI mouse or a similar device will reduce the problem of low
resolution and will allow the use of the terminal for computer graphics. Also,

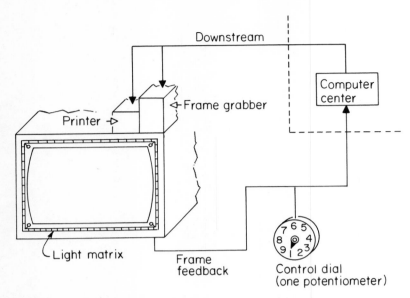

Figure 21. Trunk distribution system

the mouse will ease elicitation of responses from a distance and will provide
the needed control buttons to operate both the mouse and the light matrix. The
low price of the mouse makes it a very attractive addition.

Using the "Trunk" Distribution System

Using a trunk distribution system (Figure 21) will not allow the use of the terminal for computer graphics, assuming that no channels are reserved for that
purpose. The light matrix is again recommended for the reasons stated in the
preceding section. The low resolution is not a constraint in this case since the
resolution could be high enough to allow the projection of a full keyboard on
the screen. (Remember that the resolution of the light matrix is not related to
the resolution of the TV screen.)

A cheap control dial could be added to the light matrix. This dial could provide the control functions as well as allow a limited remote selection among
alternatives. (The remote selection will require the coding of the alternatives.)

These two alternative terminals both compare favorably with the other techniques in most design aspects, though the missing information outlined at the
beginning of this section is required for a more meaningful choice. Meanwhile, price differences among the techniques merit a careful study of all of
them.

The potentials of two-way communications for both evil and good are immense. Technology offers a great variety of hardware to facilitate those communications. An integral part of this system is the home terminal, which will
constrain and shape future communications. The choices of home hardware as
well as the decisions about what type of communications are desirable could
have a lasting effect on social values and life-styles. Careful research in both
communications technology and the use of this technology is vital and urgent.

Notes

1. Malcolm Macaulay, A Low Cost Computer Graphic Terminal, Fall Joint Computer Conference, 1968.

2. Development of a Generalized Interactive Display and Control Concept, Internal report, Apollo Guidance, Navigation and Control, Draper Laboratory, MIT, July 1970.

3. SAC—Graf/Pen, Advertising report published by Science Accessories Corp., Southport, Connecticut.

4. D. C. Englebart and W. K. English, A Research Center for Augmenting Human Intellect, SRI, AFIPS, Conference Proceedings, Vol. 33. (Palo Alto, Cal.: SRI).

5. J. T. Locascio, G. L. Karanza, and J. J. Dalton, Light Pen Versatility, Information Display, Nov./Dec. 1967.

6. ERICON, advertising pamphlets by Shintron Electronics, Cambridge, Mass.

7. Michel Guité, New Technology for Citizen Feedback to Government, unpublished report prepared for the Computer Task Force, Dept. of Communications, Ottawa, Canada, Dec. 1971.

8. John E. Ward, "Present and Probable CATV/Broadband-Communications Technology" in Sloan Commission Report on Cable Communications, June 7, 1971 (reprinted in this book).

CATV in the form in which it exists today is one-way communication, just like the other mass media. Even in that form it is important because it ends the scarcity of channels on the broadcast spectrum. It makes possible a television of abundance with channels available to small groups and minority tastes.

CATV is also important because it provides technical prerequisites for a new kind of two-way communication. At the time of writing we are on the verge of that change. The FCC is requiring cablecasters to install two-way hardware, and the cablecasters all over the country are actively planning for what they will do.

In this section, we turn to the social implications of two-way communications. The next two chapters consider the use of electronic feedback in political discussion. The last two chapters deal with commercial services that could be provided with the aid of two-way cable.

The last chapter also moves further into the future by dealing not only with audience responses to fixed downstream programs (all that any CATV system is now actively considering) but also with the kind of interaction that allows the audience member to control what gets sent to him downstream. When that kind of truly interactive on-demand communication comes, as it will, the social consequences for business, for education, for politics, for culture, and for our whole way of life may be dramatic indeed.

12. TECHNOLOGY FOR GROUP DIALOGUE AND SOCIAL CHOICE

Thomas B. Sheridan

INTRODUCTION

Usually the best way to discuss and resolve the choices that arise within groups of people is face-to-face and personally. For this reason, city planners and educators alike are calling for new kinds of communities for working, living, and learning, based more on familial than on contractual relationships. When people get to know one another, conflicts have a way of being accommodated.

Beyond the circle of intimacy the problem of communication is obviously much greater; and while social issues can still be resolved more or less arbitrarily, it is more difficult to resolve them satisfactorily.

The "circle of intimacy" is constrained in its radius. One analyst has estimated that the average person in his lifetime can get to know, on a personal, face-to-face basis, only about 700 people—and surely one can know well only a much smaller number. The precise number is not important: the point is that it is dictated by the limitations of human behavior and is not greatly affected by urban population growth, by speed of transportation and communication, by affluence, or by any other technologically induced change in the human condition.

Indeed, these changes underlie the problem as we know it. Although the number of people with whom we have intimate face-to-face communication during a lifetime remains constant, we are in close proximity to more and more people.

We are, moreover, a great deal more dependent on one another than we used to be when American society was largely agrarian. We are all committed together in planning and paying for highways and welfare. We pollute each other's water and air. We share the risks and the costs of our military-industrial complex and the foreign policy it serves. Technology, while aggravating the selfishly independent consumption of common resources, has made communications beyond the circle of intimacy both more awkward and more urgent.

Beyond the circle of intimacy, what kind of communications make sense? Surely most of us do not demand personal interactions with "all those other people." Yet in order to participate realistically in the decisions of industry and commerce and in government programs to aid and regulate the processes that affect us intimately, we as citizens need to communicate with and understand the whole cross section of other citizens.

Does technology help us in this? Can it help us do it better? We may now dial on the telephone practically anywhere in the world, to hear and be heard with relatively high fidelity and convenience. We may watch on our television sets news as it breaks around the world and observe our president as though he were in our living room. We can communicate individually with great flexibility; and at our own convenience we can be spectators en masse to important events.

But effective governance in a democracy requires more than this. It requires that citizens, in various ways and with respect to various public issues,

----

The research at M.I.T. described herein is supported on National Science Foundation Grant GT-16, "Citizen Feedback and Opinion Formulation," and a project, "Citizen Involvement in Setting Goals for Education in Massachusetts," with the Massachusetts Department of Education.

can make their preferences known quickly and conveniently to those in power.
We now have available two obvious channels for such "citizen feedback." First,
we go to the polls roughly once a year and vote for a slate of candidates; second, we write letters to our elected representatives.

There are other channels by which we make our feelings known, of course —
by purchasing power, by protest, etc. But the average citizen wields relatively
little influence on his government in these latter ways. In terms of effective
information transmitted per unit time, none of the presently available channels of citizen feedback rivals the flow from the centers of power outward to
the citizens via television and the press.

What is it that stands in the way of using technology for greater public participation in the important compromise decisions of government, such as
whether we build a certain weapon, or an SST, or what taxes we should pay
to fund what federal program, or where the law should draw the line that may
limit one person's freedom in order to maintain that of others?

Somehow in an earlier day decisions were simpler and could involve fewer
people — especially when it came to the use of technology. If the problem was
to span a river and if materials and the skills were available, you went ahead
and built the bridge. It would be good for everyone, and thus with other blessings of technology. There seemed little question that higher-capacity machines
of production or more sophisticated weapons were inherently better. There
seemed to be an infinite supply of air, water, land, minerals, and energy.
Today, by contrast, every modern government policy decision is in effect a
compromise — and the advantages and disadvantages have to be weighed not
only in terms of their benefits and costs for the present clientele but also for
future generations. We are interdependent not only in space but in time.

Such complex resource allocation and benefit-cost problems have been attacked by the whole gamut of mathematical and simulation tools of operations
research. But these "objective" techniques ultimately depend upon subjective
value criteria — which are valid only so far as there are effective communication procedures by which people can specify their values in useful form.

THE FORMAL SOCIAL-CHOICE PROBLEM

The long-run prospects are bright, I think, that new technology can play a
major role in bringing the citizenry together, individually or in small groups,
communicating and participating in decisions, not only to help the decision
makers but also for the purpose of educating themselves and each other. Hardware in itself is not the principal hurdle. No new breakthroughs are required.
What is needed, rather, is a concerted effort in applying present technology to
a very classical problem of economics and politics called "social choice" —
the problem of how two or more people can communicate, compare values or
preferences on a common scale, and come to a common judgment or preference
ordering.

Even when we are brought together in a meeting room it is often very awkward to carry on meaningful communication due to lack of shared assumptions,
fear of losing anonymity or fear of seeming inarticulate, etc. Therefore, a
few excitable or most articulate persons may have the floor to themselves
while others, who have equally intense feelings or depth of knowledge on the
subject, may go away from the meeting having had little or no influence.

It is when we consider the electronic digital computer that the major contributions of technology to social choice and citizen feedback are foreseen. Given the computer, with a relatively simple independent data channel to each participant, one can collect individual responses from all participants and show anyone the important features of the aggregate—and do this, for practical purposes, instantaneously.

Much of technology for such a system exists today. What is needed is thoughtful design—with emphasis on how the machine and the people interact: the way questions are posed to the group participants; the design of response languages that are flexible enough so that each participant can "say" (encode) his reaction to a given question in that language, yet simple enough for the computer to read and analyze; and the design of displays that show the "interesting features" or "pertinent statistics" of the response data aggregate.

This task will require an admixture of experimental psychology and systems engineering. It will be highly empirical, in the same way that the related field of computer-aided learning is highly empirical.

The central question is, How can we establish scales of value which are mutually commensurable among different people? Many of the ancient philosophers wrote about this problem. The Englishmen Jeremy Bentham and John Stuart Mill first developed the idea of "utility" as a yardstick, that could compare different kinds of things and events for the same person. More recently the American mathematician Von Neumann added the idea that not only is the worth of an event proportional to its utility, but that of an unanticipated event is proportional also to the probability that it will happen.[1] This simple idea created a giant step in mathematically evaluating combinations of events with differing utilities and differing probabilities—but again for a single person.

The recent history of comparing values for different people has been a discouraging one—primarily because of a landmark contribution by economist Kenneth Arrow.[2] He showed that, if you know how each of a set of individuals orders his preferences among alternatives, there is no procedure that is fair and will always work by which, from this data, the group as a whole may order its preferences (that is, determine a "social choice"). In essence he made four seemingly fair and reasonable assumptions: (1) the social ordering of preferences is to be based on the individual orderings; (2) there is no "dictator" whom everyone imitates; (3) if every individual prefers alternative A to alternative B, the society will also prefer A to B; and (4) if A and B are on the list of alternatives to be ordered, it is irrelevant how people feel about some alternative C, which is not on the list, relative to A and B. Starting from these assumptions, he showed (mathematically) that there is no single consistent procedure for ordering alternatives for the group that will always satisfy the assumptions.

A number of other theoreticians in the area have challenged Arrow's theorem in various ways, particularly through challenging the "independence of irrelevant alternatives" assumption. The point here is that things are never evaluated in a vacuum but clearly are evaluated in the context of circumstance. A further charge is a pragmatic one: while Arrow proves inconsistencies can occur, in the great majority of cases likely to be encountered in the real world they would not occur, and if they did they probably would be of minor significance.

There are many other complicating factors in social choice, most of which
have not been, and perhaps cannot be, dealt with in the systematic manner of
Arrow's "impossibility theorem." For example, there is the very fundamental
question of whether the individual parties involved in a group-choice exercise
will communicate their true feelings and indicate their uncertainties, or
whether they will falsify their feelings so as to gain the best advantage for
themselves.

Further difficulties arise when we try to include in the treatment the effects
of differences among the participants along the lines of intensity-of-feelings
versus apathy, or knowledge versus ignorance, or "extended sympathy" versus
selfishness, or partial versus complete truthfulness; yet these are just the
features of the social-choice problem as we find it in practice.

To take as an ultimate goal the precise statement of social welfare in mathematical terms is, of course, nonsense. The differing experiences (and consequently differing assumptions) of individuals ensure that commensurability
of values will never be complete. But this difficulty by no means relieves us
of the obligation to seek value commensurability and to see how far we can go
in the quantitative assessment of utility. By making our values more explicit
to one another we also make them more explicit to ourselves.

POTENTIAL CONTRIBUTIONS OF ELECTRONICS

Electronic media notwithstanding, none of the newer means of communication
yet does what a direct face-to-face group meeting (town meeting, class bull
session) does — that is, permit each participant to observe the feelings and
gestures, the verbal expressions of approval or disapproval, or the apathetic
silence that may accompany any proposal or statement. As a group meeting
gets larger, observation of how others feel becomes more and more difficult;
and no generally available technology helps much. Telephone conference calls,
for example, while permitting a number of people to speak and be heard by all,
are painfully awkward and slow and permit no observation of others' reaction
to any given speaker. The new Picturephone will eventually permit the participants in a teleconference to see one another; but experiments with an automatic
system that switches everyone's screen to the person who is talking reveals
that this is precisely what is not wanted — teleconferees would like most to observe the facial expressions of the various conferees who are not talking!

One can imagine a computer-aided feedback-and-participation system taking
a variety of forms all of which are more or less characterized by Figure 1.
For example:

1. A radio talk show or a television "issue" program may wish to enhance its
audience participation by listener or viewer votes, collected from each participant and fed to a computer. Voters may be in the studio with electronic
voting boxes or at home where they render their vote by calling a special telephone number. The NET "Advocates" program has demonstrated both.

2. Public hearings or town meetings may wish to find out how the citizenry feel
about proposed new legislation — who have intense feelings, who are apathetic,
who are educated to the facts, and who are ignorant of them — and correlate
these responses with each other and with demographic data, which participants
may be asked to volunteer. Such a meeting could be held in the town assembly
hall, with a simple push-button console wired to each seat.

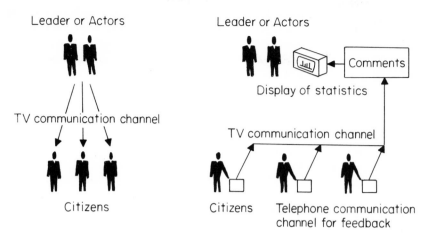

Figure 1. General paradigm for citizen feedback (right) added to top-down communication (left)

3. Several PTAs or alternatively several eighth grades in the town may wish to sponsor a feedback meeting on sex education, drugs, or some other subject where truthfulness is highly in order but anonymity may be desired. Classrooms at several different schools could be tied together by rented "dedicated" telephone lines for the duration of the session.

4. A committee chairman or manager or salesman wishes to present some propositions and poll his committee members or sales representatives who may be stationed at telephone consoles in widely separated locations, or may be seated before special intercom consoles in their own offices (which could operate entirely independently of the telephone system).

5. A group of technical experts might be called upon to render probability estimates about some scientific diagnosis or future event that is amenable to before-the-fact analysis. This process may be repeated, where with each repetition the distribution of estimates is revealed to all participants and possibly the participants may challenge one another. This process has been called the "Delphi Technique," after the oracle, and has been the subject of experiments by the Rand Corporation and the Institute for the Future[3] and by the University of Illinois.[4] Their experience suggests that on successive interactions even experts tend to change their estimates on the basis of what others believe (and possible new evidence presented during the challenge period).

6. A duly elected representative in the local, state, or national government could ask his constituency questions and receive their responses. This could be done through radio or television or alternatively could utilize a special van, equipped with a loudspeaker system, a rearlighted projection/display device, and a number of chairs or benches, which could be set up rapidly at street corners prewired with voter-response boxes and a small computer.

These examples point up one very important aspect of such citizen feedback or response-aggregation systems: that is, that they can educate and involve the participants without the necessity that the responses formally determine a decision. Indeed the teaching-learning function may be the most important. It demands careful attention to how questions are posed and presented, what operations are performed by the computer on the aggregated votes and what operations are left out, how the results are displayed, and what opportunity there is for further voting and recycling on the same and related questions.

Some skeptics feel that further technocratic invasion of participatory democracy should be prevented rather than facilitated—that the whole idea of the "computerized referendum" is anathema, and that the forces of repression will eventually gain control of any such system. They could be correct, for the system clearly presupposes competence and fairness in phrasing the questions and designing the alternative responses.

But my own fear is different. It is that, propelled by the increasing availability of glamorous technology and spurred on by hardware hucksters and panacea pushers, the community will be caught with its pilot experiments incomplete or never done.

THE STEPS IN A GROUP FEEDBACK SESSION

Seven formal steps are involved in a technologically aided interchange of views on a social-choice question:

1. The leader states the problem, specifies the question, and describes the response alternatives from which respondents are to choose.

2. The leader (or automated components of the system) explains what respondents must do in order to communicate their responses (including, perhaps, their degree of understanding of the question, strength of feeling, and subjective assessment of probabilities).

3. The respondents set into their voting boxes their coded responses to the questions.

4. The computer interrogates the voting boxes and aggregates the response data.

5. Preselected features of this response-aggregate are displayed to all parties.

6. The leader or respondents may request display of additional features of the response aggregate or may volunteer corrections or additional information.

7. Based upon an a priori program, on previous results, and/or on requests from respondents, the leader poses a new problem or question, restarting the cycle from Step 1.

The first step is easily the most important—and also the most difficult. Clearly the participant must understand at the outset something of the background of any specific question he is asked, he must understand the question itself in nonambiguous terms, and he must understand the meaning of the answers or response alternatives he is offered. This step is essentially the same as is faced by the designer of any multiple-choice test or poll, except that there is the possibility that a much richer language of response can be made available than is usually the case in machine-graded tests. Allowed responses may include not only the selection of an alternative answer but also an indication of intensity of feeling, estimates of the relative probability or importance of some event in comparison with a standard, specification of numbers (for example, allowable cost) over a large range, and simple expressions of approval ("yea!") or disapproval ("boo!").

The leader may have to explain certain subtleties of voting, such as whether participants will be assumed to be voting altruistically (what I think is best for everyone) or selfishly (what I think is best for me alone, me and my family, etc.). Further, he may wish respondents to play roles other than themselves (if you were a person under certain specified circumstances, how would you vote?).

He may also wish to correlate the answers with informedness. He may do this by requesting those who do not know the answer to some test question to refrain from voting, or he can pose the knowledge test question before or after the issue question and let the computer make the correlation for him.

Ensuring that the participants "play fair," own up to their uncertainties, vote as they really feel, vote altruistically if asked, and so on, is extremely difficult. Some may always regard their participation in such social interaction as an advocacy game, where the purpose is to "win for their side."

The next two steps raise the question of what equipment the voter will have for communicating his responses. At the extreme of simplicity a single on-off switch generates a response code that is easily interpreted by the computer but limiting to the user. At the other extreme, if responses were to consist of natural English sentences typed on a conventional teletypewriter — which would certainly allow great flexibility and variety in response — the computer would have no basis for aggregating and analyzing responses on a commensurate basis (other than such procedures as counting key words). Clearly something in-between is called for; for example, a voting box might consist of ten on-off switches to use in various combinations, plus one to indicate "ready," plus one "intensity" knob.

An unresolved question concerns how complex a single question can be. If the question is too simple, the responses will not be worth collecting and will provide little useful feedback. If too complex, encoding the responses will be too difficult. The ten switches of the voting box suggested above would have the potential (considering all combinations of on and off) for 2^{10} or 1024, alternatives, but that is clearly too many for the useful answers to any one question.

It is probably a good idea, for most questions, to have some response categories to indicate "understand question but am undecided among alternatives" or "understand question and protest available alternatives" or simply "don't understand question or procedures," three quite different responses. If a respondent is being pressured by a time constraint, which may be a practical necessity to keep the process functioning smoothly, he may want to be able to say, "I don't have time to reach a decision"; this could easily be indicated if he simply fails to set the "done" switch. Some arrangement for "I object to the questions and therefore won't answer" would also be useful as a guide to subsequent operations and may also subsume some of the above "don't understand" categories. Figure 2 indicates various categories of response for a six-switch console.

The fourth step, in which the computer samples the voting boxes and stores the data, is straightforward as regards tallying the number of votes in each category and computing simple statistics. But extracting meaning from the data requires that someone should have laid down criteria for what is interesting; this might be done either prior to or during the session by a trained analyst.

It is at this point that certain perils of citizen feedback system arise, for

Identification of self (note: if one of 1, 2, 3 not switched, assume unregistered or other party; if one of 4, 5, 6 assume other or none)	(1) Republican (2) Democrat (3) Independent (4) Protestant (5) Catholic (6) Jew
Expressions of feeling and experience	(1) Am intensely interested (2) Am mildly interested (3) Am uninterested (4) Daily experience (5) Occasional experience (6) No experience
Four numerical categories Two administrative categories	(1) Less than 10% (2) 10 to 30% (3) 30 to 60% (4) Greater than 60% (5) Don't know (6) Don't understand
Three alternatives Three administrative categories	(1) I want plan 1 (2) I want plan 2 (3) I want plan 3 (4) Undecided as to plans (5) Object to available plans (6) Confused by procedure
Rank ordering of three alternatives, A, B, C first choice second choice	(1) A (2) B (3) C (4) A (5) B (6) C
Response to interpersonal communication of actors as to I	(1) Miss Adams (2) Colonel Baker (3) Doctor Crank (4) Agree (5) Disagree (6) Am bored

To select one of 8 on each of two questions
    Question 1
 	
(dots under your answer indicate switches to be thrown)
    Question 2
	A B C D E F G H
	(1) . . . .
	(2) . . . .
	(3) . . . .
	(4) . . . .
	(5) . . . .
	(6) . . . .

Figure 2. Sample categories of response for a six-switch console

the analyst could (either unwittingly or deliberately) distort the interpretation of the voting data by the criteria he selects for computer analysis and display. Though there has been much research on voting behavior and on methods of analyzing voting statistics, instantaneous feedback and recycling poses many new research challenges.

That each man's vote is equally important on each question is a bit of lore that both political scientists and politicians have long since discounted — at least in the sense that voters naturally feel more intensely about some issues than about others. One would, therefore, like to permit voters to weight their votes according to the intensity of their feeling. Can fair means be provided?

There are at least two methods. One long-respected procedure in govern- ment is bargaining for votes — "I'll vote with you on this issue if you vote with me on that one." But in the citizen-feedback context, negotiating such bargains does not look easy. A second procedure would be to allocate to each voter, over a set of questions, a fixed number of influence points, say 100; he would indi- cate the number of points he wished the computer to assign to his vote on each question, until he had used up his quota of 100 points, after which the computer would not accept his vote. (Otherwise, were votes simply weighted by an un- constrained "intensity of feeling" knob, a voter would be rather likely to set the "intensity of feeling" to a maximum and leave it there.)

A variant on the latter is a procedure developed at the University of Arizona[5] wherein a voter may assign his 100 points either among the ultimate choices or among the other voters. Provided each voter assigns some weight to at least one ultimate alternative an eventual alternative is selected, in some cases by a rather complex influence of trust and proxy.

Step 5, the display of significant features of the voting data, poses interest- ing challenges concerning how to convey distributional or statistical ideas to an unsophisticated participant quickly and unambiguously.

The sixth step provides an opportunity for nonplanned feedback — informal exposition, challenges to the question, challenges to each other's votes, and verbal rebuttal — in other words, a chance to break free of the formal con- straints for a short time. This is a time when participants can seek to influ- ence the future behavior of the leader — the questions he will ask, the response alternatives he will include, and the way he manages the session.

EXPERIMENTS IN PROGRESS
Experiments to date have been designed to learn as much as possible as quickly as possible from "real" situations. Because the mode of group dialogue just discussed introduces so many new variables, it was believed not expedient to start with controlled laboratory experiments, though gradually we plan to make controlled comparisons on selected experimental conditions. But the initial em- phasis has been on plunging into the "real world" and finding out "what works."

Experiments in a Semilaboratory Setting within the University
In one set of experiments in the Man Machine Systems Laboratory at MIT the group-feedback system consists of fourteen hand-held consoles, each with ten on-off switches, a continuous "adjust" knob, and a "done" switch. The consoles are connected by wire to a PDP-8 computer with a scope display output. Closed- circuit television permits simulation of a meeting where questions are being

posed and results aggregated at some distant point (for example, a television station in another city) and where respondents may sit together in a single meeting room or may be located all at different places. Various aggregation display programs are available to the discussion leader, the simplest of which is a histogram display indicating how many people have thrown each switch. Other data reduction programs are also available, such as the one I have described permitting voters to give a percentage of their votes to another voter. A variety of small group meetings, seminars, and discussions have been held utilizing this equipment.

Two kinds of leadership roles have been tried. The first is where a single leader makes statements and poses questions. Here, among other things, we were concerned with whether respondents, if constrained to express themselves only in terms of the switches, can "stay with it" without too much frustration and can feel that they are part of a conversation. Thus far, for this type of meeting, we have learned the following:

1. Questions must be stated unambiguously. We learned to appreciate the subtle ways in which natural language feedback permits clarification of questions or propositions. Often the questioner doesn't understand an ambiguity in his statement — where a natural language response from one or two persons chosen at random only for the purpose of clarifying the question is often well worth the time of others, though this by no means obviates the need to have some "I don't understand" or "I object" categories.

2. The leader should somehow respond to the responses of the voters. If he can predicate his next question or proposition on the audience response to the last one, so much the better. Otherwise he can simply show the audience that indeed he knows how the vote on the last question turned out and freely express his surprise or other reaction. In cases where the leader seemed as though he was not as interested in the response and simply ground through a programmed series of questions, the audience quickly lost interest.

3. Anonymity can be very important, and, if safeguarded, permits open "discussion" in areas that otherwise would be taboo. For example, we have conducted sessions on drug use, in which students, faculty, and some total strangers quite freely indicated how often they use certain drugs and where they get them. Such discussions, led unabashedly by students (who knew what and how to phrase the key questions!) resulted in a surprising freedom of response. (We made the rule that voters had to keep their eyes on the display, not on each others' boxes, though a small voting box can easily be held close enough to the chest to obstruct others' view of which switches are being thrown.) It was found especially important, for this kind of topic, not to display any results until all were in.

In the same semilaboratory setting we have experimented with a second kind of leadership role. Here two or more people "discuss" or "act" and the audience continuously votes with "yea," "boo," "slow down and explain," "speed up and go on to another topic" type response alternatives. Voters were happy to play this less direct role but perhaps for a shorter time than in the direct response role. Again it proved of great importance that the central actors indicate that they saw and were interested in how the voters voted.

Experiments with Citizen Group Meetings Using Portable Equipment

As of this writing five group meetings have been conducted in the Massachusetts towns of Stoneham, Natick, Manchester, Malden, and Lowell to assist the Massachusetts Department of Education in a program of setting educational goals. In each case cross sections of interested citizens were brought together by invitation of persons in each community to "discuss educational goals." Four similar meetings were conducted with students and teachers at a high school in Newton, Massachusetts. (A similar meeting was also held in a church parlor in Newton to help the members of that church resolve an internal political crisis.) All groups ranged in size from 20 to 40, though at any one time only 32 could vote since only that number of voting boxes have been built.

The portable equipment used for these meetings, held variously in church assembly halls, school classrooms, and television studios, features small hand-held voting boxes, each with six toggle switches, connected by wire to substations (eight boxes to a substation, each of the latter containing digital counting logic), which in turn are series-connected in random order to central logic and display hardware. The display regularly used to count votes is a "nixie tube"-type display of the six totals (number of persons activating each of the six switches). The meeting moderator, through a three-position switch, can hold the numbers displayed at zero, set it in a free counting mode, or lock the count so that it cannot be altered. A second display, little used as yet, is a motorized bar graph to be used either to display histogram statistics or to provide a running indication of affective judgments, such as agree with speaker, disagree, too fast, too slow, etc.

The typical format for these meetings was as follows. After a very brief introduction to the purpose of the meeting and the voting procedure itself, several questions were asked to introduce members of the group to each other (beyond what is obvious from physical appearance), such as education, political affiliation, marital status, etc. An overhead projector has been used in most cases to ask the questions and record the answers and comments (on the gelatin transparency) since, unlike a blackboard, it need not be erased before making a permanent record. Following the introduction, the meetings proceeded through the questions, such as those illustrated by Figure 3, and those posed by the participants themselves. The categories of "object to question" or "other" were used frequently to solicit difficulties or concerns people had with the question itself—its ambiguity, whether it was fair, etc. Asking persons who voted in prepared categories to identify themselves and state, after the fact, why they voted as they did, was part of the standard procedure. Roughly twenty questions, with discussion, can be handled in 1 1/2 hours.

After the meeting, evaluations by the participants themselves suggested that the procedure does indeed serve to open up issues, to draw out those who would otherwise not say much, and generally to provide an enjoyable experience—in some cases for three hours duration.

EXTENDING THE MEETING IN SPACE AND TIME

The employment of such feedback techniques in conjunction with television and radio media appears quite attractive, but there are some problems.

A major problem concerns the use of telephone networks for feedback. Un-

---

How are preschool children best prepared for school?

	(as school now exists)	(as school should be)
(1) lots of parental love	9	11
(2) early exposure to books	2	1
(3) interaction with other kids	14	8
(4) by having natural wonders and aesthetic delights pointed out	1	5
(5) unsure	4	3
(6) object	0	1

Salient comments after vote ("as school now exists" and "as school should be" not part of question then): One man objected to "pointed out" in (4), as it emphasized "instruction" rather than "learning." Discussion on this point. Someone else wanted to get at "encouraging curiosity." Another claimed, "That's what question says," and another "discover natural wonders." Consensus: "leave wording as is." Then a lady violently objected that the vote would be different depending on whether voter was thinking of school as it now existed or as it should be. Others agreed. Two categories added. Above is final vote.

Student attendance should be:

(1) compulsory with firm excuse policy	11
(2) compulsory with lenient excuse policy	0
(3) voluntary, with students responsible for material missed	12
(4) voluntary, with teachers providing all reasonable assistance to pupils who miss class	2
(5) unsure	3
(6) object	0

Comments centered on the feeling that some subjects require attendance more than others do. (Note the 0 vote on category (2), which is inappropriately self-contradictory.)

Figure 3. Typical questions and responses from the citizen meetings on educational goals

fortunately telephone switching systems, as they presently work, do not easily permit some of the functions one would like. For example, one would like a telephone central computer to be able to interrogate, in rapid sequence, a large number of memory buffers (shift registers) attached to individual telephones, using only enough time for a burst of ten or so tone combinations (like touchtone dial signaling), say, about 1/2 second. Alternatively one might like to be able to call a certain number, and, in spite of a temporary busy signal, in a few seconds have the memory buffer interrogated and read over the telephone line. However, with a little investigation one finds that telephones were designed for random caller to called-party connections, with a busy signal re-

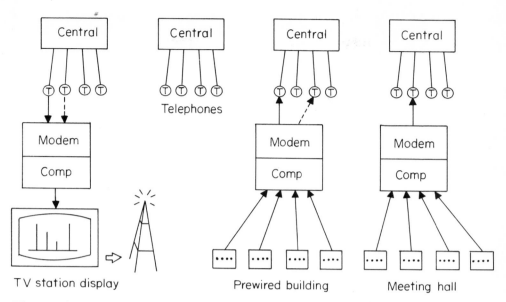

Figure 4. Multigroup arrangement for television audience response

jecting the calling party from any further consideration and providing no easily employed mechanism for retrieving that calling party once the line is freed.

For this reason at least for the immediate future, it appears that sampling a large number (much more than 1,000) on a single telephone line in less than 15 minutes, even for a simple count of busy signals, is not practical.

One tractable approach for the immediate future is to have groups of persons, 100 to 1,000, assembled at various locations watching television screens. Within each meeting room participants vote using hand-held consoles connected by wire to a computer, which itself communicates by telephone to the originating television studio. Figure 4 illustrates this scheme.

Ten or more groups scattered around a city or a nation can create something approaching a valid statistical sample, if statistical validity is important, and within themselves can represent characteristic citizen groups (for example, Berkeley students, Detroit hardhats, Iowa farmers, etc.). Such an arrangement would easily permit recycling over the national network every few minutes, and within any one local meeting room some further feedback and recycling could occur that is not shared with the national network.

Cable television, because of its much higher bandwidth, has the capability for rapid feedback from smaller groups or individuals from their individual homes. For example, even part of the 0-54 MHz band (considered as the best prospect for return signals[6]) is more than adequate theoretically for all the cable subscribers in a large community, especially in view of time-sharing possibilities.

These considerations are for extensions in space. One may also consider extensions in time, where a single "program" extends over hours or days and where each problem or question, once presented on television, may wait until slow telephone feedback or even mail returns of an IBM card or newspaper "issue ballot"[7] variety come in.

Development of such systems, fraught with at least as many psychological, sociological, political, and ethical problems as technological ones, will surely have to evolve on the basis of varied experiments and hard experience.

Notes

1. J. von Neumann and O. Morganstern, Theory of Games and Economic Behavior, 2nd ed. (Princeton, N.J.: Princeton University Press, 1947).

2. K. Arrow, Social Choice and Individual Values (New York: John Wiley, 1951).

3. N. Dalkey and O. Helmer, "An experimental application of the Delphi Method to the use of experts," Management Science, No. 9 (1963).

4. C. E. Osgood and S. Umpleby, "A computer-based system for exploration of possible systems for Mankind 2000," Mankind 2000 (London: Allen and Unwin), pp. 346-359.

5. W. J. Mackinnon and M. K. Mackinnon, "The decisional design and cyclic cooperation of SPAN," Behavioral Science, Vol. 14, No. 3 (May 1969), pp. 244-247.

6. The Third Wire: Cable Communication Enters the City, Report by Foundation 70 Newton, Massachusetts, March 1971.

7. C. H. Stevens, "Citizen feedback, the need and the response," MIT Technology Review, Cambridge, Mass., pp. 39-45.

13. CITIZEN FEEDBACK IN POLITICAL PHILOSOPHY

Ithiel de Sola Pool

Electronic devices for citizen feedback may be new, but the philosophical is-sues posed are old. New gadgets may make possible instantaneous polling or national town meetings, but the question has been with us since Plato of where, when, and which citizens should be heard.

It is easy to debunk the more naive notions of how to use electronics in pol-itics. One science-fiction fantasy has the public engaged in a national town meet-ing on cable TV with various issues being debated and then decided by an in-stant push-button vote. War could be declared on Monday, canceled on Tues-day, and declared again on Wednesday, depending not only on which demagogue was most effective but also on who happened to be home and was not tired of politics from last night's session. If full-time Congressmen with substantial staffs cannot keep up with the details of a thousand bills and thousands of pages of appropriations, even though they do most of their work in specialized com-mittees, it is clear that part-time citizens cannot do the job.

But that is putting up a straw man. No one who has given the matter even a few minutes' thought proposes such a scheme. More modest approximations to the national town meeting are either schemes providing for occasional referen-da on major issues or purely advisory polls. These may be useful devices, but these, too, have their problems.

Referenda may be useful in a democracy, but California's experience illus-trates what happens when referenda are too frequent and too easy to put on the ballot. Year after year appealing crackpot notions go on the ballot, debated not in full text but by bumper stickers and billboards whose total context is often no more or no less than "yes on 9" or "no on 14." Changing the polling place for such referenda from a precinct firehouse to a button in the living room is probably a step away from giving the vote which the citizen casts that singular importance that might lead him to take adequate time for thought. It is hard to see what is gained by voting from the home, other than keeping the citizen dry if it rains.

As for opinion polling, the accuracy of a poll depends above all on drawing a truly representative sample. The votes of those self-selected citizens who choose to watch a political program and then choose to push a response button tells us little about how the rest of the public feels.

Push-button voting in Congress is an equally dubious idea. During the time taken for a roll call vote, Congressmen are scurrying back to the Chamber from whatever they have been doing, getting themselves informed as to what the vote is about, what its parliamentary consequences will be, and how those informed fellow-members whose general views they share intend to vote.

In short, casual push-button voting has little to recommend it.

But if we stopped at this point we would miss the truly profound significance of the new technology of electronic citizen feedback. There are better things to do with it than encouraging ill-considered votes and unreliable polls. What the future may hold is only caricatured by such proposals.

To put things in perspective, let us consider five issues that a political the-orist might raise about any system of political representation or political par-ticipation, including the new electronic ones.

First, there is the issue of whether the prime justification for encouraging citizen participation is the psychic satisfaction that it gives to the participating citizen or the achievement of better governmental performance. In a forth-

coming paper, Charles Murray raises that issue in regard to rural development programs in Thailand. The community development literature, Murray
points out, contains many pleas for increasing the degree of participation in
the planning and development of projects for rural communities. But the justification (when any is adduced) is most often in the form of evidence that if
people are given some voice they will feel more committed to the projects,
feel better about them, and have better morale. But whether participation will
produce more miles of irrigation ditch, more latrines — or less — is a question
generally left unanswered. Murray meets this issue directly. He shows by
careful statistical analysis that at least in Thai villages local participation pays
off in results but to a degree that differs with the dependency of the content of
the particular project on the knowing of local facts about which the villagers
are more expert than the outside advisors.

In the early 1950s the same issues were raised regarding task group organization. A classic study by Lippett and White[1] found that democratic groups
had higher morale than authoritarian ones but did not find that they produced
more. So too, Bavelas,[2] in experiments on message transmission in groups
with different structures, found that diffuse equalitarian structures produced
better morale than did centralized authoritarian ones, but they did not generally do more work.

These results raise a fundamental value issue about the purpose of government. How does one weigh the trade-off between the public welfare that arises
from being treated with respect and equality and that which arises from delivered outputs?

Closely related is the issue between two criteria for political action: civic
involvement in the polity versus a felicific calculus of individual benefits. Two
hundred years ago, within fifteen years of each other, two men wrote the classic treatises on either side of that issue. Jean-Jacques Rousseau in the Social
Contract argued that in a world of conflict the only way that men could be free
without tearing society apart was if each assumed as his own desire that the
general will of the society should prevail. Only when each citizen civic-mindedly
and freely subordinated his personal interests to the good of the whole would
each be free in that act of subordination. From this Rousseauian paradox
stemmed the concepts of freedom as being a willing subordination to society,
held by Hegel, Marx, and the totalitarians.

But Rousseau was not unanswered. His contemporary Adam Smith also wrote
of order in a world of conflict. The theory of laissez faire developed in the
Wealth of Nations was a demonstration that there are circumstances in which
each may freely strive for his individual advantage and the outcome nonetheless may be order and public benefit. Economists, from Adam Smith on, sought
to define ever more precisely the circumstances and conditions in which mutual self-seeking would have stable and beneficial or unstable and harmful outcomes.

One of these conditions for a stable society clearly is, as Rousseau, too,
recognized, that the competitors each personally value the rules of the game
and value the society that enjoins obedience to those rules. Business competitors have to believe in enforcement of contracts, respect for property, and
honest bargaining, or competition becomes mayhem. Football players have to
love the game and respect the umpire, or it will no longer be football.

Thus there is no polar opposition between the role of the participating citizen as fighting for his own interest and his role as a duty-bound participant in a civic effort. Society requires a complex pattern of these two elements. When, how, and in what relation these elements must appear for society to be both stable and free is a complex matter at the heart of political theory. An example of that complexity is the issue of whether the fostering of vigorous citizen feedback will produce commitment to the polity or active pursuit of self-interest, or both, or neither.

Depending upon how one feels about the goals of government as being some organic general welfare on the one hand, or payoffs to the individual citizen on the other, one may view the potential contribution of new information technologies as being primarily to the leader or primarily to the led. This is the third issue to which I turn.

Over-the-air television has given the President the capability of talking directly to the American people in one type of giant town meeting. In all communities personal leadership is one dominant fact of politics. Very few families, or clubs, or offices, make decisions by equalitarian discussion among all concerned. Whether we like it or not, the empirical fact is that in such environments personal loyalty of the led to leaders gives legitimacy to a process in which a few individuals actually dominate decision making.

Mass society tends to weaken the bonds of personal attachment between leader and led. The power holder is more often an unloved, unknown person. The boss in the factory or the administrator in a university, the union boss or the high public official—all are chronic targets of disaffection.

For heads of nations, radio and television have provided some means to partially restore a personal tie between the leader and his followers. Franklin D. Roosevelt did it in the United States when he introduced the fireside chat. Adolf Hitler at the same time, but in a very different way, used radio to establish his direct relationship with the German people. Television renews this bond even better. Richard Nixon, hardly one of the most charismatic of politicians, has nevertheless increasingly chosen to talk directly to the American people on all major occasions and to hold fewer and fewer press conferences for the reporters. And it has worked well for him.

Thus the electronic national town meeting may work far better for the leader than for the led. The gadgets that enable one person to communicate to many are far easier to use and make effective than those that enable millions of scattered individuals to somehow answer back.

And while it is popular these days to deplore political "manipulation" and "elitism," an objective treatment of the issues has to point to both sides of the coin. The very existence of a stable, happy society depends upon the population forming a positive cathexis to a leader who represents their aspirations and values. This, Freud argues in Group Psychology and the Analysis of the Ego, is the crux of the formation of any group. So, too, Rousseau argued that the great lawmaker, a respected national hero, might be necessary to help individuals to find the general will. A Washington, Lincoln, Roosevelt, or Churchill becomes a symbol of faith around whom the nation can cohere.

However, the argument for electronic citizen feedback is that technology can reverse this process and help the citizens make their leaders listen to them. Is that possible, or is it inevitable that in the electronic forum the citizens, even if they push feedback buttons, are really responding to leaders made more effective by the same electronic devices?

A fourth fundamental issue concerns the time dimension in deliberation and decision making. Democratic principles require that the public have the government they want — but over what time period: every minute, every day, every four years? The system of checks and balances about which every American schoolchild learns is in large part a system for slowing up decision until everyone has had a chance to be heard and until second thoughts have had a chance to jell. Bills go through committees, two houses, and then to the President and sometimes cycle back again.

Finally, one may ask whether the function of the democratic political process is the airing of all views so that they can be considered fully by whoever reaches decisions, or whether democracy means the actual making of decisions by majority rule. The latter is the more common view, but there is a strong case for the former.

In point of fact, the U.S. Congress devotes a very large part of its efforts to investigations, to hearings, and to resolutions that do not have the force of law. In the great debate on Vietnam, for example, there has hardly ever been a major vote on a decision. The appropriation bills for it are passed without much problem. The great dramas have been on resolutions such as that on the Gulf of Tonkin or on setting withdrawal policy, or in investigations of atrocities, refugee care, defoliation, official secrecy, and responsibility for the decisions made.

A critic might denigrate that Congressional behavior, alleging that a process that avoids decision making is a charade to fool the public while changing nothing. But a very good case can be made to the effect that political debate in the Congress, as in election campaigns, is properly a process of airing issues rather than one of reaching decisions. Very few policy decisions are all-or-none matters. Most are necessarily incremental and concern degree and direction of some continuing activity. Research on budgets has shown, for example, that the best predictor of next year's budget is this year's budget. At most there is a change of a few percentage points here or there. Thus, for example, if the Congress of the United States were to decide this year that highway construction is despoiling the countryside and that public transport is what is needed, it could not make that decision promptly. Some roads are already under construction. Others need feeders to be useful at all. For such reasons alone a decision now would make only small changes in the next budget. Furthermore, suppose that in a radical mood all road construction were stopped and railway construction started; the next election might bring in a Congress with opposite views. Probably any policy on either side would be better than the scattered incomplete starts that would exist if each Congress thought that it should legislate de novo, in spite of the prospect of another Congress reversing its decisions. Road construction versus public transport is not an all-or-none issue. It is a matter of how much to spend on each of them and where. On such matters of detail the general public has no clear view. Each individual wants some roads built and others not. Majority voting can only indicate the general direction of movement.

Thus, the case can be made that the important thing about the democratic process is that it allows the venting of all considerations. It allows issues to be formulated and coalitions to form around them. The actual decisions are reached in many ways: by passing general laws, by executive decisions, by changing the responsible officials, by bargains and deals, by judgments in the law courts, by letting events take their course, or by some combination of these. But in a participant society all of the modes of decision are under the pressure of full and free comment by those who would like to affect the process.

What has all this got to do with electronic feedback?

Perhaps if we keep these issues in mind we may visualize uses of interactive cable television in politics more plausible than a town meeting on a national scale. There is no reason to believe that an interactive terminal in every home will make ordinary men replace private concerns with civic ones as their priority activity or make them regard the "vibes" of politics as "their thing" in preference to having efficient delivery of services. Nor is there any reason to believe that just because the communication is electronic, political leaders will or should listen passively to instructions from the cable rather than use the same communications devices to exercise leadership. It is nonetheless true that the new communications technologies can be used to make democracy work better. Let us consider how. One may well ask why it makes any difference that the complaint box is electronic. One could write a science fiction story about a society on another planet that had expensive electronic devices for recording complaints and in which an inventor came along and created a new device consisting of a cheap wafer-like object called paper and a short marking stick costing about 29¢ called a pen, and a system called the mails whereby for eight cents one could have a complaint delivered to the responsible person in a permanent record form called a letter. What a vast step forward in participatory democracy that would be.

In fact, letters are important, but even though we use them we are dissatisfied with how that system works. What the parable tells us is that no mechanical device will by its mere existence solve the problem of making bureaucracies responsive. It tells us that it is not enough that a device for registering complaints exists. One must consider how it is used and how its use can be made attractive to the public. So for the electronic complaint box one must ask how that technology can be used to do things a little better than we now do them with the mails and the telephone.

The mails are cheap but do not give that instant feedback that modern psychology has shown to be so valuable as reinforcement of any behavior. Also, writing letters requires, for a good half of the population, skills in which they are deficient; therefore the process is discouraging. The telephone call does give instant feedback and does not call on writing skill, but it leaves no permanent record for the bureaucracy to act upon. Also, it is often hard to find the proper place to which to address one's call. And it is a little more frightening to say nasty things to a live person than to a blank piece of paper. In short, each present available medium has its limitations.

New technology can be designed to overcome these problems. One can design an electronic complaint box to avoid that frustrating aspect of phoning, not knowing how to get the right person at the other end of the line. One may not know who he is, or, if one does know, he may not be easily reachable. One can design an electronic complaint box to avoid the frustration of letters that, although they may ultimately be bucked to the right person (or at least the trusting citizen believes they have been), still provoke no feedback, perhaps for weeks, that anything is happening.

An oral message can be recorded on a tape to provide a permanent record on which someone has to act. There can be immediate human or machine acknowledgment of receipt. A continuous computer record could be kept of where the complaint is in the processing path, and periodically a message could be printed out on the home console as to where it has moved and how it is being handled. That mode of handling of spontaneous citizen-generated complaints might be sufficiently more effective than what exists today to encourage a considerable increase in citizen activity in registering their dissatisfactions with the authorities.

For sensitive issues that generate much complaint, CATV could be used in another way at the initiative of the authorities. A public official could schedule citizen feedback sessions not unlike the appearances of candidates. For example, every week at a certain hour the school superintendent could get on the cable for dialogue with the public. Visible on the screen, he could answer questions phoned in. That can be done today on over-the-air radio or TV for high priority matters. What cable adds is enough channels to make that worthwhile if only 50 concerned parents choose to watch and question.

Such a system would be undermined and its use discouraged if only the same few self-selected busybodies tuned in every week. A canny superintendent would focus each program at a different school or grade and would notify at least a number of people in the special population that he wanted to reach, thus assuring himself meaningful participants in his telemeeting.

Better communication technologies that create more efficient, more extensive, and more intensive interaction between public figures and their constituents may reduce the sense of alienation by making the public figures better able both to respond to their constituents and to influence them. There is no electronic difference between these two processes. They are both enhanced by efficient two-way communication. The specter of electronic manipulation is simply the other side of the coin of the hope of electronic democracy.

Closer interaction between citizen and official is not necessarily either a good or a bad thing. There is no healthy politics without leadership any more than there is manipulation without it. He who is disturbed at this thought should read once more the classic essay on this subject, Max Weber's Politics as a Vocation. Weber in 1919 wrote presciently about the specter of totalitarianism looming then on the horizon. That was the direction organized mass movements could and did go in Europe between the wars. Weber's prescription against that trend was better mass political leadership. He knew that the dangers inherent in mass mobilization could not be met by having less of it but only by affecting its character. The narrow path between mobilization of the people by demagogic populist hysterias (whether of left or right) on the one hand and the political silence of bureaucratic reaction on the other can be trod only by popular movements led subtly by politicians morally committed to liberal values.

This is an overly simple summary of a very profound and complex essay, but for our purposes it will do. The point here is that there is no distinction between the electronics that make demagogy easier and those that make responsible politics easier. It is the men who use them that make the difference. Modern society needs better modes of communication between the people and those in power. It needs to lessen the citizen's sense of alienation, his sense of powerlessness and isolation. To do so it must provide communications that are faster, more individualized and more responsive than are the mass media today. Electronic, computer-controlled, broadband, two-way facilities can make such communications possible. But whether the outcome of that new technology will be the "dark night" of ideological manipulation or the responsible politics of liberal discourse depends not on the technology but on men.

Let us then, in closing, look at the new communications technology from a committedly liberal point of view. Let us ask ourselves how the increased communication capabilities can be used in ways that promote open discourse and pluralism in society.

The first thing to note is that increased citizen feedback via electronic consultation may well make it harder, not easier, to make high-level decisions. It may well increase the cost of the political process. The decisions that get made with wide public consultation may conceivably be better ones, but one cannot avoid the fact that bringing everyone into the discussion is a costly, time-consuming process that often generates stalemates. Consider civil rights, for example. It was a widely recognized conclusion of social scientists working on civil rights in the 1950s that the way to integrate places of work, restaurants, parks, or swimming pools was just to do it without discussing it at all. If not discussed, the physical presence of blacks tended to be taken for granted as natural, but once the issue was raised, polarization, dispute, and deadlock followed.

On the other hand, it is also true that to move the race relations struggle beyond the winning of small battles required the forcing into national consciousness of the moral issue in the Negro's status in America.

This example illustrates one conclusion that needs to be underlined. Citizen feedback in large communities like a nation is primarily useful as an educative process that serves to bring issues into the public consciousness and to get them defined. It is likely to make the reaching of decisions on the matters on which it operates more costly, more time-consuming, and more difficult; but at the same time it makes for deeper commitment and understanding when the decision is made. What follows is that highly politicized national decisions must be small in number. A nation can debate seriously perhaps half a dozen issues at a time, but not more.

It follows that if electronic citizen feedback is, as it can well be, an instrument for increasing the politicization of the populace, then the scope of high-level decisions on which it is brought to bear must be severely restricted. A national or other large community that tried to politicize everything would tear itself apart in endless conflict.

Worse still, the problem is not that attempts at total citizen participation lead to chaos. Worse than that, movement and reforms that proclaim as their goal the securing of massive direct democracy end up producing the contrary. So-called people's revolutions become totalitarian. The dictatorship of the proletariat becomes the dictatorship of the Party. The supposed voice of the Volk is really the voice of the Führer. It cannot be otherwise, for with or without electronic devices, the citizen body has neither the time nor capacity to

handle everything in public affairs at once. If a pretense is made at comprehensive direct democracy, then the participant populace must be organized into a disciplined structure for anything at all to get done. The demand for total
citizen participation is thus invariably in reality a demand for authoritarian
citizen organization.

A liberal democratic policy, then, must be predicated upon the fact that
there is a price for increased citizen involvement and participation. If the price
is not to be less freedom and more authoritarianism, then the price must be a
reduction in the scope of political action. Direct citizen participation is clearly
desirable and possible, and the newer communications technologies make it
more possible, but only for a tolerable portion of the citizens' time, over well-
organized topics, and with due process — otherwise it becomes a pretense.

The restrictions on citizen action in a liberal democracy are many. The
Bill of Rights restricts political action that is at the expense of the freedom of
other individuals. Other rights, such as privacy, also restrict political actions by one set of citizens against another. Some topics, such as religion in
the United States, are barred as topics of public policy. Nonetheless, democracy certainly implies a very wide latitude of subject matter for proposals,
complaints, and issues that a citizen may inject into the political arena. Liberal democracy requires, however, that the issue once raised be subject to a
procedure that allows time and opportunity for rebuttal from those who care to
rebut; it further requires notice to those who might be affected before action
is taken and also protects the citizen body from being subject to the constant
harassment by political controversy that is of significance to only a few.

In this respect, as in others, the hallmark of liberalism is concern for
procedures. It is a concern that whatever is done be done in accordance with
established rules designed to assure fairness. So, too, must it be with electronic feedback. A communications system that allowed any self-designated
group to inject itself at any time into organized decision processes is a violation of the freedom of those who do not choose to be vigilant and politicized all
the time. Specifically, the notion that every city council meeting and school
board meeting or Congressional hearing should be on the air with electronic
feedback influencing its processes is an absurdity whose consequences would
be the opposite of the intent of those who propose such processes. Just as today there are scheduled occasions for citizen participation with limited agendas
(which we call election campaigns, public hearings, and demonstrations), so,
too, with electronic feedback. The more intense and real the involvement that
electronic feedback creates for the citizen in public affairs, the more crucial
it is to limit the scope of its operation and what is affected. If citizens are
brought, by effective personal participation, to the point of caring very deeply
about political outcomes, then there had better not be too many important political decisions, for every time one is made there are losers as well as winners.

It is a fact of politics that the more active citizens are more partisan and
care more.[3] The young McCarthy-McGovern activists, for example, repeatedly
tell us that if they do not win this next time around, their "generation" (or,
more accurately, those among them who share their views) will be turned off
on the political process because society does not choose to listen. They have
identified a very genuine dilemma. The more politicized a minority like them-

selves becomes, and the more vital the sphere of politics, the more disillu-
sioned they will be at society's failure to accept their version of the truth.
Thus the price for having a politically active citizenry in a free society is a
sufficient devaluation of political decisions so that losing is not intolerable to
the losers.

To understand how this may be, let us introduce a distinction that students
of small group behavior and organization have found useful, namely, the distinction among democratic, authoritarian, and laissez faire structures.[4] Both
democratic and authoritarian groups reach collective decisions as to what the
group as a whole will do, though by very different means, one by consensus
and one by elite fiat. On the other hand, laissez-faire, or individualistic, organization lets each person go his way. That is neither democratic, nor undemocratic; it is on a different dimension.

A liberal society de-emphasizes political decisions in that way. It reduces
as far as possible the areas where there must be consensus. It does so in various ways. It does so in part by delegating as many decisions as possible to
small subunits, for example, state and local governments and corporations
(which are also chartered public bodies). These subordinate units may each
act differently, thus avoiding the trauma of all-or-none political decision. A
politicized population living in such a pluralist system was the Jeffersonian
ideal. It is in contrast to the Rousseauian version of the democratic ideal (with
its totalitarian implications) that requires a collective decision by all for the
good of all.

The newer technologies of communication promise to be favorable to pluralism. In contrast to today's mass media, the new technologies work best in
providing means for linking small specialized populations and in providing feedback from them. There are few ways other than a gross opinion poll in which
electronic feedback mechanisms can create interaction on the scale of millions
or hundreds of thousands of persons. Those feedback devices are likely to work
best on the scale of tens or hundreds or low thousands. Such technologies as
graphic display of cumulated audience reactions or storing, forwarding, ab-
stracting, and retrieving of questions and comments from the participants can
make seminar-type procedures possible for what are now lecture-size audiences (see Chapter 12 in this volume). The economics of teleconferencing makes
feasible interactions among mini-audiences scattered over a distance at savings
in both cost and effort over travel. At the same time CATV is inherently a
neighborhood device, at least in its current stage of development as less than
a fully switched system and is therefore favorable to community organization.

Perhaps, then, the most important way to think about citizen feedback in
the CATV era is not as a device that the President or the Congress will find
easy to use, nor as a device that will give citizens much more voice in those
top-level decisions, but rather as a device that will promote grass-roots interactions among citizens with special interests.

Is this a good thing or a bad thing? Like all trade-offs, it is some of each.
It makes meaningful and intensive citizen participation possible — far more
possible than it can be at the national level, except sporadically. But on the
other hand, a devolution of the focus of politics to the local and interest-group
level may make this already inchoate nation even harder to unify and govern
than it is now (see Chapter 5).

We are now in the realm of prediction. How the conflicting forces will balance out is uncertain. What it will mean to American society is that each citizen can participate more effectively in his own community and perhaps feel that he is heard more than he is today, and that at the same time the resulting special-interest groups will become more powerful and better organized.

Perhaps it is not so much a matter of prediction, but more a matter of planning. Perhaps if we understand these processes and the trade-offs involved, we can create feedback devices and practices that will both let people feel more efficacious in their own groups and also somehow strengthen our national consensus. Whether we can do that I do not know, but understanding what is at stake is the first step toward trying to achieve both those seemingly antithetical goals.

Notes

1. Roland Lippitt and Ralph K. White, "The Social Climate of Children's Groups," in Roger G. Barker, et al., Child Behavior and Development (New York: McGraw Hill, 1943).

2. Alex Bavelas, "Communication Patterns in Task Oriented Groups," in Daniel Lerner and Harold Lasswell, eds., The Policy Sciences (Stanford, Calif.: Stanford University Press, 1951).

3. Bernard Berelson, Paul Lazarsfeld, and William McPhee, Voting (Chicago: University of Chicago Press, 1954), pp. 246 ff.

4. See Kurt Lewin in Gardner Murphy, ed., Human Nature and Enduring Peace, 3rd Yearbook, Society for the Psychological Study of Social Issues (Boston: Houghton Mifflin, 1945), pp. 303 ff; and Margaret Mead, Cooperation and Competition among Primitive People (New York: McGraw-Hill, 1937).

14. COMMERCIAL USES OF BROADBAND COMMUNICATIONS

Ephraim Kahn

INTRODUCTION

Demand for business communications will grow markedly during the decade of the 1970s and thereafter. This demand can be satisfied by broadband communications systems — cable and microwave, almost certainly multiplexed to increase their capacity to reach different addressees simultaneously. This paper will not try to discuss the complexities of ownership of these broadband facilities. But it appears that, in the absence of a quantum jump in the availability of high-quality lines from the Bell System, other telephone companies, or Western Union, one or more new communications suppliers will be needed. For many purposes, existing Bell System or WU lines provide satisfactory service. But large-volume transmission of electronic data is far more simply and economically handled by a broadband (or at least a broader band) system.

For business use, two-way communications capability seems essential — and it is precisely for this reason that cable, as distinguished from over-the-air information dissemination, has an advantage that may be crucial. There are new methods of over-the-air broadcasting under development that can compete with cable in terms of providing broad channels through which information may be pushed in large quantities at high speed. This is fine, as long as the only real consideration is getting the information from a central place to receivers — either as a group or individually addressed. But if the person who receives the information has to act on it and himself make a contribution to a return flow of information, cable seems the most economical broadband method.

The costs of the terminals for two-way transmission over cable are less than the costs of transmitters and receivers capable of carrying the quantity of information that takes full advantage of the capacity of the cable. Obviously, this is of major significance only in connection with the overall economics of a communications system, and the presumption is that the primary economic justification of a CATV system is to provide entertainment TV, and that the provision of other services is a desirable by-product of cable's great information-carrying capability.

The full information-carrying capacity of a TV channel probably would not be needed for most business-oriented information. A TV channel can carry a moving picture; far less bandwidth is required to transmit a facsimile of a document. If cable communicators make their facilities available only in increments of a full channel, a great deal of capacity will inevitably be wasted in business use. On the other hand, it will take a broad bandwidth to accommodate a number of business users: those interfaced to the same storage and retrieval computer, those requiring simultaneous display of a wide variety of data on a cathode tube screen, or those using facsimile printout.

At a minimum, revenues from use of broadband cable channels for non-entertainment uses should help directly to put into practice some of the desirable principles set forth in the 1968 report of the President's Task Force on Communications Policy and make it easier for the others to be realized.

As long as telephone companies are committed to a basic cost-per-call system of computing basic charges, an important competitive opportunity will be available to cable systems: renting a channel (or a fraction of a channel) to volume users. This evidently could be done at a cost significantly lower than the charge made by telephone companies for open lines.

A dedicated signal path is implicit in automatic checking of credit verification devices, for example. Such a path is also desirable for the volume user of data communications. The foreseeable cost advantage to two-way cable communications may be enhanced if the Bell System generally adopts a policy of imposing a surcharge on computer-connected "Information System Access Lines." This ISAL tariff would require almost all companies that have telephone lines going to computers to pay an additional fee for the privilege of computer access. Computer users protested the imposition of a surcharge when hearings on its ISAL tariff proposal were held in Illinois. To the extent that telephone company charges for computer access exceed those made by a CATV system, users of computers through distant terminals will have an incentive to use the CATV's facilities. Although the Illinois application was withdrawn at the request of the local regulatory agency, ISAL tariff applications are currently pending in Florida, Georgia, and the areas served by Cincinnati Bell.

Today, demand for data transmission capacity is not being adequately met by existing suppliers of communications lines. Complaints have frequently been made that the information-carrying capacity of existing telephone-type lines is not great enough to permit users of data-processing equipment to take full advantage of the capabilities of today's computers. Complaints are also made that existing service is too costly, service and maintenance inadequate, and that the public utilities offering data transmission lines to users do not seem to care about users' problems.

Present common carriers of communications are, of course, aware of these complaints and criticisms. The Bell System, for example, plans to have a separate network for digital data transmission installed by the middle of this decade. This new network will serve 50 to 60 major U.S. cities. Western Union also has a data network coming up. It should also be noted that the capital needs of telephone companies are great—Bell estimated $100 billion by 1980 in a 1968 forecast—and they are hard pressed just to meet growth in telephone demand and upgrade their existing systems. If this consumes all of the money available to them, then CATV entrepreneurs should be able to develop the commercial services that they are able to provide relatively free of immediate competitive pressures. Eventually, it must be assumed that telephone companies will compete in any area that requires two-way switched communications. Faced with this prospect, it seems possible that some CATV operators could elect (regulation permitting) to install only rudimentary two-way communications capability. Still others might opt to use the telephone company's lines to provide a channel from the subscriber to the CATV head end.

New cable companies conceivably could be in an advantageous situation to attract the capital they will require. For one thing, installation of local cable systems will not be inordinately costly. This will present potential investors with an opportunity to own a portion of a system instead of using the same amount to buy an insignificant fraction of Bell. For another, the investors in the new system will not be in the position of having to service large quantities of outstanding senior securities or of seeing their equity diluted through bond conversions. To the extent that "getting in on the ground floor" is in investment incentive, cable companies will be able to offer it.

Growing demand for broadband communications will very likely be satisfied

at first by microwave radio and telephone company facilities for long distances
and by cable over shorter distances. These present the lowest costs for digital
data, the prime business use. At present, the telephone companies and Western Union offer the widest range of switched facilities. Existing cable installations typically offer primarily entertainment TV on a one-way basis to individual subscribers. Eventually, they will become more sophisticated, and
the public policy issues involved in permitting the existence of what would essentially be a second national telecommunications network for individual subscribers (business and nonbusiness) will have to be faced.

CURRENT COMMERCIAL CAPABILITIES
The cost of two-way capability is relatively modest. The NCTA has estimated
that it would cost about $180 to $200 per mile to include a pair of twisted
wires (usable on a time-sharing or party-line basis by all subscribers) when
a new system is built. This would provide a 4-KHz message channel. Adding
this capability to an existing system would cost between $300 and $400 a mile.
Additional costs would be incurred for subscribers' terminal devices and for
switching equipment at the CATV head end. A broader channel — 1 MHz — could
be included in the two-way capability. This would be adequate for Picturephone-
type transmissions. No estimates of cost for including this are available.

The Subscription Television system, devised by Subscription Television,
Inc., in South Pasadena, California, "has been specially designed to transmit
and receive digital information over today's existing coaxial cable networks,"
the company says. It requires 5 MHz bandwidth — 2.5 MHz in the mid-band of
the VHF spectrum going out to the subscriber and 2.5 MHz in the sub-channel
region below channel 2 returning to the STV interface equipment. Maximum
capacity of the system is 500,000 bits per second to and from all subscribers,
but within that ceiling it is infinitely flexible.

The STV system is controlled by a central processor, which not only receives and transmits information but acts as a traffic control device so that
information will not be put into the system when it is overcrowded. This central processor interfaces with the CATV transmitter. Each subscriber on that
system has an STV service selector, which can communicate with the central
processor. The central processor can send messages to any subscriber individually and identifies the source of information any subscriber sends back.
Coded signals can be sent at regular intervals to all subscribers to find out
what services offered by the system are being used, whether purchases have
been made from a shop-at-home channel, whether pay TV is being watched,
etc. Since it takes only 40 bits for the central processor to make such a query
and 40 bits for the subscriber to reply, a single transmit-reply cycle requires
only 160 microseconds. This means that each of 10,000 subscribers could be
queried once every 1.6 seconds. Something like this would appear to be very
well adapted to credit card checking, where the in-store terminal would be in
an "open" or "no response" position at all times other than when a card is
placed in the device for reading. The flexibility of the STV system enables it
to handle high-density data at high speed to a limited number of subscribers or
low-density data at lower speed to a large number. This implies, too, that
high density data could be transmitted at adequate speeds during times when
the system would otherwise be underutilized.

STV estimates that a single central processor can serve a maximum of
50,000 subscribers. It notes, however, that smaller "slaved" processors adequate for about 10,000 subscribers can be introduced into the lines to act as
store-and-forward modules. The company also says that a number of separate,
smaller CATV systems could share a central processor — five systems of
10,000 subscribers each, for example.

DATA TRANSMISSION

Estimates of the growth in demand for nongovernment pure analog digital data
transmission for business (not residential) users have been made in connection with applications to have the Federal Communications Commission establish a specialized data transmission network. Those given here were submitted
by Data Transmission Co. (DATRAN) to the FCC as part of its application.
Volume of voice, video, telemetry, telegraph, and teletypewriter traffic is not
estimated.

It should be borne in mind that DATRAN at the time was "selling" the FCC
on permitting such a network to be established. But even if the estimates were
to be discounted by 50 percent as an allowance for euphoric optimism, the projected growth would be impressive.

DATRAN's estimates cover seven industries, not all of which will be noted
in detail here. Overall in 1970 these industries engaged in 14 billion transactions carried out through 3.7 billion data calls. By 1974 there will be 50 billion transactions that will generate 12 billion data calls. In 1980 these industries are expected to make 250 billion transactions through 32 billion calls.
This would represent an increase of 750 percent in the number of calls over
the decade and cumulative growth of 1,650 percent in the annual volume of
transactions.

Demand for data transmission facilities will also grow as a result of a decision by the Federal Reserve Board to permit bank holding companies to undertake — for businesses not related to the holding company — "storing and
processing other banking, financial, or related economic data, such as performing payroll, accounts receivable or payable, or billing services." Linking
the bank computers that do the data processing with the customers who are
having this work done could well be a function of local CATV systems.

SECURITIES INDUSTRY

At present, the securities industry is a heavy user of data telecommunications,
and it will undoubtedly continue to be. The industry is spread all over the country, but most transactions take place in New York City.

Typically, the customer of a brokerage house wants fast action: if he places
an order to buy, he wants confirmation of execution in a matter of minutes. The
seller is similarly impatient. Because of brevity of content, sales and purchase
orders, requests for quotations, size of the market in individual securities, and
the like, can be handled adequately with the facilities now made available by the
Bell and other telephone systems.

In the future, this may not be true. Although stock quotations are now available in many brokerage offices through visual display machines (Quotron, for
example), there may well come a time when customers want more.

With a two-way broadband communications system, customers could obtain

research data and information on the status of their accounts and make other
one-to-one inquiries. Answers would be displayed on TV-type screens or as
a hard-copy facsimile of the material transmitted by the broker.

With a fully switched broadband system, customers could communicate with
any broker and obtain the fruits of his research. But such inquiries would take
up computer time and they could well be more costly to brokers than mailing
a prospect some sheets of paper. Furthermore, since this information would
be accessible to anyone willing to expend about as much energy (and perhaps
less money) as it now takes to make a telephone call, many more investors
(and the idly curious) might choose to shop the research departments of several brokerages before making a purchase or a sale.

If the inquiry load becomes too heavy, this could impel brokers to either
(1) limit access to their research to their existing customers, which they might
be reluctant to do, or (2) impose a fee for research services — in which case
they might market them on a fee-for-service basis, giving their customers
either a preferential price or a full rebate when they undertake a transaction
through the firm, based on its research. Payments and receipts, by the time
this develops, may well be handled electronically so there would be little or
no need to have branch offices where customers can make payments or which
can mail checks to customers.

Indeed, in the fairly distant future it is easy to conceive of a fully computerized stock market in which sale and buy orders are placed electronically
direct by the customer and matched by the stock exchange computer. Capacity
of a broadband system would be adequate to support such a setup, and growth
in the number of investors may demand it.

BANKS

A strong spur to the expansion of electronic transfers is the adoption by the
Federal Reserve Board of a policy statement stressing that it is "a matter of
urgency" to set up a nationwide, direct, fast, and economical system to transfer funds and settle balances; to reduce the volume of items banks must handle;
expand at least some Federal Reserve Bank facilities "to include high speed
tape transmission, and computer-to-computer communications"; and to reduce
the "float," or amount of money in transit.

K. A. Randall, former chairman of the Federal Deposit Insurance Corporation, now chairman of a large bank holding company, pointed out in 1970 while
still a federal banking regulator that "technology particularly is going to force
institutions to be generally larger in size." He added that with this, there will
come a reduction in the number of financial institutions in the United States,
suggesting that there would be a shrinkage from a total of around 22,000 financial institutions of all types "maybe to around 8,000 in another 20 years."
This would be accompanied, he said, "very likely" by "branching in some
form in every state."

Bankers are likely to be receptive to new ways of doing business. They have
been hearing about the "checkless society" for years, and they are expansion
minded. Governor William W. Sherrill of the Federal Reserve, for example,
says that the time has come for bankers to reorient their concept of banking
and to recognize that banking's "function no longer is predominantly lending but
must become a concept of greatly expanded financial service to its customer."

Governor Sherrill foresees "a time when financial advice, bookkeeping, budgeting, and financial management information provided to the customer may be much more important in the customer's eyes" than the funds that banks make available. And, he adds, these services would probably be "much more profitable" to banks.

Bankers are aware of the ever-growing flood of paperwork that will confront them. Early in May 1971 the Monetary and Payments Committee of the American Bankers Association reported that banks must start at once to seek the advantage of "paperless" debiting and crediting. This is significant not only because it takes a coherent look at the problems of the future but because the men on the committee that made the report all are high-level, policy-making bankers — the only people that can push the institutions they work for in the direction of anticipating technological change and getting ready for it in time.

The ABA's MAPS study does not envisage a sudden change to a magical "checkless society." The bank group took a practical approach: it looked at the growth that has taken place in check handling in recent years, projected it into the future, and found that by 1980 banks may be processing 43 billion checks a year. Since this aspect of banking is labor-intensive, this raises a number of problems.

The MAPS committee did not foresee the end of the check as it now exists. In fact, its chairman, Richard P. Cooley, president of The Wells Fargo Bank, N.A., San Francisco, states flatly that "paper checks will be around for as long as we can envision," and that "banking can continue to operate efficiently with a paper check system."

But banks are already taking steps to cut down the volume of checks they must handle. Many banks have arranged for preauthorization of certain payments that must be made regularly. This takes the form of customer authorization of monthly withdrawals to pay such debts as mortgage loans or insurance premiums. Some creditors give customers who preauthorize these payments a discount.

Less common, but on the increase, is the multiple check. When a multiple check is used, a customer makes a list of his creditors and the amount each is owed and makes out a single check for the total. The check goes to the bank, which parcels out the money to the individual creditors by adding the funds to the creditor's account in the bank.

Within the banking community, there is broad recognition (though far from universal approbation) of a trend to bigness. In part this is because banks have to be bigger to be economically viable today. This is reflected in persistent moves to liberalize the branching laws in states that are now highly restrictive.

Many smaller banks have refrained from installing their own data-processing equipment because they are too small to justify it economically and do not believe that they can sell enough outside data-processing services to make data processing a paying business. Virtually all larger banks have their own data-processing equipment, either owned directly or owned and operated through a subsidiary. Still others use independent data processors.

With more liberal branching and a tendency for banks to grow larger and to affiliate with holding companies, the need for communication among the bank's entities will expand. Today, many banks have their computers located at the main office. It works for the main office and the branches. But the means of

getting data from a branch to the computer are relatively primitive. Within a
city, a truck picks up a day's transactions from a branch or a subsidiary and
takes them to the main office for processing. This is subject to delay, documents may be lost, and is relatively costly. There is little on-line interconnection among branch banks and a central computer, so that transactions are
simultaneously handled at a remote location and recorded by the central computer. In some banks, branch transactions are batched and sent to the computer several times a day by data-phone.

Many banks already have limited branch communication with their central
computer. Typically, this permits a teller to use a Touch-Tone R telephone
to ascertain a customer's current balance.

As broadband communications become increasingly available, it will become easier and possibly less costly for banks to install on-line intercommu-
nications. Even if a bank with several branches, or a holding company with a
number of banks, does not want to operate this kind of system, it could well
choose to use a channel (or fraction) of an existing CATV network to transmit
its data to the computer at a time of day when entertainment use is minimal.
Unlike stock brokerages, banks do not now depend on immediate transmission
of data. They can easily move their traffic at off-peak hours, since the computer will in any case start the next business day fully up-to-date. This delay
will not be tolerable when the "checkless society" comes into being, although
a continuing lag in transfers of funds is not invariably a disadvantage.

Although the checkless society may be a long way off, it is far closer for
financial institutions than for the general public. Banks and other financial institutions trust each other. They will accept electronic confirmation of transfers of funds from another financial institution. But they require a piece of
paper when they deal with their everyday customers. In the absence of a facsimile system, they would not get a signed document on an electronic funds
transfer by one individual for the account of another.

The fear of crime will in certain respects be a spur to the development of
electronic transfers. Small merchants, for example, would welcome the idea
of general electronic transfers triggered by credit cards since it would free
them from the loss of cash they suffer when held up and from the possibility
of personal injury at the hands of the hold-up man as well.

This additional security could well be worth money to the merchant—the
money that would go to pay for the communications system needed for the
credit card—or "currency card"—that buyers will use when electronic transfers are widespread.

Some resistance is likely from the general public—in part because electronic transfers mean instant reduction of the bank account. Clearly, some
method will have to be found to combine "cash" and "credit" authorizations in
a single card. This could be done by having the merchant press a "cash" button or a "credit" button on his in-store verification and authorization terminal.
Conceivably, when a purchase is made on a 30-day credit account, the transfer
system computer could store the debit and forward it for action at the proper
time.

A significant step toward larger-scale interbank electronic transfers has
been made by the New York Clearing House, whose Clearing House Interbank
Payments System (CHIPS) is currently handling between $15 billion and $20

billion a week. The CHIPS setup handles about 15,000 transactions weekly on
behalf of about 4,000 accounts that foreign banks have in eight large New York
commercial banks.

The system is based on a large Burroughs computer at the Clearing House,
with 42 smaller Burroughs terminal computers at the banks. The terminal
computers are linked to the main one by two leased telephone lines; each bank
has an additional dialed line that can be used if the leased lines are not available. The computers store payment messages and release them as they are
authorized. Their memory capacity is sufficient to enable the banks to keep
workload fluctuations from being too wide.

CHIPS saves the banks time and money by dispensing with checks for each
transaction and the cumbersome settlement procedure they involve. But a hard
copy is made at both ends of each link so that the accuracy of transmission can
be verified. With the CHIPS system, a computer terminal operator enters a
payment order, for example, upon receipt of a cable from abroad. Each bank—
and each account of foreign banks—has a code number. The appropriate numbers plus the amount of the transaction are sent to the computer. The computer
adds control numbers to the message and returns a hard-copy printout of the
data to the sending terminal. When this hard copy is approved by an officer of
the bank that originated the transfer, the terminal operator releases the message and the computer automatically sends it to the proper destination, where
the terminal prints it. The computer's codes are standard; all banks are familiar with them, and the use of a bank's code in a message results in a full
printout of its account name and other relevant data.

At the end of the business day, the main computer balances all transactions
and prints reports that show the status of each participating bank with respect
to the accounts it handles. Each of these banks gets full daily reports from the
computer. The CHIPS system does not yet handle all interbank payments in
New York, but its operators anticipate that its workload will speedily climb to
5,000 a day, and that eventually it will eliminate about 40,000 payment checks
a week. Banks in Southern California have inaugurated an even larger electronic transfer system.

The executive vice-president of the New York Clearing House Association,
John F. Lee, believes that bigger networks and greater geographic spread
"will have to evolve...as costs, capacity and needs are evaluated. The system has been developed," he notes, "and we have the computer equipment and
programs available to make expansion, in terms of single or multiple systems, entirely feasible."

The Federal Reserve Board is concerned about the future adequacy of its
electronic funds transfer network. At present, this system is based on telephone lines that handle about 150 words per minute. In the foreseeable future,
the FRB system will be upgraded through equipment changes to handle at least
10 times as much. Currently, the board apparently sees no need to investigate
the potential of broadband systems—but it may do so in the future. It seems
clear, however, that FRB technicians believe that by the time demand would
be ready to overwhelm the present system, Bell will have more capacity available.

FACSIMILE

Facsimile transmission is one area in which the advantages of broadband can not be seriously disputed. But even here, it is not necessary to use a full TV channel-width to obtain satisfactory definition on the final copy. Facsimile transmission over the ordinary telephone line is slow — between 4 and 6 minutes per page. This may be reduced as data compression techniques are applied to facsimile and as the telephone company makes available wider lines that will be able to exploit this capability.

The cost of high-grade, speedy facsimile machines — and the unavailability of wider-band lines to accommodate this traffic — have interacted to inhibit the widespread use of facsimile by the public.

It has been suggested that wider-band networks be used to supplement — and eventually to supplant — the mails as we know them today as a carrier of information. If current growth rates persist, the U.S. Postal Service will have to handle more than 108 billion pieces of mail (excluding checks) in 1980.

A proposal to substitute facsimile via satellites for intercity mail distribution has been filed with the FCC by General Electric. Local distribution would be handled by CATV systems, at an eventual cost of about 10 cents per 150 words, with delivery to terminals equipped with alphanumeric typing devices. William B. Gross, of GE, envisages three phases in the development of a facsimile-printout mail system. In the first, Bell System long lines handle intercity traffic and deliver to the local CATV. In the second, long-distance traffic moves by microwave. The third phase encompasses both satellite and microwave long-distance transmission. Costs would decrease as the mail system matures; at first it would very likely be used only for business letters that would otherwise go special delivery, via airmail. The existence of such a mail transfer system would be particularly helpful when next-day delivery of business mail is sought, since this traffic could move through the system at times when other demand is slack.

Long-distance broadband transmisssion (using the equivalent of 10 or 12 telephone lines) is already in use within the U.S. government, where Long Distance Xerox moves facsimiles of documents at the rate of six to eight pages per minute. The State Department, in addition to using LDX for traffic within the United States, uses Optical Character Recognition systems to transmit overseas. This system involves the use of an OCR reader interfaced to the department's communications computer, which automatically sends communications to the proper destination. This bypasses manual tape-punching completely and enables information to be delivered at improved speed and accuracy even to places with only a telegraphic-grade communications system.

It should also be noted that there are a number of alternatives to facsimile transmission that provide a hard copy at speeds satisfactory for most business purposes, certainly faster than common carrier printed messages, including Telex. The IBM magnetic tape Selectric typewriter, for example, transmits at 180 words a minute over telephone lines and produces a second "original" — or a duplicate magnetic tape — at the receiving end. This could well be satisfactory to many users who do not have the volume to justify an OCR system or who wish to take advantage of some of the machine's other capabilities, such as typing with a justified right-hand margin.

Very recently, a device was demonstrated that makes it possible to use an ordinary telephone line to transmit documents while simultaneously using the line to carry a conversation. At present, the machine takes several minutes to transmit a page. If it can be speeded up, this could present a practical alternative to broadband facsimile transmission for many users.

The Fair Credit Reporting Act should be a spur to the development of facsimile transmission, since all grantors of credit must — in specified circumstances — provide persons who have been denied employment or credit, or for whom the cost of credit has been increased on the basis of a consumer credit report, certain data in written form. If a bank refuses credit — or increases its cost — on the basis of a consumer credit report, the dealer must disclose to the customer the name and address of the bank, and the bank in turn must give the customer the name and address of the agency that made the report. Facsimile transmission of the required data to the merchant would provide an adequate record of compliance.

CREDIT CARDS

Credit card authorization is expected to be a major new business use for data communications, particularly as low-cost card readers become available. The American Bankers Association Bank Card Standardization Task Force has urged the credit card industry to adopt a system of magnetic stripe encoding to make credit cards machine-readable. It proposes that the magnetic stripes, to be located on the backs of cards, be adequate for dual-density encoding. Airlines want the stripe on the face of the card. In the future, the Task Force plans to recommend a standard message format for authorizing credit. The standard is expected to meet the needs of all credit card issuers, not only bank cards. Adoption of cards that have data coded on them will lead to "zero-balance" credit cards, which must be verified electronically before each purchase.

Adoption of the standard would, of course, make it possible for a single terminal to provide the information required by a variety of credit card issuers. It would also enhance the probability of development of not only a national but an international credit card authorization system.

A system has also been patented that enables a merchant to place a customer's credit card in a sensor that generates a picture of the rightful card holder on a Picturephone of closed-circuit TV receiver display. The picture is kept in a microfilm file at a central location. If a picture is not desired, a description of distinguishing physical characteristics of the card holder may be sent, or other personal data that could be used to establish that the person presenting the card is the authorized user.

Banks and other issuers of credit cards are vitally interested in having these cards used only by authorized people. While banks are hopeful of eventually turning a handsome profit on their credit card operations, many are still marginal or losing ventures. (In part, this is a problem of the banks' own making — they voluntarily sent credit cards to people who turned out to be cheats and deadbeats.)

In this field, too, the Bell System is moving. Its credit card verification system is clearly more cumbersome than the automatic card-reading that will be possible if the proposed ABA credit card magnetic standard is adopted—but it's better for merchants than manual credit verification.

The Bell device is an adaptation of its automatic card dialer. When credit is to be verified, a card punched to make a call to a computer is inserted in the device. Once a link has been established, a card identifying the merchant is put in the machine, followed by a card that identifies the customer. If a purchase would exceed predetermined credit limits, the merchant indicates the amount by pushing buttons on the device's Touch-Tone keyboard. The computer then makes a vocal response authorizing (or denying) the proposed transaction. In areas where Bell central offices are not equipped for Touch-Tone calling, auxiliary equipment is available so that the system can be used with a regular dial telephone. The Bell device is also believed adaptable to magnetic tape encoding.

Rental is expected to come to about $5.00 per month for the device. If the computer is accessible through a Wide Area Telecommunications Service (WATS) lines, a charge is made only for the time the line actually is in use—expected to be less than one minute per call since the Bell automatic card-dialer takes only about one second to make the connection.

The range of business-related information that can be disseminated through a broadband system is limitless, particularly in a two-way system with rapid facsimile or fast printout at each terminal. Conceivably, the day might come when much white collar work is done at home, with the worker receiving the data he needs by cable and despatching the work he does similarly.

RETAILING

In retailing, there should be substantial demand for use of broadband cable. Like banks, retailers are vitally interested in credit cards. At present, verification of credit card validity is often lax—the retailer knows that the credit card issuer will pay him for the goods even if they are "paid for" with a lost or stolen card. As point-of-sale verification terminals become available, it seems reasonable to expect the credit card issuers to change their policy in this regard. If the credit card verifying terminal is combined with a charge slip imprinter, card validity could be checked automatically. Using a multiplexed cable channel and an open line, it is estimated that each retailer's credit card device could be checked automatically each 1.6 seconds. Insertion of an invalid card would be signaled to the retailer; if no signal was received the card would be presumed to be valid.

Multiplexed cable attached to the cash registers in stores could transmit sales data automatically to a computer—by individual items if necessary since the clerk would ring up the stock number of each product sold at the time of sale. (This is already done in some stores, where cash registers produce machine-readable tapes.) This gives management a chance to improve inventory control and to check actual sales against projections very quickly.

The more glamorous use of cable in selling is, of course, display of goods on the TV screen at home, with the customer able to make a purchase by signaling the seller through his (or her) home communications console. Over-the-air broadcasting has clearly shown that TV is an effective sales tool. With a sophisticated two-way cable system, a customer could call a store, arrange

for TV display of goods he is specifically interested in purchasing, and perhaps discuss the products with a salesperson. It seems reasonable to imagine,
however, that first efforts in CATV sales will involve dissemination of products chosen by the seller to whatever audience happens to be viewing a merchandising channel. Buyers would then order by telephone.

METER READING

Remote reading of utility meters (electric, gas, and water) over a CATV channel has been touted as a possibly remunerative service that might be offered
by a wired broadband communications service. In theory, this is correct and
attractive, especially to CATV operators who currently charge their subscribers about $5.00 per month. If each utility were to pay $0.50 per month for its
CATV meter-reading service, the resulting income would be a welcome increment.

It has been suggested that agreement by utilities to buy meter-reading services (which, at $0.50 per meter would cost them just about as much as their
current expenditures to get data from the meter into machine-readable form)
could interest CATV operators in wiring areas completely since the return
from a fully wired area would be comparable to that now derived from a $5.00
charge to the 40 percent of homes connected that many CATV operators use as
their basis for amortization. Meter-reading revenues alone would be $150 per
month per 100 homes, and the 40 homes that would take TV service would provide an additional $200 per month. To be sure, it would take a longer time for
the meter-reading service alone to pay off—but it seems likely that some
people who would not otherwise subscribe to CATV service could be persuaded
to do so "since the cable's already in the house." To be sure, the data from
utility meters could as well be transmitted over a narrowband telephone line—
if the telephone company's price was right.

There are a number of other things that must be taken into consideration,
however. At a fee of $0.50 per month, the utility makes no immediate saving—
though it probably would over a period since meter readers' pay is likely to
rise. Water systems might lose money; they often read meters semiannually.
The utility would gain in customer convenience and efficiency. Each meter
would be read on schedule, and meter-readers would not have to enter customers' homes. (Many utility meters are already located on an outside wall.)
The utility meters now in use are, however, very long-lived, and any utility
would think twice before putting a lot of money into replacing equipment that
has many years of adequate service left in it.

It might well be that utilities could effect economies by turning to a different method of reading their existing meters—something like identifying each
meter with a sticker, then using a special microfilm camera to take a picture
of the meter's dials and the identification. The microfilm, when developed,
could be machine-read and the bills automatically processed. In CATV areas,
utilities would find it more economical, when technically feasible, to add a
shaft encoder to their existing equipment.

On the other hand, when meters are installed—particularly in new communities that are being planned and built with two-way CATV—routine installation of utility meters that can be read remotely would seem desirable.

This implies, of course, that meter-reading represents an area of rela-

tively slow growth, not a bonanza. Similarly, when utilities replace worn-out
meters in areas that are logical candidates for CATV penetration, or which
already have systems that contemplate two-way operation, they might well be
advised to use equipment that can be read remotely.

MARKET RESEARCH
Automatic market research represents another potential use for two-way
CATV. At a fairly simple level, viewers at home could indicate a preference
or opinion on a question by pressing a button on the TV set. CATV systems
could conceivably serve as testing grounds for new entertainment programs,
with viewers giving a "good" or "no good" signal.

On a technically more sophisticated level, TV receivers connected to cable
systems could be so wired as to present a constant record of the channel they
are tuned to at all times that the set is on. Such a system is now working in
Santa Maria, California.

This system can be programmed to supply the names and addresses of all
viewers who are watching a specified channel during a given time period. This
would make it possible for market researchers to limit their follow-up efforts
to those households that were watching a test program or commercial, for example. Consequently, the total cost of this reserach would be greatly reduced.

OTHER USES

Consumer Uses
Federal sponsorship of data banks stocked with information relevant to consumers has been proposed. Indeed, the Federal Trade Commission is currently
conducting a study, undertaken at President Nixon's request, of the possibility
of establishing a nationwide consumer information data bank. Currently, the
study is limited to determine the requirements for such a data bank. Sen.
Philip A. Hart (D., Mich.) and consumer advocate Ralph Nader (as well as
others) have also advanced this proposal. To be of maximum use to consumers,
it will have to be accessible to them. Access can be had through telephone lines,
of course, but it would very likely be preferable to have the data bank's computer provide written data in response to consumer queries. If, or when, consumer data bank programs become available, CATV systems could obtain them
and provide this information to their subscribers through their own computers,
presumably on a fee-for-service basis.

Reservations
Computerized reservation services are already in use by many airlines, car
rental agencies, and hotel chains. But they are accessible only by making a
telephone call to one of the company's reservation desks. With widespread use
of point-of-sale terminal devices, it would be possible for travel agents, for
example, to obtain for their customers immediate information on seating availability for air travel, rooms and their rates for hotel patrons, and vehicle
availability for car renters. Required data could be displayed automatically
on a cathode tube, and reservations made on an associated keyboard.

Real Estate

In real estate businesses, even smaller companies would find it useful to be able to generate, from their own or remote computers, pictures of the properties they currently offer for sale, maps showing the location of each, and a printout of descriptive material.

Insurance

Companies could arrange to give their agents access to all policy information at the touch of a few buttons. This would be helpful in settling claims quickly, in bringing accounts up to date, and in checking on the insurance planning status of customers and prospects. If a system of no-fault auto insurance is enacted, a central information bank accessible to insurance agents may be a necessity.

CABLE'S COMPETITORS

Advancing technology will certainly have an impact on purveyors of broadband communications services. In this area, it is impossible to approach the Bell System with anything other than the highest respect and to assume that it will continue to keep pace in those communications fields in which it does not lead. But the future of cable will also be affected if new methods of broadcasting that provide sharply increased spectrum capacity are successfully developed. Obviously, the incentive to lay cable (FCC says a "working figure" for cost is $4,000 per mile) is diminished if the number of TV channels available to the public can be increased by cheaper broadcast methods. Since commercial uses of cable depend heavily on two-way communication, use of new broadcast methods would tend to keep merchants and others dependent on telephone company facilities, which are (or will be) adequate for most commercial transactions.

The fact is that the Bell System (and other telephone companies) have lots of problems other than broadband communications. Under current FCC policies, telephone companies are barred from owning CATV companies in those areas where they operate as telephone utilities, and they must make their poles available for stringing CATV lines at reasonable cost. This, obviously, encourages independent CATV systems. More significant, perhaps, is the fact that in many parts of the country the telephone utility is hard pressed to upgrade its existing services and to make them fully satisfactory to present and anticipated subscribers.

As long ago as 1968, the Bell System publicly recognized that huge demands were going to be made on it for new equipment, and that usage of its lines was going to rise very sharply. At that time, it projected that by 1980 there would be an increase of 70 percent (70 million) in the number of telephones in use. It expected a tripling of long-distance calls from about 5 billion a year to 15 billion a year and a tenfold increase in the number of overseas calls made from the United States, from 12 million to 120 million. The typical day would bring 500 million calls in 1980, Bell said, up from 300 million in 1968. Bell added: "By 1980, as much information may be exchanged by data transmission as by voice communications." Actually, this point has already been reached. Bell also asserted that the 1968 "interstate network represents only about 15 percent of the plant-in-service that may be needed" as the 1980s begin.

Evidence to support this view is readily visible. The Touch-Tone telephone is really a 12-button device that gives access to a computer — and hence to all the data that can be stored in a computer. In May 1967 the Bell System's public relations department asserted that "your telephone may be able to do almost anything you want it to by the year 2000." Today, this type of telephone is being sold to the public as "the fastest thing since the wheel," and its promise of future delights goes untrumpeted. Since Touch-Tone telephones cost more, one can not help but admire a business strategy that seems to be succeeding in persuading the public to pay in advance for the use of a device that will, for most users, confer significant new benefits only at some unspecified future date — and even then almost certainly at an additional charge for each service.

To be sure, some of the potential of Touch-Tone telephones is being exploited. Devices are available that enable telephone users to "dial" frequently called numbers automatically. A variant on this card-dialing system is in use for inventory control and stock replenishment systems.

So far, the CATV industry has spawned a number of technical advances, and more can be expected as demand for cable-related devices increases. The needs of cable have been fairly simple up to now, so that the economic incentive to develop two-way response capacity has been small. Businessmen have been reluctant to try to market equipment for which there is hardly any demand. This will surely change as wired broadband systems become more widespread and as use of its two-way capabilities becomes a real, rather than a potential, profit maker.

Cable systems that exist today are small markets, and the suppliers of equipment have tended to be small as well. In common with most other parts of the electronics industry, supply of basic CATV components is a business that can be entered fairly readily. The Bell System has an in-house monopoly supplier, Western Electric, which also has considerable expertise in cable-type communications and Bell Labs as a powerful research arm. Experience so far with cable communications implies that (1) the Bell System will continue to use Western Electric and (2) that existing suppliers of cable equipment to CATV operators will be joined by others as demand expands. Some of these small suppliers will be technologically innovative. Thus, prospects are good that technical advances will leapfrog from Bell to the independents and vice versa. A long-range cost advantage may lie with CATVs, since their suppliers will be able to import equipment, whereas Bell is tied to Western Electric.

It should be borne in mind that the Bell System operates on a vast scale. When it adopts a design for equipment, this design tends to remain basically stable for many years. Installation of equipment costs vast sums. Cable installations also have to be paid for — but a small city or neighborhood can be wired for a total cost far smaller than setting up a new long-distance cable from Boston to Miami. Thus, CATV operations may be able to avail themselves of technically advanced equipment more rapidly than the Bell System.

New methods of broadcast transmission promise to become available in time. FCC has already authorized over-the-air Community Antenna Relay Systems, which use microwave radio, but their use is limited. Short-haul, microwave, Local Distribution Systems, the FCC says, could be cheaper than laying CATV trunks. Development of these technologies, and others, poses a definite threat to the future development of CATVs that are primarily oriented to supplying entertainment signals to subscribers on a one-to-many basis.

Another development, the Amplitude Modulated Link, is said to be useful for relaying off-the-air signals from the site of the AML transmitter to a CATV head end.

FCC has also approved another system said to be useful for transmitting CATV signals. Although called a "quasi-laser," it is actually based on use of the infrared portion of the spectrum. It is claimed that a single quasi-laser transmitter can carry up to 18 TV channels on a single beam and provide service covering a 15-mile radius.

Still under development is the "FM laser," invented by Dr. William J. Thaler, head of the physics department of Georgetown University and originator of over-the-horizon radar. FM laser is said to be capable of doing as much as microwave, to be substantially cheaper, and to function within a very narrow beam.

Dr. Thaler says that at present he is operating a 1.2-mile link transmitting "black-and-white color TV information." In the laboratory, he has succeeded in transmitting color video information. And, he adds, "we're just about on the verge of improving the modulator to the point where we should be able to multiplex color TV channels over this system, hopefully up to something like 15 channels."

Dr. Thaler believes that a major advantage of his system is that it is frequency modulated, thereby avoiding interference from amplitude-modulated atmosphere noise. Still in the works at Dr. Thaler's laboratories is another device that uses electro-optics to produce frequency-modulated laser light. In the laboratory this device is already "in excess of a Gigaherz," but its full development is obviously still some time away.

The Bell System will not be standing idly by while independent CATV and microwave systems pre-empt its long-distance specialized traffic. Bell was a pioneer in coaxial cable; it began to study coaxial structure in the 1920s and applied for a patent on coaxial cable in 1929. The first Bell long-distance coaxial system was demonstrated in 1936, TV was transmitted over coaxial cable in 1937, and Bell's first commercial coaxial service (capacity, 480 circuits) was introduced in 1941. Currently, Bell is installing a long-distance cable system in calls L-4, estimated to cost $2 per circuit mile, with the capability of carrying some 32,000 voice messages simultaneously. This cable's successor — L-5 — is being worked on. Estimated to be capable of carrying over 90,000 voice messages, it could cost as little as $1 per circuit mile. Furthermore, Bell says that L-4 cable can be upgraded with L-5 electronics. In addition, Bell is working on its planned high-capacity cable for transmission of digital messages, including computer data and Picturephone signals.

As the Bell System sees it, starting around 1980, it is likely that circular waveguides — with a digital message capacity of over 250,000 circuits — will be installed when new lines are needed in areas where usage is heavy. In the more distant future is the possibility of using laser beams as message carriers — and one of these optical channels would be able to carry more than 10 million voice channels.

Bell's plans for digital data transmission imply that it expects to keep
highly competitive in this field. At present, Bell is using a system it calls
T-1, a short-haul (less than 50 miles) pulse-code modulated system. It has
a field trial of a T-2 system under way at Willow Grove, Pennsylvania. The
T-2 system is a 400-mile system. By 1975, Bell expects to have a T-5 system capable of spanning the continent.

Bell's Picturephone system, which currently uses a bandwidth of one MHz,
is currently offered on a very limited basis, and it is finding few takers. At
present, the resolution it offers on its display tube is inadequate for typewriter type. But, interfaced to a computer, the Bell device can provide display of a great deal of business-related information, and it permits the user
to manipulate the data. A self-contained, non-Bell system could do as much,
and a two-way cable to a central computer could provide not only data relevant
to a particular business but access to a variety of other materials as well.

For the present, a moderately skeptical view of Bell's expansion plans must
be taken. Complaints about ordinary telephone service are frequent both for
Bell and independent companies. Late in April 1971 the New York State Public
Service Commission denied a New York Telephone Co. application to set up an
experimental Picturephone service for intercom use only in lower Manhattan.
The modest setup envisaged by the utility would have required 10 employees
for installation and servicing.

But the PSC denied it, saying that "any diversion of company assets and resources to other than essential services, no matter how miniscule, is contrary to the Commission's objectives and policies." The regulatory body agreed
to allow the utility to resubmit its application—"whenever the level of basic
telephone service has improved sufficiently to permit favorable consideration
of this new and important service offering."

Nevertheless, it must also be borne in mind that for the purposes of most
business-oriented users of communications, the telephone company's network
provides adequate facilities. To be sure, some users of existing data lines
pay a penalty because they can not run their computers and associated equipment at optimal speed and must suffer the expense of modulator-demodulator
equipment. Bell points out that when it offered a service capable of 50,000 bits
per second, "people didn't beat our doors down." The volume of traffic is such
that it can be accommodated by telephone lines and move fast enough to keep
users fairly well satisfied. Since the Bell System has a substantial leg up in
creating a wider-band network, it may be able to satisfy its customers fully
before competitive broadband networks are built.

Nevertheless, data users seem eager to grasp any alternative to telephone-
grade lines. Remote batch equipment—a fast growing area—operates optimally
at about 12,000 words per minute. But limitations imposed by the availability
of service from telephone companies force these devices to work at speeds of
about 3,000 words a minute.

A two-way cable network in a small area could very well move fast enough
to have local facilities available before the Bell System and Western Union do.
Certainly, such a system could be in operation sooner than a microwave link
network, since it is estimated that three to five years will elapse between May
1971, when the FCC authorized such specialized communications networks, and
the time national data transmission service can be offered.

Bell's role at present seems to be to sit tight and improve its service. Within a few years, it expects to have ample, technically excellent, data-carrying networks available to volume users. If FCC requires cable opera-tors to provide two-way communication capacity, Bell might elect to have these systems interface with their long-lines data networks. Independent spe-cialized data networks could well be inclined to use CATV facilities to carry their traffic from their point of origin to the microwave transmitter and from the receiver to their destination, although one such network has proposed using CATV facilities only when it is impossible to provide direct network-to-addres-see links by optical laser transmission.

In all likelihood, the Bell System would be happy to devote itself to handling the traffic generated by large-volume users of data. Small users — individual credit card terminals in stores and restaurants, for example — can be serv-iced adequately by ordinary telephone lines, although probably at higher cost than through an open-channel automatically scanned CATV link.

OTHER ISSUES

Initial Congressional response to future developments in the use of two-way cable could well be hostile. When an entrepreneur obtained the TV and other rights to a heavyweight championship fight so that it was visible only in the arena or over closed-circuit TV, complaints were vociferous. A bill cospon-sored by 32 members of the House and several members of the Senate was in-troduced to force sports events to be shown on free home TV by banning the use of closed-circuit TV whenever a sports event is of sufficient public inter-est for radio or TV stations to want to broadcast the event.

Two-way cable communication, with sales made direct from the TV set, will have a major impact on business. To the extent that it has an adverse ef-fect on small business, it is all but certain that efforts will be made in Con-gress to impose restrictions on the use of cable systems for consumer pur-chases.

The extent to which cable entrepreneurs will be allowed to engage in non-communications activities has yet to be settled. The FCC has already ruled that common carriers with annual revenues of $1 million or more can provide data processing services only through affiliates that are completely separated from the parent company. General adoption of the principle underlying this FCC rule could put serious restrictions on the diversification permissible to cable companies.

In turn, this could be an inhibiting factor for expansion of the cable systems or networks. Costs to consumers might well be higher if cable system opera-tors are allowed to do nothing more by way of business-related activity than supply a channel to an advertiser or other user.

PRIVACY

The need for privacy in financial transactions carried over a cable or other broadband system can hardly be overstressed. In one way, this may be a per-suasive argument in favor of cable rather than over-the-air broadband links, which are more susceptible to unperceived interception.

When CATV is used for financial transactions, steps will have to be taken not only to ensure privacy but to preclude unauthorized persons from gaining access to computer data so they can make improper transfers of funds. Some computer technicians believe that "you can get all the privacy you're willing to pay for," and this may well be the case. Right now, most people get privacy they believe adequate for their confidential transactions by using the mails.

Nevertheless, the need for privacy, security, and absolute accuracy in the transmission of transactions argues that all of the devices used in electronic data transfer—both from the home and into the home—will have to be thoroughly reliable and of high quality. This implies greater cost than is now experienced in buying entertainment-grade electronics. Once business operations are integrated with a communications system, failures become very costly, and highly reliable equipment is essential to the functioning of the whole.

The need for privacy extends far beyond financial transactions in any society that is rushing headlong toward computer assistance in such quantity that some people think in terms of computer domination. The National Science Foundation in April 1971 funded a two-year study for ground rules that will protect privacy in the computer age. The primary objective of the study, NSF says, is to find an accommodation between the need for confidentiality of information and the desire to gain greatest possible usefulness from computer technology and data bases. This may result in drafting legislation aimed at protecting individual privacy without sacrificing the social benefits of computer technology. The study will also extend to the technological or administrative safeguards that will have to be adopted to provide maximum confidentiality and to minimize the possibility of improper intrusion, but this is expected to be done in the second phase of the study.

ALTERNATIVES

By the time computer-interconnected two-way CATV for business activities is in use on a large scale, competitors will develop, and some people may find them more attractive. To ensure privacy, businesses and families may choose to use minicomputers for their accounts instead of using a central one. Some computers adequate to this task are already available for about $4,500, and they could no doubt be leased for a smaller current outlay or bought on the installment plan. Home ownership of a computer could provide substantial economies to fairly heavy users who have to pay a fee-for-service whenever they use a central computer. And it would, of course, be a great status symbol.

15. PROSPECTS FOR ON-DEMAND MEDIA

Ithiel de Sola Pool,
Charles Murray, and
Kenneth Dobb

INTRODUCTION

The coming decades may see mass communication widely displaced by media that are interactive rather than one-way. Among the interactive media, some may be described as "on-demand."[1] What are these new media, and are they really coming? That is the subject of this essay.

What Is "On-Demand Communication"?

On-demand communication is communication characterized by (1) audience control of the timing and content of the messages received and (2) the production of the messages transmitted to the audience by mechanical means.

To illustrate: A person phoning for weather information and hearing a tape recording is engaged in on-demand communication. A person hearing a weather report on radio or TV is not, because the broadcaster, not the receiver, controls the timing of the transmission. A person phoning the weather bureau and talking to someone on its staff is also not engaged in on-demand communication, because he has tapped a human, not a mechanized, source.

A person in a library getting a listing of publications from a computerized information retrieval system on a topic that he has specified is engaged in on-demand communication. A person talking to a reference librarian is not, nor is the reader of a published bibliography.

Ordinary phone usage is two-way communication but is not on-demand communication, because the source of the reply is human. It takes little ingenuity to think of borderline cases, but the basic distinction is clear.

Can We Forecast Communication Systems?

To estimate the 1980-1990 demand for on-demand communication is technological forecasting at its hardest. We are asking about a technology that is not yet deployed. We are asking not only whether it is technically feasible and at what cost but also how far ordinary people will accept innovations that will change their way of life. We do not believe that a categorical forecast is possible. People themselves seldom know how they will act and what they will desire in a totally new situation. This is testified to by the experience of all social research on hypothetical behavior, such as new-product testing. What people say they will do has very little relation to what they actually do under new circumstances.

On the other hand, total defeatism would be just as wrong as exaggerated confidence. Social research is often able to predict what people will do even though they are quite unsure themselves.

A basic methodological principle on which we operate in this study may be described as "reducing the possibility space." In the social sciences it is seldom that the situation is so well controlled that one can achieve that paradigm of scientific rigor: the prediction that if x then y. More often, the social researcher is forced into a tendency statement that "other things being equal, x tends to promote y." Often y is an "either-or." For example, a child denied warm personal contact will either grow up to crave affection or withdraw into frigidity. It is common, but unjustified, to dismiss such statements as scientifically worthless. In fact, the statement is significant, for it excludes at least one possibility, namely a high incidence of normal middle-of-the-road adults coming from that background. The possiblity space has been reduced.

Most social prediction is of that character. It cannot forecast a specific outcome. It can nonetheless be useful in reducing the infinity of possible outcomes.
     Note that this is not a procedure that eliminates subjective judgment at the
end of the forecast process. Infinities, as any mathematician knows, can be
of different sizes. Eliminating some outcomes from an infinity of alternatives
still leaves a smaller infinity of alternatives. There can still be no analytic
solution or firm prediction. But reducing the possibility space makes very
much easier and more reliable the intuitive judgment of the human estimator
as he looks at the situation.

Summary of Conclusions

In this paper we conclude that of all the on-demand services that might use
two-way telecommunications facilities, the one that may attract enough dollars
from clients to pay for its development is remote marketing from the home.
Other services that would use the same hardware, such as educational applications, would then, but only then, become economically feasible. We also
conclude that there is evidence of consumer interest in on-demand communication services, the evidence including trends in expenditure for communication facilities and also a trend of preference for a way of life that we call the
"insular family," that is, a trend toward doing more and more things in the
home. However, we also conclude that the cost of on-demand communication
is such that discretionary consumer income will not pay for such facilities before the 1990s at the earliest. Communication systems become both economical and desirable only when they are diffused to a large part of the population.
The commercial introduction of on-demand communication facilities is therefore likely to come in a rush, but only long after the technical capability and
the potential demand for on-demand communication services have existed for
a long time. Then, when a tip point is reached, the system may grow with explosive speed. One way to avoid protracted delay in development is for the government to provide 100-percent penetration of homes by broadband cable for
educational and other public services; such an installation would greatly accelerate the addition of other services that would use the cable. However, the
more probable course in this country is reliance on commercial processes for
introduction of cable. That means that services dependent on close to 100-per-
cent penetration will have to wait for many years. Cable, it is true, is only a
part of the hardware for an on-demand communication system. Telephones
and cassettes will also be part of a comprehensive system. But cable is required to provide adequate bandwidth for real-time video elements of an interactive system. The development of on-demand communication facilities in the
United States thus depends upon the growth of cable television to that point
where there will be not only a telephone but also broadband cable in virtually
every home, schoolroom, and office. That will be a communications revolution of major social impact.

ON-DEMAND COMMUNICATION FOR RETAILING AND OTHER SERVICES

Most kinds of human communication can at some cost be transmitted electronically. Audio and video signals, both of which we know how to reproduce from
electronic signals, cover the great bulk of symbols humans use. Smell, taste,
and touch are also used, but much less.

The issue is, therefore, not what kinds of on-demand two-way communication can be made electronic. Almost all of it can be. The issue is what kinds of devices would be sufficiently useful and economical in an electronic mode to be widely adopted and paid for by society.

One principal example already exists: the telephone, on which Americans spent $19 billion in 1971. In some ways it is unique in modern communication technology. Most of the communications revolution of the past century has consisted of the emergence of mass media, that is, ways in which one or a few originators can talk at great economy to a very large audience. The success of the newspaper, magazine, billboard, radio, movie, bulk mail, and TV are accounted for by ability to deliver messages for pennies by dint of the economies of mass production and distribution.

In contrast to the mass media, there are three widely diffused technologies of individualized person-to-person communication: the mails, telegraph, and telephone. Of these the mails are old, although as an efficient and economical system they go back only to the 19th century. Mail and telephone, like the mass media, have been able to get costs per message down to pennies, and they have therefore been universally adopted in developed countries. Telegraph, which in its earlier form served to cut long distance transmission time from that of ships and trains to instantaneous transmission, but at a cost of dollars, never became a pervasive communication system, though important for specialized uses. Newer technologies that, like telegraphy, transmit written text but bring it into the home electronically at low costs may become as universal as telephone and mails in the future. Up to now, however, telegraphy has succeeded only in institutional uses, like press services.

This brief review of historical cases may help orient our thinking about future prospects. There are new communication technologies that may find their way into every home, but probably only if the services cost very little. There are also more expensive technologies that businesses, governments, and other institutions, though not households, may adopt.

It is our judgment that only one new application of on-demand electronic communication to the home may meet a sufficiently universal need at a low enough price to have a chance of diffusing widely by itself in the near future: that is retail marketing.

Other home uses of on-demand communication, such as teaching, information services, and entertainment, may ride the coattails of retail marketing. Many of these services may reach the home in mass media form, such as CATV or over-the-air TV, but the cost of the added wrinkle of allowing the user to dial up what he wants when he wants it is likely to be prohibitive if the users of just one such service have to pay the full cost. These additional services may become economic, however, if use is made of equipment already in place for retail marketing, telephone, radio, and TV purposes.

Thus retailing seems likely to be the driving factor in the development of any universal on-demand communication system. Predictive efforts should therefore focus on the retailing prospects.

On the other hand, businesses, educational institutions, government, and other such organizations will use two-way on-demand communication systems extensively. Closed-circuit applications of that kind may develop independently of the spread or nonspread of demand communications into the home.

A number of previous efforts have been made to catalogue possible on-demand services that would be potentially marketable or socially useful. For example, other listings of two-way services—not necessarily on-demand—will be found in Baer and in Baran.[2] Another listing is presented in the appendix to this chapter.

The possible future applications noted there range from such simple ones as no-school announcements, through surveillance, citizen feedback, and community services, to such complex applications as remote medical care. Almost all of them depend on some degree of technical innovation.

BARRIERS TO THE ADOPTION OF ON-DEMAND TECHNOLOGY

The Problem of Step Functions

The speed with which innovations can be introduced and the accuracy with which predictions can be made about their introduction depend upon the stage at which they are located on the R and D sequence. We might order technologies in the sequence from basic science through development to deployment:

1. Scientific understanding exists. Clues to the technology exist, but both research and development problems remain to be solved.
2. Technological solutions are within the state of the art, but many alternative solutions remain to be evaluated for their relative effectiveness. Development problems remain to be solved.
3. R and D has been done. Prototypes exist but not implementation of the system in the field. Investment problems remain.
4. Equipment is on the shelf ready for deployment. Production and marketing problems remain.

The technologies we are examining are largely at stage 2. It is hard at this stage to see clearly into the investment and deployment prospects, for the technological alternatives are still unresolved. When technological issues get resolved they may suddenly open up the sluices to extensive investment and marketing activities.

Eleven groups of hardware configurations are listed in the appendix, arranged roughly, but not exactly, in an order of rising sophistication. There is a breakpoint at Group VIII, a configuration that provides (1) specified still pictures on-demand (2) two-way text and (3) two-way voice transmission. That combination of communication resources gives the user enough power to begin to do radically new things that he cannot do with presently deployed hardware or to do old things very differently than the way he does them today. That configuration perhaps offers the user enough to get him to pay the costs of new facilities. Many simpler services, while cheaper to introduce, do not offer enough to sell themselves. Some more powerful services are impossibly costly. Whether the costs of that configuration are likely to be acceptable even by the end of the 1980s is hard to know. That is a critical question.

The significant thing about that configuration, which makes for a step function in utility, is the inclusion of some individualized video capability. Video bandwidth is so much greater than that for text or audio that provision of pictures on-demand represents a major challenge. Is video necessary? And how little on-demand video capability can one get away with and still have an attractive service?

Certain message services and institutional services would not require video to be acceptable. Certain entertainment and political services can be provided with ordinary cable television and individualized digital feedback, without sending on-demand individualized pictures downstream. Many of the most interesting services, however, and notably both marketing and education in the home, depend critically, we believe, on being able to provide a picture to the user on his demand.

For many products, unless people can see the particular object they are ordering it seems unlikely that people will buy from the home, rather than going to the store. Consider several different levels of picture that might be offered.

A standard advertising picture of the generic class of the object being sold might be useful in selling a new game or a toy that the buyer had never seen before. Each exemplar of the manufactured product is presumably the same. Thus no individualized picture is needed. What is required is much like current TV advertisements. Also it is like current practice in that motion is presumably important, with a demonstrator showing how the article is used and turning it to its various positions. On-demand (but not immediate) communication of such pictures could be provided by recording the picture in the home when it is broadcast and replaying it.

A still black and white picture of the actual article to be bought would be the minimum necessary to sell meat, vegetables, flowers, or clothing by remote means. Also, the selling of most small standard items that the buyer selects from among many alternatives would require this facility, for example, greeting cards, books, cosmetics, gifts, hardware. For this facility there must be enough bandwidth for the store clerk to place the article under consideration in front of a camera, so as to deliver at least a still picture within a few seconds.

A further refinement would be a color picture. And finally, a moving picture of the actual article to be bought clearly has great advantages for demonstration and rotation but at prohibitive costs.

The relative costs and benefits of color and motion is an interesting issue to be explored. It varies by class of product. Color may be desirable for video purposes, but unfortunately it makes the TV set a less good CRT for text.

For education, what can be done without any pictures is very limited, but different subjects need pictures of different levels. Algebra, music appreciation, and languages can perhaps be taught well without pictures, but not history, or cooking, or musical technique, or science or sociology.

1. Preplanned stills recorded off the cable onto a cassette in nonpeak hours can be linked to audio tape and to interactive digital exercises, so that a student can do his course at his own pace and convenience.

2. Stills called interactively out of a library in a branched teaching program would be still better and for many purposes are essential but place more burden on peak load capacity. That is the video capability that the Plato system at the University of Illinois and also the Mitre frame grabber will provide. That is also what the critical configuration (VIII) would make possible.

3. Movies provided by cassette for preplanned use in a course can be very useful. Electronic transmission of them may be a luxury; the mails may be a cheaper method of delivery and fast enough.

4. Movies called promptly on-demand over cable can hardly be justified eco-

nomically unless it be at the call of a teacher in a classroom, for there the
movie will be seen by many people at once.

It seems clear that one critical test of the usefulness and economy of a system is what level of video it can provide, with what speed, and at what costs.
This is one of the major bottlenecks to the development of effective on-demand
systems. What are the others?

Potential Bottlenecks

For the implementation of on-demand systems a number of resources must be
provided, each of which costs money. Let us list some of these.

1. The user must be provided with a directory service so that he can learn and
locate what is available to him[3] (for example, a directory channel on CATV,
the phone book, 555-1212, TV Guide, book reviews).

2. Downstream capacity: How many channels of what capacity each? How much
and how fast can text and pictures be transmitted? How many services can run
at once without interference (for example, fire alarms, phone, facsimile, and
TV all running simultaneously)?

3. Upstream capacity

4. Terminals, for example, CRT, typewriter, facsimile, tape recorder, video
tape recorder, cassette player, camera, speaker, sensors. In dollar volume
terminals are likely to be by far the largest part of the hardware costs of the
system. If the average interactive terminal costs only $500, a 50-million-
home system will cost $25 billion for terminals alone. That is about double
what the public has invested in TV. It is about half of the over $50 billion that
is the value of the current investment in the American telephone system. Such
an investment will not come about overnight. Clearly, there are strong economic considerations in favor of making maximum use of the TV sets and telephones that are already present in almost all homes, and which represent sunk
costs. Noam Lemelshtrich summarizing some work by Malcolm Macanly has
made the point regarding the use of TV sets as two-way terminals: "Using a
standard home TV as a functional part of the home terminal has the following
advantages: (a) economy — it makes use of the huge capital society has already
invested in TV sets; TV receivers are much cheaper than conventional computer CRT terminals (due to relative volumes of production) and consume less
power (random scanning which is used in computer CRT terminals requires
more power), (b) maintenance — component parts are cheap and well distributed
in neighborhoods, trained servicemen are available, (c) universality — TV receivers can be used with inputs from TV cameras (color signals can be easily
synchronized), unlimited number of feedback displays can be projected (like
keyboards, adding machines or engineering as well as artistic design), (d) human factors — users can construct their own "keyboards" or display for feedback; TV is a familiar device to people and can be expected to be more accepted
and understood than adding complex terminals to the TV for communications."[4]

5. Switching system capacity: The present telephone switching system will not
meet needs that require extensive hour-after-hour connections.

6. Cumulator and computer input capacity: What are the peak loads of feedback
that the head end of the system can accept?

7. Computer controlled bulk storage capacity: For information retrieval systems and all applications requiring large libraries, one economic constraint

is the cost of bulk storage and retrieval. While trillion-bit memories that are
as cheap as storing paper are clearly on the horizon, the problem of economical conversion to machine-readable form remains acute.[5] Electronic and microform storage and combinations of them constitute a rapidly evolving technology.
8. Remote interconnections of local systems: Problems of cost, standardization, privacy, and method are involved in, for example, allowing local physicians' record systems to access records from other cities.
9. Production of program material: The cost and difficulty of producing effective news, entertainment, educational and other material for on-demand services is likely to be greater than all the technical efforts put together.

This last point cannot be stressed too much. Most pricing exercises, our
own included, focus on the costs of getting bits to receivers. Making these bits
into meaningful information that a human receiver will want is, however, an
even more expensive proposition, and someone has to pay for that, too.

The telephone system has the blessing that the production of messages is
an uncharged item for which AT&T has no responsibility. On-demand communications are not like that. Even if regulation separates companies, as it
may, into carriers and originators, both must be compensated for their services. Thus hardware investment roughly equal to what had gone into the telephone system is dependent upon other massive investments in software production.

Television when it was introduced had the advantage of having on hand an
enormous reservoir of old motion picture films. "That reservoir gave the
medium something to market in its early days without heavy production costs."[6]

On-demand communication starts without the advantages that have helped
either telephone or TV to fill their channels. Software will be a critical problem.

ESTIMATING THE PROSPECTS ON THE SUPPLY SIDE
In the previous section we have identified the facilities needed for possible
future on-demand communication services. Whether and when such facilities
may be disseminated is an economic matter. At what price can they be offered?
Will they be bought at that price? Elementary economics teaches that such estimates are not point estimates. How much will be offered and bought is a function of price.

Technological Assumptions as to What Is Available
A supply curve specifies how much of the service would be offered at any given
price. In the long run that is determined by the cost of production at different
levels of volume. But note that a particular technological solution or several
alternative ones must be specified before one can estimate cost of production.

The cost of any technological solution is a function of date. The long-run
supply curve in 1980 will not be the long run supply curve in 1990. Thus we
must predict a shifting family of curves (see Baer).

On-demand communication services are not independent of each other nor
of nondemand communication services. Many services can use the same hardware. If two-way communication terminals are in the home for marketing services, then it becomes much cheaper to provide educational services to the same
homes. If broadband cable enters the home for CATV then perhaps facsimile

services can be supplied more cheaply than otherwise. One looks for simpli-
fying assumptions about the basic extant hardware on which new on-demand
services will build. One useful assumption would seem to be that in the period
of the 1980s almost every American home and office will have available to it
two types of local wired connections, telephone pairs and CATV cables, and
generally no others. Among the costing exercises done to date are that of Baer,
that by John Ward for cable TV, and that of Michel Guité and Noam Lemelshtrich for a two-way terminal. They provide a starting model for continuing
exercises in this direction.

Maintenance

One set of considerations that cannot be disregarded in a supply-demand analysis is the allocation of responsibility for providing and maintaining the various
parts of the communication system. The telephone system, for example, is a
completely rented system, owned throughout by the phone company. TV sets,
on the other hand, are the property and responsibility of the viewer.

The more complex the terminal equipment required in an ordinary user's
premise, the more desirable an arrangement like that of the telephone company. Dependence on unskilled and unscrupulous repairmen and whopping bills
for repairs may make otherwise good systems untenable. Thus one step toward making on-demand services feasible is the movement of CATV companies into television set rental and repair. The majority of complaint calls to
cable companies turn out to be for set malfunctions. Also, multichannel CATV
requires the company to install set adapters. It thus makes sense for the cable
company to move in the direction (common in England) of providing sets and
maintenance for a rental fee. Current opposition by the manufacturers and repairmen to this natural trend will slow down the adoption of on-demand services.

If new equipment worth perhaps $500 is to be installed in private homes,
the rapidity of adoption will depend very much upon the ease of citizen operation and on the establishment of satisfactory systems of repair and maintenance, and also upon creation of financing arrangements that do not require
lump sum purchase decisions.

System Installation

In forecasting the extent of implementation of a technical innovation, one must
consider not only its technical feasibility (which for on-demand communications
is clearly positive) or the cost/benefit trade-offs (which for on-demand communication is more problematic). One should also consider the ease or difficulty of initial implementation. Systems that may be equally useful and profitable if once established may vary greatly in the ease of getting them established.

The assumption underlying the discussion is that, other things being equal,
systems are more likely to take hold if they can be installed incrementally. We
consider three types of increment: increments in service, in users, and in
sponsors.

Incremental Improvements in Service

We are asking here if the whole package
must be bought before any benefit is derived.

The problem of using computers in libraries is an apt example. One of the reasons that general-purpose libraries have been so slow to move to on-demand computerized catalogs despite their obvious advantages to the user is that there is no convenient first step. The economics of scale can become highly attractive as computer storage costs fall and traditional cataloging costs rise. Even now, it is possible that entering the second book into a computer-based catalog could be cheaper than cataloging the same book by conventional methods. But the cost of getting the first book into the system is enormous. The capital investment is large, and sunk costs in the existing system cannot be applied to implementing the new one.

Similar problems may be expected to affect the development of on-demand household systems. Once the hardware is bought to provide one two-way service, the sponsor can add a set of others at relatively little additional cost. But there is no cheap way to provide that first service.

Incremental Increases in Users For virtually all new systems, costs fall as the number of users rise. But the shape of the decreasing cost function is critical. For some services, costs are more nearly constant with number of users; for some, the fall is sharp. Most communications services are of the latter type. It is expensive to start small.

That is the consideration from the point of view of the supplier. A similar consideration applies on the side of the user. In estimating the value to him, the prospective subscriber to a service must ask whether he is dependent on others participating in the system. Communications facilities are peculiarly sensitive to this problem. When a consumer buys an ordinary product, it is often a matter of indifference to him how many other people have the same, or he may even want to be unique. If the service is on-demand access to a library of films, it may make no difference to the individual user whether anyone else is on the system or whether there are other libraries elsewhere. This does not mean that it is economical for the supplier to provide the service to only a few consumers. It does mean, however, that the selling job will be relatively simple if the service itself is attractive.

However, when a person subscribes to a phone system it is vital to him that other people are doing the same. If the on-demand service being considered is home shopping, the consumer is clearly dependent on widespread acceptance. Unless a variety of stores are hooked into the system, there is no reason for him to acquire an expensive terminal; and unless there are a large number of consumers with home two-way consoles, there is no reason for any one store to join. Getting the system off the ground requires a highly coordinated organizational effort.

Sponsor Independence This characteristic applies primarily to business and government on-demand systems and the problems involved in coordinating major new programs among several institutions.

At one end of the scale is the system that can be installed and used internally by a single institution. The airlines' computerized on-line reservations systems are a case in point. No industry-wide effort was needed to make the system desirable for any one airline. The largest trunk carriers bought their system independently; indeed, one of the competitive incentives to do so was that other airlines did not have them. Plans for linking the systems waited until after several independent systems were in operation. A coordinated effort was not required to make the basic decision to go ahead.

A more difficult problem of coordination arises when the utility of the service depends on several institutions installing the system nearly simultaneously. The degree of difficulty depends in turn on subsidiary factors, such as the relationship of the proposed service to competitive pressures among the companies. Paperless credit transfers among banks would be an example of a relatively low-conflict service. In the retail and consumer goods industries, the room for cooperation would be somewhat more restricted.

Perhaps the most difficult area for cooperation would be among institutions in entirely different fields. For banks to join with retail firms to supply an automatic bill payment service would require accommodations among institutions working in disparate environments and with quite different priorities in the design of the system.

Investment in Two-Way Services

If our interest lies in forecasting the likely tempo of the introduction and acceptance of on-demand communications services, we must assess the magnitude and types of economic hurdles on-demand communications must clear before such services achieve widespread use.

One plausible assumption regarding the provision of on-demand services is that they will evolve in large part from CATV technology. According to the Sloan Commission, CATV systems service about 5.9 million subscribers or about 9 percent of the television viewing market. CATV systems have been installed in over 2,750 communities. New FCC rules (March 31, 1972) require two-way capability on all new CATV systems.[7] However, the FCC has left to the future any definition of the level of two-way service that they have in mind. Several two-way terminals and system head ends have already been designed with experiments in seven communities presently underway. (These are described in Ward,[8] Baer,[2] and Guité.[9]) In addition, other two-way systems are being devised and developed. (See Ward and Lemelshtrich in this volume.) But despite this froth of activity and investigation, no sense of the likelihood of the provision of two-way services by the present CATV industry or of their demand by consumer households has yet emerged.

The most critical factor influencing the provision of these services is the availability of entrepreneurial skill and venture capital. Likely sources of the necessary fixed capital are (1) federal, state, or local government, or some combination of the three and (2) the CATV industry.

There is a possibility that government subsidies may play a role in financing two-way services. The strongest case for government initiative is the advantages that accrue from 100% penetration. To serve 20% or 50% of the homes, cables still need to pass down every street in front of every home. The additional cost for putting drops in every home is small. The advantages of 100% penetration are considerable. For example, if every home has cable, meter reading by cable becomes highly economic. If meter readers have to call at some houses, it saves much less to have automatic reading at some of the houses on their routes. If every family has cable, schools can use cable for homework. If some families do not have it, schools cannot organize homework

that way. Thus there are real social benefits to be served by public action to
provide 100% cable penetration. Some advocacy of this cause is currently
taking place. Also, it might prove practicable to perform some intragovern-
mental functions more efficiently by means of a cable system. However, strong
national norms against governmental control of communications media weigh
against subsidization.

The willingness of the private decision makers to risk capital is predicted
on their expectations of return on investment. They must estimate the size of
the initial fixed-capital, the cost of venture capital, the sources of subscriber
revenues, and the responsiveness of two-way subscription to changes in price.

Given an existing and operational cable system, the initial investment costs
are a function of expenses for upgrading the head-end equipment, converting
the laid cable and amplifiers to accommodate new capabilities, and installing
or persuading consumers to install more sophisticated attachments as terminals. First, let us try to give a sense of the direction of head-end equipment
costs.

Perhaps the first thing to note is that head-end costs relative to transmission and receiver costs are less sensitive to system size. In mass communication systems the head-end and production costs are almost totally insensitive to system size. In on-demand systems, that is not completely true but remains largely so. Head-end costs are more sensitive to the type of system
(subscriber response, subscriber initiated, point-to-point, etc.) and the number of services it offers than to its size. For subscriber response systems
minicomputers are an estimated $40,000-$120,000, the variation being made
up largely of memory capacity and peripheral equipment variables. Vicom
Co. currently uses a DEC PDP 8 to service a five-subscriber two-way hookup
at a cost of $10,000. They estimate capital costs on a computer servicing
4,000 subscribers at about $20,000-$60,000. Fully developed, that is fully de-
bugged and tested, software systems for the same system might range from
$15,000 to $25,000 per system. Development costs for the innovator, however, are estimated by companies now developing software systems at between
$150,000 and $250,000. Subscriber response display facilities in the form of
real time video or hard copy terminals might add another $10,000-$20,000.

Transmission plant costs are highly sensitive to the number of households
passed per system mile and to market penetration rates, that is, the share of
households passed who are subscribers. CATV operators today estimate that
they need a 35 percent penetration rate to break even.

For two-way systems, the costs vary with the type of transmission medium
required. Five options of varying technical complexity for providing the upstream return link from subscriber to head end may be isolated. These are (1)
wire pairs, (2) two-way transmission on a single cable, (3) return transmission on a separate cable, (4) separate upstream-downstream trunk cables with
common feeder lines, and (5) a switched distribution system of the Rediffusion
type. At present one-way CATV systems average $4,000 per mile above ground
and $15,000 per mile underground.[10] The incremental cost per mile for each
of these five options in a 200-mile cable system, assuming aerial distribution,
has been estimated by Baer at (1) $390 per mile, (2) $830, (3) $1,625, (4)
$1, 045, and (5) not relevant.[11] Thus he concludes that two-way systems would

cost 16 to 32 percent per household more than one-way systems <u>before</u> considering terminals.

Table 1 presents Michel Guité's estimates for a variety of levels of one- and two-way systems — estimates that are in basic agreement with the others.[12] John Ward estimates the total costs of providing two-way service in the sophisticated configurations we have been discussing at between $200 and $500 extra per subscriber and simple two-way capability at less than twice the cost of one-way CATV systems. Table 2 gives Guité's estimates for conversion of existing systems to two-way; it considers four alternatives that seem to be available to Canadian developers.[13]

Empirical research has indicated that contrary to expectations and classical economic theory, cost of capital in terms of interest rates seems to have little impact on investment decisions. Rather, according to J. R. Meyer and E. Kuh,[14] the relevant financial consideration is the availability of internal liquidity. "Finance from currently generated funds will permit a rate of accumulation of fixed capital consonant with the residual quantity of financing remaining after these prior claims (inventories and dividends) have been met. The major flows of current funds are net retained income and depreciation expense." An additional finding is that the unavailability of liquidity adversely affects the investment performance of smaller firms. The problem of firm size and available liquidity is a recurrent one in the economics of innovation. Led by Schumpeter and Galbraith, one school contends that the costs of innovation are so great that only large firms can now be involved and, in addition, that projects must be carried out on a large enough scale so that successes and failures can balance out. A second group contends that there is no evidence to support these theoretical hypotheses. The results of a preliminary investigation indicate that size of firm correlates with the rate of technological innovation in some industries but not in others. Three factors were observed to hold where the largest firms did account for a disproportionately large share of innovations. This phenomenon was observed where (1) the innovation required a large investment relative to the size of the potential users of the innovation, (2) the minimum size of firm to which the innovation would apply is large relative to the average size of the firm in the industry, (3) the average size of the largest four firms in the relevant industry is much greater than the average size of all firms that are potential users of the innovation.[15]

The size of firm issue is one that is important for future CATV innovations. A quick glance at Table 3 suggests that the capitalization of the average CATV firm is very small.

The investor in cable communication services has a variety of potential sources of revenue. The most obvious source is subscriber fees. By expanding services he may justify higher charges to the consumer. An additional source of revenue will probably be commercial enterprises that find that they can reap substantial savings or other benefits (advertising, market information, etc.) by communicating with their customers by cable. The same kind of considerations would apply to government utilization of cable systems. (It is not yet clear how much of potential profits will be siphoned off by regulation or requirement to provide public service channels free.) New sources of revenue might arise by expanding the firm's functions from simply providing in-

Table 1. Costs for Several One- and Two-Way Systems (in dollars; 10,000 subscribers; 50 percent market penetration

| | Hardware Costs | | | | Operating Costs | Total Costs |
	Cable and Head End	Home Terminal	Total	Cost per Month	Cost per Month	Cost per Month
Standard one-way 12-channel	125		125	2.65	3.24	5.89
Standard one-way 24-channel	145		145	3.06	4.78	7.84
New two-way 24-channel with four-button pad	170	85	255	5.42	6.00	11.42
New two-way 24-channel with modified keyboard and printer	170	200	370	7.86	6.25	14.11
New two-way 24-channel with modified keyboard and printer and voice	170	225	395	8.38	6.50	14.89
New two-way 24-channel with modified keyboard and printer, voice, and frame grabber	170	415	585	12.43	7.00	17.43

Source: Michel Guité, New Technology for Citizen Feedback to Government (mimeo), MIT and Stanford, December 1971, p. 24.

Table 2. Costs for Conversion of Existing Systems to Two-Way (in dollars; 10,000 subscribers; 50 percent market penetration)

	Initial Cost	Cable Modification	Head End Modification	Total
Typical 12-channel one-way cable	125	25	20	170
Converted (from 12- to 24-channel) one-way cable (for 300 MHz cable only)	145 (includes converter)	30	20	195
New 24-channel two-way cable	150	—	20	170
Switched 24-channel two-way cable	350	—	20	370

Table 3. CATV Systems by Subscriber Size (February 7, 1969)

Size	Number
20,000 and more	8
10,000-19,000	50
5,000-9,999	144
3,500-4,999	123
2,000-3,499	279
1,000-1,999	423
500-999	427
50-499	730
49 and under	46
Not available	260

Source: TV Factbook # 40.

formation transmission facilities to other sorts of duties. For instance, the
cable operator might take over the billing function from retail outlets using
two-way services and, in the manner of contemporary credit cards, charge
a commission on every sale.

Finally, in making an innovative decision, a CATV operator must anticipate
the response of other actors to his decision. A firm normally takes other
firms and governmental regulatory agencies into account in these strategic
considerations. The CATV industry, though highly populated with individual
firms and displaying little market concentration—the top seven members of
the CATV industry account for only 27 percent of the total subscribers to
cable—is nevertheless characterized by local monopoly. One consequence, in
some communities, has been shoddy cable equipment or installation, which
has impaired the reliability or quality of the picture signal. An additional consequence, given the lack of local competition, might be a high aversion to in-
novate risk.

ESTIMATING THE PROSPECTS ON THE DEMAND SIDE

Functional Equivalences

Most proposed two-way services provide not new services but rather existing
services in a more convenient form. Telemeetings or teleshopping are substitutes for travel that save time, labor, and capital expenditures. Such proposed services can be analyzed by comparison with the system they replace,
noting where savings are generated for either the customer or business firm.
However, sometimes the latent function of an activity is quite different from
its avowed function. The latent function of marketing or politics may sometimes be social contact. That fact must be taken into account in estimating
whether a cable system replacing traditional functions is an attractive proposition.

Communication can be face to face, with the communicators physically
present, or it can be mediated, with an artificial medium used to convey the
message from source to receiver. As each artificial medium has developed,
there have been trade-offs between using it versus face-to-face communication. One could describe this process for each new medium starting with the
development of writing. An illustrative and very pertinent modern example is
the effects of media on watching professional football.

The trade-offs between being physically present or electronically present
at a pro football game center around (1) the amount of information about what's
happening; (2) the visual vividness of that information; (3) the intangible qualities of "being there," for example, crowd excitement; (4) money costs; (5) the
desire to stay home or get out of the house; (6) convenience and comfort. As
the media have developed, the balance between the advantages of direct versus
mediated communication have steadily moved toward the latter. With radio,
the only benefits to the listener were cost and convenience; when television
came in, the viewer approached parity with the spectator in the amount of information he received; with color TV he approached parity in visual vividness;
and with the development of split screen and instant replay techniques, he
moved clearly ahead in the amount of information. The sophistication of the
media is now such that there is a real argument for watching pro football on

television even if the tickets to the game were free. And there is at least one
case in point: the 13,000 empty seats in ordinary weather (mostly already paid
for) at the 1971 American Football Conference championship game in Baltimore, when the customary television blackout was compromised by the telecast of a station in Washington, D.C.

That example indicates that media can even improve upon a formerly direct
mode of communication. Obviously, most substitutions of mediated for direct
communication are not so clear-cut. Even in the case of pro football, television viewing is not exactly equivalent to being there (though it is in some ways
superior), and the analysis eventually boils down to specifying the competing
advantages of using direct or mediated communications and the relative costs
of each.

The most universal cost associated with face-to-face communication is the
cost of getting the source and the receiver of the message together. The increases in those costs over the next two decades can be expected to be large.
A strong and apparently irreversible trend is that of the growth of geographic
areas over which people operate. Retail markets have expanded from the village marketplace and neighborhood store to the metropolitan shopping center
with fruits and vegetables from 3000 miles away. Commercial markets are
increasingly continent-wide and worldwide. Politics moves from the country
courthouse to the state to the nation. Scientific and cultural invisible colleges
increasingly operate by meeting, Xerox, and telephone among functional spe-
cialties without reference to geography. At the same time our urban conglom-
erations are spreading out rapidly as people seek elbow room.

"Because of the transportation revolution brought on by the automobile and
the superhighway ... the population per square mile in urbanized areas will
have decreased from 6,580 in 1920 to around 3,800 in 1985."[16] Despite urbanization, there is taking place an increased dispersion of the population: geographic dispersion in terms of distance and effective dispersion in terms of
time and money.

The Changing Ecology of American Life: Dispersion

The Insular Family   The last fifty years have seen a major shift in the dispersion patterns of the U.S. population. At the broadest level of aggregation, the
proportion of people living in urban areas has risen steadily. But within that
is a subsidiary trend that developed during the late 1940s: the move from the
central city into the suburbs. It is this phenomenon that can be expected to
have important effects on the relative desirability of face-to-face and electronic communications in the 1980s. Many of the effects are already emerging.

The critical factor is not only the physical distance that separates the suburbanite from his job, friends, shopping, and leisure facilities, but the effective distance in terms of the time, money, and trouble that it costs him to get
to those people and places. Reasonably detailed estimates of the trends in time
and money costs can be obtained. The elementary fact that they are increasing
is part of the day-to-day experience of most people in both suburbs and cities.

These costs are highly sensitive to both the specific communications function in question and the persons involved. The profile of the current movie
audience is a case in point. One need not look to only the generation gap to ex-

plain why college youth constitute so high a proportion of the market. A part of the explanation is that college students generally do not have to hire baby-sitters, fight highway traffic, and pay parking fees in order to get to a theater. For a family with children in the suburbs, the money costs alone of going out of the house often dominate the solution.

Even when the money costs are not decisive, the time and trouble costs may be changing the citizen's calculation of when face-to-face communication is sufficiently desirable to warrant a trip out of the home. AT&T's Management Science Department is reported to have come to this conclusion in their analysis of the sudden increase in New York telephone traffic during 1969. They noted that telephone usage had shown the most dramatic jump in low-income areas and theorized that "as New York's traffic snarls intensified, as its subways grew uglier and as its streets grew more dangerous people stayed home and telephoned more."[17] In light of the experience in 1969, AT&T has concluded that their traditional trend line projections must be modified to include this type of changing human calculation.

Analysis of the role of <u>costs</u> in determining the future of different types of face-to-face interaction has its counterpart in the analysis of the future <u>demand</u> for such interaction. For what purposes do people want physical presence? Does the housewife find a social value in going to the supermarket that would make her unreceptive to a home-shopping service?

The trade-offs involved tend to fall into a few broad classes — those of substitution of human-to-machine interface for human-to-human interface, and those of substitution of audio or visual transmission for face-to-face interaction. Because the trade-offs do bear these family resemblances, it is plausible to expect that the various on-demand services will elicit similar patterns of acceptance or resistance. Cost factors aside, it is unlikely that shopping at home would become a great success while on-demand access to news stories would be a total failure. The degree of popularity may well vary from service to service, but the basic yes or no questions are likely to be answered consistently across the board.

One of the most prominent of the factors that will be pervasively influential is the course of development of what we have called "the insular family" — insular not only out of necessity through cost factors, but insular out of choice as well.

The evidence we can bring to bear at this point is impressionistic. It is not enough to even begin to make a solid case one way or the other, but it does point up the potential utility of a more thorough investigation.

The role of the family house as a center of entertainment is suggestive. Figure 1 is a graph of annual expenditures per capita for six categories of leisure activity from 1940 through 1968. Movie ticket and spectator event expenditures have been normalized for changes in ticket prices. Lacking specific data, the others have been normalized for changes in the overall consumer price index.[18] All six categories use 1958 as the base year.

Thus in the 22 years from 1945 to 1967 the expenditure by the average American for movies, for other spectator events, and in taverns — in short, for entertainment outside the home — dropped by more than half, from $61 to $26 per year. At the same time his expenditures for TV, radio, recording equipment, books, magazines, and bottled liquor — in short, for amusement in the

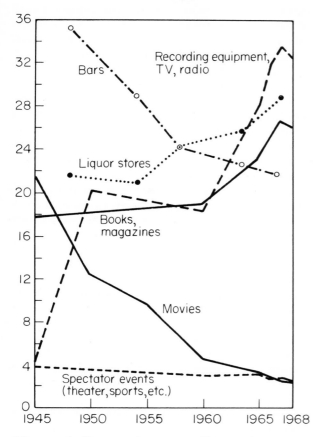

Figure 1. Per capita expenditures in constant dollars for selected leisure activities.

home — more than doubled from $44 to $89 per year. In 1945 the average American spent most of his entertainment budget away from home. By 1967 (except on vacations) he spent more than three-quarters of it in his own home.

Other data could be applied to the question of in-home/out-of-home activity. It would appear from a survey of newspaper accounts that the function of restaurants is undergoing a significant change. Owners of recently closed carriage-trade restaurants — the Colony and "21" in New York, for example — seldom attribute their problems solely to increased labor costs or changes in expense account rules. A consistent theme is that people do not approach dining out in the same way that they used to. Elegant public dining has lost its cachet. The growth industry in the restaurant field has shifted to the fast-food take-out franchise, which in effect does nothing for the customer but allow him to buy cooked food and eat it in private.

In addition to the changing role of the home as a center of activity, investigation should be undertaken into the hypothesis that people are increasingly averse to face-to-face interpersonal contact and increasingly drawn to electronic or servo-mechanical substitutes. Is it a phenomenon with deeper roots than just the change in the nature of American cities?

Again, we must rely on impressionistic evidence for formulating the hypothesis. But the anecdotal evidence is abundant, and the implications for on-demand communications, potentially significant.

For example, it has become almost a truism in the sports pages that it is better to watch sports on television than to see them in person, because of the capability of television to isolate, replay, and slow down the action. There is also the less frequently cited advantage of being able to instantly "leave" a one-sided game. These comments place a relatively low value on the excitement of being there, but evidence suggests that the public is beginning to share that opinion, as in the Baltimore AFC championship game referred to earlier. That game is not an isolated case — similar effects have consistently resulted from leaks in the television blackout at other, less important games in other cities.

The relationship of customers to salesmen (even soft-sell salesmen) is another fruitful source for the hypothesis. Advertising campaigns offer interesting evidence on this point, most prominently in the recent series of commercials by American Motors promising that salesmen will not approach anyone who enters the showroom and who, if asked for assistance, will limit themselves to answering questions — a sort of human substitute for an on-demand system. Perhaps most provocative in this respect is the occasional television commercial asking the viewer to write for information on a service or product while promising that "no one will call on you." That it is considered normal for a person to be interested enough to take the trouble to write and mail a letter, but be deterred by unwillingness to listen to a personal sales pitch, says something worth investigating about American attitudes toward interpersonal contact.

Finally, it should be inquired whether the desirability of interpersonal contacts is being seen differently even when friends are involved. The change over the last hundred years in the manner of making and maintaining friendships is suggestive. An obvious institutional change may be seen in the disappearance among the middle and upper classes of the concept of being "at home" to visitors. With that disappearance has occurred a marked limitation in the range of people who are expected to call without a specific invitation, from a circle that in 1900 included not only friends but people who wanted to become friends through a series of calls, to a very small circle in 1970 that includes only a few of the closest personal friends. And even when these few do drop in, the visitor is likely to phone ahead; moreover, either he or the host is likely to feel obligated to create a "reason" for the visit, even if it is something as trivial as giving the children a chance to play together. The casual conversation with no justification except its own pleasure is more likely to occur on the telephone, if at all.

Growing social inwardness of the household, corresponds to a larger trend that sociologists have observed in commenting on American society, and particularly on the working class strata within it. The trend is toward an increased familial orientation in general with do-it-yourself projects at home displacing the tavern or lodge and material family advancement being one of the highest values. A noted sociologist writes: "Working-class families now frequently orient themselves toward "giving" their children the good things of life.... .

This is a relatively new conception of parental responsibility for the working class. It accompanies a more suburbanite orientation and a greater involvement of the husband in the home as opposed to his own male peer group. Working-class people themselves often comment on the change, contrasting their own behavior with their parents' more standoffish and removed style when they were growing up.... .

"The lion's share of the income of the working class family, particularly in the family-building years, is spent for things that in one way or another can be regarded as devoted to the well-being of the family as a whole. The standard consumer package which concretizes the good American life style allows only as the last frosting on the cake expenditures that are seen as personally indulgent by only one member of the family ... .

"Most of the expenditures that make up the standard package are of course for food, the house and its furnishings, and the automobile."[19]

Dispersed Businesses The incentives for business to develop networks of on-demand services are enhanced by current trends in geographical dispersion among and within businesses.

The move of corporate offices from a few major centers to suburban and small-city locations has been widely publicized during the last two years, but such relocations have been occurring steadily for more than a decade. In 1960, headquarters of 264 of the 500 largest U.S. corporations were located in the inner cities of the ten largest Standard Metropolitan Statistical Areas (SMSA); by 1970 that number was reduced to 226 — a drop of more than 14 percent in ten years.[20] If that rate persists, by 1990 only one-third of these large corporations will be headquartered in the ten largest SMSAs. A quick survey of business directories suggests that a similar trend exists among smaller corporations, and that within companies the number and dispersal of office locations has been increasing.

According to New York's Regional Plan Association, in 1910, 80 percent of that region's office jobs were in the city. By 1940 it was still 70 percent but is now 56 percent. Of the 2.4 million jobs of all kinds they expect to see created in the region by 1985 they expect 2 million to be in the suburbs.[21]

There is no reason to expect that the trend toward greater dispersal of business will stop. On the contrary, there are persuasive arguments that the rate of dispersion will accelerate. Many of the reasons that prompted the relocations — the rising costs of urban sites and the deterioration of inner-city life, to name two — still exist, and there is a circularity of cause and effect that could make them even more compelling in the future. As businesses move out, the city steadily becomes less important as a center of activity for those that remain. The tax base through which the city could otherwise try to regain its attractiveness is reduced. The upper-income population is diminished and with it the number of customers for expensive urban amenities, which, in turn, have been one of the city's major selling points in its competition with the suburbs.

In the absence of massive federal expenditures or tax incentives to reverse the trend, continued dispersion of corporate business at the current rate would seem to be a conservative prediction. Such a dispersion of business activity will increase the attractiveness of on-demand systems for inventory control, information retrieval, logistic planning, and labor allocation. More generally,

it will increase the incentives to substitute electronic means for a wide variety
of business communications that formerly could be conducted face-to-face for
the price of a short taxi ride.

Corporate dispersion has its parallel in retail trade. Even when the increases in suburban population are discounted, the city is losing popularity as
a place to shop. As an illustrative statistic, during the period from 1963-1967,
retail trade in the suburbs of the ten largest SMSAs increased by 35.5 percent.
In the inner city, the increase was less than half as great — 15.3 percent.[22] In
eight of those ten cases, suburban retail sales already account for more than
half of the total. What is left to the downtown stores are the least affluent customers with the lowest variety of demands, as Table 4 indicates.

An obvious first effect of retail trade dispersion is to increase the complexity and expense of distribution systems. The department stores are particularly aware of this problem. Because the population is less concentrated, a single
large store in the inner-city can no longer adequately tap a metropolitan market. The logical response is to build a number of branch stores in the suburbs.
But at this point the department store encounters a dilemma: to maintain a large
and varied inventory of merchandise at several branches, each of which has a
relatively low volume of traffic compared to the old one-store system, means
a very low profit margin; yet the popularity of department stores depends on
having a large selection. Montgomery Ward's solution, and one that is being
adopted by other firms, is to build what it calls "modular stores," which maintain very small inventories on most merchandise and only floor models for big
ticket items.[23] The system hinges on responsive store-to-warehouse communications and rapid, flexible distribution systems to keep the store stocked with
take-home items and to make home delivery of the others.

Table 4. Shopping Done Downtown by Cleveland Women, by Social Class (in
percentages)

Proportion of Downtown Shopping*	Social Class					
	L-L	U-L	L-M	U-M	L-U	U-U
High	68	50	42	33	22	18
Low	19	33	37	50	59	64
None	11	15	19	15	16	18
Don't know	2	2	2	2	3	--
Total	100	100	100	100	100	100
Number of cases	132	346	265	206	36	11

*High downtown shoppers shop downtown half or more of the time; low down-
town shoppers, one-quarter or less of the time; none means women who do
not shop downtown.
Source: Stuart U. Rich and Subhash C. Jain, "Social Class and Life Cycles as
Predictors of Shopping Behavior" in Steuart Henderson Britt, Psychological
Experiments in Consumer Behavior (New York: John Wiley, 1970).

The economics of this type of retailing deserves close monitoring by anyone interested in the feasibility of home shopping systems. A logical next step for a modular store system is point-of-sale and warehouse terminals linked to central computer storage of inventory status. A substantial fleet of delivery vehicles is required. Both conditions facilitate hookup with a system serving home consoles that are capable of displaying merchandise and of taking orders. For a company with the internal facilities to make a modular store organization work, much of the capital investment that would otherwise be required to set up a home shopping system would already have been defrayed in the form of sunk costs on adaptable equipment. Thus the development of household on-demand services will interact with changes in the business world. Many household services that appear impractical when seen in isolation could well come about as adjuncts to communications and distribution networks developed for business use.

Conditions of User Acceptance

We come now to another key problem. Predicting the adoption of "on-demand" communications systems depends, as we have seen, on a myriad of circumstances including price, characteristics of the system, trends in the society, and so on. But of all of them, the least easily predictable is the attitude and behavior of potential users. This emerges as a critical variable.

Among the variables that affect system acceptance are:

The relationship of new to established behavioral patterns

Degree of voluntarism involved in the change

Amount of learning required

Cost

Social resistance or reinforcement

Changes in social relations induced

Associations of the new practice with positively or negatively valued symbols

Propensity to imitation; visibility and character of models

Reinforcement pattern from use of the new practice.

Relationship to established behavioral patterns   A new device may involve no change in behavior, or it may require great changes indeed. A newspaper delivered by facsimile, if it looked the same, would require little change on the user's part. An on-line computerized news information retrieval system would require him to learn entirely new patterns of news acquisition.

The new behavior may supplement existing ways of life, or it may displace old behaviors. A traffic avoidance advisory service, for example, does not displace any presently valued activity. On the other hand, in-the-home shopping displaces much of the housewife's present social and physical activity. Resistance will, of course, be much greater to any innovation that displaces a valued activity. Since as long as there is no alternative, people do not know how strongly they value an established behavior, it is extremely difficult to esti-

mate how attached to the status quo they will prove to be once an alternative
appears.

Degree of Voluntarism   Closely related to the preceding considerations is the
degree of voluntary choice involved in making the change. If a newspaper in a
single-newspaper town decided to go from home delivery to facsimile, wide
acceptance could be predicted because the individual user has little choice. On
the other hand, a single grocery that went exclusively to broadband selling in
competition with others would be offering its customers a choice, perhaps to
its regret. Pay television sports events in full competition with continued free
sports events might not succeed, but if they succeeded enough to dry up sponsorship for major free broadcasts of sports events, then in the absence of
choice they would succeed still more. In that last example the degree of voluntarism is so critical as to make the whole situation metastable, with all the
difficulties of analysis that that poses.
    Voluntary choice operates fundamentally differently in the home and in in-
stitutional applications of on-demand communication. In jobs, people do what
they are told. In return for their wages, employees accept the requirement to
be trained for and do rather extraordinary things, such as going door-to-door
and telling complete strangers rather ridiculous stories to persuade them to
part with their money or walking along high scaffoldings hundreds of feet above
the ground. Because of the lack of voluntarism in the work situation, quite exotic systems may win acceptance in an institutional setting that would not win
acceptance in the home.

Amount of Learning Required   People can learn to do extremely difficult things,
such as reading and writing. In practice, however, the social situation severely
affects how much learning they are willing to do. In job situations where, as
we have just noted, voluntarism is low, a rather considerable amount of training may be successfully required. People are trained to be physicians, for example. On the other hand, as far as home use is concerned, the general principle seems to be that almost nothing is adopted by adults that takes more than
an hour or so of learning. Children are made to devote many years to acquiring
such skills as reading, writing, arithmetic, playing an instrument, learning a
foreign language, and learning athletic skills. An occasional adult voluntarily
subjects himself to the same kind of discipline. There are those who take classes in hobbies, who learn a foreign language, or who take up a new sport. They
are a tiny minority. The most difficult-to-learn innovation that has been universally adopted by consumers in modern times is driving a car. The automobile is the interesting exception in this, as in a number of other respects.
Elsewhere in this paper we note that it is the most expensive innovation in
recent times. American families have been willing to commit more than a
thousand dollars a year to have this device. No other universally adopted device comes anywhere near to that in cost. The same thing applies to learning
difficulty. While today most people learn to drive as adolescents, in the period
of the auto's original diffusion many people had to learn to drive as adults.
    Except for the automobile, almost all the devices that consumers have fairly
universally adopted, and which are not taught in youth, can be mastered in an
hour, and most of them in five minutes. Turning on a radio or TV set, dialing

a telephone, running a washing machine or vacuum cleaner, are examples of
the things people do learn. Relatively few people make the transition from pen
to typewriter in adulthood. The motion picture projector was never widely
adopted, though it gave a better picture than TV, primarily because it put too
much burden on the user. He had to select and acquire a film, run the tape
through the sprockets, and rewind afterwards. All of this says a lot about what
can be expected of consumers in systems for automated marketing or information retrieval. Many proposed systems will fall afoul of complexity.

Cost   There is no fixed sum that represents the acceptable cost for a new service. As just noted, all sorts of barriers have been broken by the automobile.
The amount that people will pay is a function of the importance of an item to
them. Most communication devices seem to fall into the category of utilities.
A good guideline of what people are likely to spend is what they spend today
for such items as telephone service, electricity, gas, monthly parking, television, etc. They may spend more if they commit themselves in small discrete units. That is one of the secrets of purchase of telephone service, for
example. On the whole, people are not willing to spend more than pennies for
single discrete acts of communication, though an occasional movie or sports
event is a different matter.
    On the other hand, it may be that people would spend somewhat more than
they now do for utility service. We do not have a conclusion on that. Perhaps
the elasticity of demand for utilities could be examined to come up with better
estimates. A reason for suspecting that people would pay more is the relative
satisfaction that they express with value currently received. The Consumer
Research Institute reports the following percentages of the public naming the
following when asked: Which of these products do you think gives you the most
value for your money?

Electricity	62%
Telephone service	50
Major appliances	21
Grocery products	16
Automobiles	13
Gasoline	11
Prescription drugs	8

    Nonetheless, the indications are that a communication service would have to
fit in with the present structure of low utility costs. In this respect, however,
institutional decision making is entirely different. Spending on institutional
facilities is likely to be much higher per unit. A successful home service must
presumably reach a penetration of between 20 and 95 percent of households, or
in other words, roughly 10 million to 50 million households, each at low cost.
A successful institutional service may serve only thousands of locations or perhaps as many as a million, that is, one to three orders of magnitude fewer,
but each with an ability to pay one or two orders of magnitude as much.

Social Resistance or Reinforcement A large social science literature demon-
strates that before the adoption of new practices, people virtually always con-
sult and seek the advice of other people.[24] The prospects for any new on-demand communication device will depend heavily upon whether it will elicit supportive or critical reactions. In many instances communication devices require widespread adoption to make them feasible at all. The extent to which
they elicit social support may be very important. An extreme example will be
the picture phone, if AT&T ever makes a serious effort to get it adopted.[25]
The strategy for making it an object of popular interest and social discourse
will be extremely critical for the success of the effort.

Changes in Social Relations Induced We have discussed the concept of the insular family. Shopping from the home, receiving an education at home, and
working at home would all drastically change the pattern of interpersonal relations in a person's life. A key issue in whether any one of these practices
gets generally adopted depends heavily on how people will feel about not having
contacts with sales personnel, classmates, and officemates. In casual conversation today, many people are horrified at the idea; others find it extremely
attractive. In the absence of experience it is difficult to predict how most people
will respond to such changes in social relations. There is, perhaps, no more
important issue in determining the success of on-demand communication.

Association with Positive or Negative Symbols Consider, for example, present concerns about computers and privacy. The symbols associated with computers in that context are "big brother," control, manipulation, impersonality.
Legal regulations on file handling arising from such concerns could affect the
rate of adoption of computerized information systems. Right now we see such
image phenomena affecting cable TV, the success of which is extremely sensitive to regulation. Images of corruption in franchise allocations, the movie
owners' slogans against pay TV, fear of monopoly, the symbolism of a "community" medium — all profoundly affect CATV prospects.
    Education in the home could mean many different things to people depending
on how it comes into their consideration: is it an economy move, a medium for
lifelong education, a quality improvement, or a supplement to entertainment?
The fact that TV is an entertainment medium is clearly an important image
factor against other possible uses for it.

Propensity to Imitation; Visibility and Character of Models Another factor in
adoption is the nature of the adoption leaders. There is a very large sociological literature on this subject, dealing most extensively with agricultural innovation and summarized by Rogers (see note 23). The pattern, observed not only
in agriculture but also among physicians, in politics, and in marketing, is for
an innovation to take hold slowly until opinion leaders in a community adopt it,
after which adoption is rapid. The specific social and personal characteristics
of the individuals who become innovators, early adopters, opinion leaders, and
late adopters have been extensively studied. Age, education, and experience
are all factors.
    The adoption of many of the hardware devices we have talked about are

clearly subject to marked social influence, for example, by fads, social role
of the early adopters, publicity, etc.

Reinforcement Patterns   Last, but by no means least, we mention the phenomenon of positive and negative conditioning by the kind of reinforcement
that physical use of a facility gives to its user. The work of B. F. Skinner is
perhaps the most prominent summarization of the general thesis that the detailed sequencing of positive and negative reinforcement from anything will
largely shape acceptance or nonacceptance of it.

   It seems clear that elementary kinetic experience while driving a car, watching TV, or having answers come back out of a teaching machine may strongly
attract or repel further use of it. Technological predictions and planning often
grossly underestimate the extreme importance of effective man-machine design, particularly given the unwillingness of people to do anything difficult.

   One of the most promising features of on-demand two-way communication
is that it can be designed to correspond to the basic psychological principles
of how people do learn. Interactive devices tend to give reinforcement almost
instantaneously, which is of critical importance in learning. They can be self-
paced. They can eliminate the experience of being embarrassed before a cen-
sorious individual—there need be no schoolmarm present.

   Yet designers often fail to realize that the burden is on them to create things
people find easy to use. The gap between presently proposed systems and what
people would enjoy using is wide indeed.

   All of these various conditions affect the degree of user acceptance of any
new system. To evaluate the ease of acceptance of a device, it should be measured against each of these considerations.

   But there are other factors, too. The ones we have just listed all bear upon
the individual user and how he responds to the prospect of a new product that
he could have. There are also social factors that affect the acceptance of a new
facility by structuring the situation, partly independently of any individual preferences. Choices are not always completely free. An illustration of the social
determination of adoption is a phenomenon that could be called the mushroom
effect.

The Mushroom Effect
"As each new medium has arrived on the scene, it was adopted more rapidly
than the ones that came before."[26] DeFleur documents the point with the graphs
of standardized diffusion curves reproduced as Figure 2.

   This does not mean that the diffusion curve for on-demand communications
will necessarily be of the next order of steepness. There are reasons to believe that likely, but a number of characteristics peculiar to on-demand services should also warn against applying past experience too literally. But past
experience does raise the need to consider those conditions that could accelerate the spread of on-demand systems by orders of magnitude beyond normal
speeds of adoption.

   The usual equation for determining the economic feasibility of an on-demand
service that would replace an existing service involves two principal cost calculations: the projected costs for installing and operating the on-demand system,

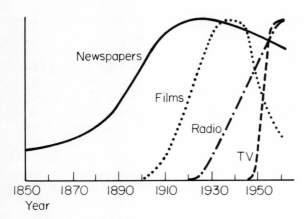

Figure 2. Standardized diffusion curves for four media. The units on the vertical axis have been "standardized" so that each curve reaches its peak at the same point. This procedure facilitates comparison of one curve with another.

against the projected costs of maintaining the old one. The validity of those projections can break down if the option of maintaining the old system is foreclosed because "nobody else is doing it that way any more." Theoretically, the costs of the labor and equipment may not have increased. But if a new norm of operation has been established, the facilities for staying with the old one may not be available.

Examples occur constantly as technology changes. The consumer no longer has a realistic option of keeping an icebox or of buying 78-rpm records, to use two homey illustrations. Given the tremendous leverage wielded by a relatively few corporations, similar phenomena are possible in the adoption of on-demand communications.

Taking one example out of many, it might be twenty years in the ordinary course of events before a small chain of banks would find it economical to buy the hardware for a paperless credit transfer system. But Chase Manhattan and Bank of America are likely to reach that point much sooner. If they install such systems the costs of the traditional paper methods for the small chain of banks could easily rise much faster than a cost projection would otherwise suggest. Even if large banks maintained nonelectronic facilities to handle transactions with its smaller colleagues, it would be in their interest to provide incentives for the others to fall in line, so that they could phase out their redundant traditional facilities. Because of the leverage of the very large bank, those incentives could be substantial. Increases in usage of new system will generally tend to drive up the costs of maintaining old ones in a steepening curve.

There are many systems that one believes to be too complex, too expensive, and faced by too many bureaucratic obstacles to be installed in the next twenty years. But if contrary to one's expectations, one is installed, it may be certain of being accepted.

Suppose, for example, that the Congress voted the several hundred million dollars to allow the Library of Congress to be fully computerized, with the contents of any book then being available for a modest fee to anyone with a terminal. The forecasting task is turned on its head—the questions about the time frame for U.S. universities to hook into the system become almost trivial; the main problem is trying to grasp all the potential ramifications of changes that would occur within a highly telescoped time period.

In short, we face a situation in which gradual growth of communication systems may not be the pattern. On-demand two-way communications may make very little progress for ten or twenty years and then suddenly mushroom in half a decade. The timing of such a turning point is very hard to predict until one is quite close to it.

Real Time Needs

One further consideration profoundly affects the demand for on-demand communication, namely, the importance of having rapid response in real time.

If real time response is not important, cassettes can be delivered by mail, books can be picked up at the public library, samples can be brought to the door by salesmen. Electronic communication may compete if it is cheaper but will lose otherwise.

On the other hand, for uses where timeliness is important, such as delivery of the day's news, daily grocery shopping, or immediate reinforcement of learning, an on-demand device may compete even when more expensive. Low-cost video tape recorders and cassettes compete significantly otherwise.

A bridge between on-demand instant-response devices and slow, hand-delivered mails may perhaps come in low bit-rate electronic delivery during off hours. Electronic mail schemes envisage business firms sending routine messages to branches, customers, and suppliers by telegraphic or facsimile means in the hours between 5:00 PM and 9:00 AM. A decade or two later one could anticipate similar devices in homes, creating the capability of transcribing entertainment and educational cassettes for use the following day.

Thus one can anticipate an interaction among three levels of communication, as shown in Table 5, with economy and need for speed determining the mix among them. The most complex services like marketing and education will combine all three.

Disposable Income

We may now try to sum up our preliminary consideration of the demand for on-demand media. We noted that they are largely more convenient substitutes for things that are now provided in other ways. We noted that they are responsive to needs created by an apparent change in pattern of American life, the trend toward urban dispersion and the insular family. We noted various problems of user acceptance of innovations, including ease of learning, reinforcement patterns, and aptness to incremental introduction.

All of these considerations get expressed in dollar and cents terms in what people will pay. But most directly their propensity to purchase is affected by how much they have available to spend. And so we come to the critical question of the money consumers will have available for communications services and how they will spend it.

Aggregate purchasing power (in current dollars) in the United States rose from about $165.9 billion in 1946 to 504.9 billion in 1965, an increase of 200

Table 5. Comparison of Three Levels of Communication

Mode of Delivery	Costs	Uses
1. Man-carried messages (for example, mails, deliveries, visiting, shopping, meetings, and video cassettes)	Cost largely borne by unpriced movements of consumer carrying a physical object, therefore, apparently low cost	Irreplaceable for delivery of physical objects and for some psychological values
2. Slow off-hour electronic delivery to transcribing devices	Cost for some uses higher, and for some, lower than man-carried messages	Requires 24-hour anticipation of wants; most important for broadband materials, for example, video
3. Instant electronic delivery (for example, phone, TV, teletype, radio)	High cost	Essential for conversational-rate interaction

percent. The aggregate figure is the sum of disposable personal income, new household credit, and credits from government insurance. By far the largest factor in this figure is the contribution of wages and salaries, counting for about 70 percent of total purchasing power in 1965.

Continued increases in aggregate consumer purchasing power are likely to expand total dollars available for discretionary purchases. The discretionary income figure is derived from aggregate consumer purchasing power by subtracting net contractual saving, essential expenditures, fixed commitments, and imputed income. Furthermore, some part of discretionary income is saved rather than spent and must be discounted. As a percentage of aggregate purchasing power, discretionary income has remained a fairly constant 35 percent in the twenty-year span from 1946-1965. Thus the 200 percent rise in aggregate purchasing power has been matched by a similar increase in discretionary income, from $55 billion in 1946 to $171 billion in 1965. The major reason for the lack of a more rapid upswing in discretionary purchasing power in that period was the postwar rise in fixed commitments (primarily housing expenses and insurance payments). Of interest to the CATV operator, then, will be the effects of other consumer commitments on the size of the discretionary dollar. At this writing the most important of these appears to be a willingness to pay for an increased supply of public goods, such as clean air, clean water, safe automobiles, etc.

For our purposes it is important to examine fairly closely what the trends in consumer income and its uses may be in the coming decades. If present trends continue, median family income will grow from about $10,000 to about $16,000 per year between 1971 and 1985, or in other words, because the growth is exponential, it will grow as much in the next 17 years as in the past 50.

Table 6. Per Capita Real Personal Consumption Expenditures (1971)

Category	1958 dollars	1971 dollars
Total	2,396	3,243
Nondurables	1,029	1,359
Durables	445	505
Services	923	1,378

Table 7. Percentage of Consumption Expenditures by Category

Category	1948	1971
Nondurables	52	43
Durables	12	17
Services	36	39

Table 8. Per Capita Annual Real Rate of Growth 1960-1970

Category	Percent Increase
All consumption	2.9
Nondurables	2.0
Durables	4.8
Services	3.2

But more important for us than the total income is how it may break down in components. Tables 6-8 present the annual rate of consumption from the third quarter of 1971 in both current dollars and separately normalized by their price trend to 1958 dollars.

What these figures suggest is that a growing portion of a growing income will be spent on the kinds of services and appliances that are represented by on-demand communication.

The increasing sense by the consumer that he is able to spend money on new things is reflected in some survey results obtained by Lee Rainwater. He found that what people say is "the smallest amount of money a family of four needs each week to get along in this community" was between 80 and 90 percent of median family income at the end of the 1940s and is now only between 60 and 70 percent.

Objectively we find that the current level of consumer spending is about $10,000 per household, of which about $1,650 goes for durable goods, of which about $55 goes for consumer electronics.

What is the trend in that figure at the present time — without the stimulus of new two-way services being offered? A Conference Board report projecting consumer spending to 1980 projects a 4.4 percent annual growth rate in consumer spending with an annual population growth of 1.2 percent (that is only a

Table 9. 1950 Per-Capita Expenditures for Recreation, Reading, and Alcoholic Beverages by Type of Product and Income Class (dollar expenditures rounded to nearest tenth)

| | Income Class | | | | | | | | | |
	Under $1,000	$1,000- $2,000	$2,000- $3,000	$3,000- $4,000	$4,000- $5,000	$5,000- $6,000	$6,000- $7,500	$7,500- $10,000	$10,000 and Over	U.S. Urban Average
20th Century Fund urban income distribution	8.9	14.2	16.5	18.4	15.2	9.5	7.7	5.6	4.0	
Per capita after taxes income	409	730	938	1,089	1,312	1,514	1,789	2,108	4,301	1,251
Recreation										
Radio, TV, and musical instruments*	4.6	5.7	11.1	18.8	25.0	26.7	31.9	28.0	32.8	18.2
Phonograph records, music	0.3	0.5	0.5	0.8	1.3	1.6	2.0	2.0	3.4	1.1
Cameras and equipment	0.1	0.5	0.4	1.4	2.3	2.5	3.1	4.0	7.1	1.7
Sporting goods	0.4	0.6	1.5	1.4	1.8	2.4	3.3	5.1	7.3	1.9
Athletic clothing	0.1	0.2	0.3	0.5	0.9	1.2	1.4	1.7	3.0	0.8
All toys	1.0	1.0	2.3	3.8	5.8	4.5	4.1	4.4	7.9	3.5
Movies	4.1	6.9	9.2	10.1	11.1	12.8	14.3	15.3	13.9	10.1
Other admissions	0.5	1.2	1.7	2.3	3.1	4.1	5.1	7.1	13.6	3.1
Dues; Social rec. clubs	1.0	1.2	1.2	1.6	2.3	3.1	4.4	5.9	21.9	2.9

Recreation away from home	0.1	1.2	1.5	1.9	2.8	3.3	5.3	5.9	15.3	2.9
Reading										
Newspapers	7.5	6.9	6.8	7.0	7.3	7.7	8.2	8.2	10.4	7.4
Magazines	1.3	1.5	2.0	2.4	2.9	3.2	3.9	3.8	7.1	2.6
Books	0.5	1.0	1.0	1.3	1.4	1.7	2.4	2.7	5.2	1.4
Alcoholic beverages consumed at home										
Beer-ale	2.0	3.9	4.8	8.1	8.4	9.4	8.3	7.4	8.7	6.6
Wine-liquor	2.0	4.2	4.2	4.5	6.5	7.7	9.6	13.6	31.5	6.8
Rented vacation housing										
Vacation home	0.1	0.8	0.7	1.1	2.0	2.6	3.7	5.2	10.8	1.7
Lodging while traveling	2.3	1.4	1.5	1.9	2.9	4.0	7.9	7.0	10.8	3.3
Flowers, seeds, potted plants										
Flowers for the house	0.2	0.3	0.4	0.5	0.7	0.8	1.2	1.4	4.0	0.7
Flowers, lawn seed, bulbs, etc.	0.6	0.5	0.5	0.7	1.0	1.3	2.0	2.0	4.5	1.1

*Includes phonograph records and music.

Source: G. Fisk, Leisure Spending Behavior (Philadelphia: University of Pennsylvania Press, 1963), pp. 92–93.

marginally more rapid predicted rate than the actual per capita growth rate reported for the last decade). The sector for which they predict the most rapid growth—8.1 percent per annum—is radios, TV sets, and records. "Next come toiletries (6.7%), foreign travel (6.5%), automobiles (6.4%), higher education (6%), and drugs (5.8%). These are followed by food at home (3.4%), alcoholic beverages (3.2%), and footware (2.9%)."[27]

If the 8.1 percent projection turns out to be correct, then applying it to the present base of $3.3 billion or $55 per household for consumer electronics, Americans should be spending about $7.2 billion on consumer electronics or about $100 per household in 1980 and $15.7 billion or $175 per household in 1990.[28] Later, when we put the supply and demand figures together, we shall consider what that means for the prospects of on-demand systems.

Elsewhere, we have dealt with the hypothesis that changes in life-styles may affect the acceptance of two-way home terminals and themselves be affected by the provision of two-way services. The relevant issue for our purposes is what part of aggregate demand will be affected by such a shift in life-styles and what implications these changes will have on the budget allocations of the family. These concepts may be subsumed under the concept of income elasticity.

Income elasticity measures the relative change in expenditures for any good, given an increment in income. The relevant computations are straightforward: the ratio of expenditures on an individual commodity to aggregate consumer purchasing power as it changes over a period of years. If aggregate income continues to increase, what increment will be devoted to communications devices?

An important factor in determining both the size of the discretionary spending dollar and income elasticity for a commodity is the shape of income distribution in society. Table 9 shows per capita expenditures for selected leisure goods by income class. Clearly, dollar expenditures on these goods increase with income. In some cases, too, these figures demonstrate income elasticity across income groups. Contrast the inelastic expenditure on newspapers, movies, and beer with the highly elastic expenditures on wine and liquor, and vacation homes. In a society with an increasing number of high-income people new exotic communication devices might well prove popular.

Between AT & T and CATV, transmission facilities for two-way communication will be available in the coming decades. But in cost, the terminal is the dominant part of sophisticated systems. Total telephone company investment today amounts to approximately $535 per phone. CATV two-way multichannel linkage, depending on population density, penetration, and kind of system, can be commercially achieved for amounts starting at about $200 per customer plus the cost of the TV set. The only way to bring that per capita cost down substantially is by government programs designed to give 100 percent penetration. The minimum two-way terminal is a TV set at about $100 plus another $100 in equipment. A terminal usable for on-demand marketing and education would require cassettes, VTR, frame grabbing, a scrambler or enabler for pay TV, and two-way voice and would cost at least $500 and maybe substantially more.

If 50,000,000 families are to be supplied, we are talking of investments of $25 billion or more for terminals. The estimates by Michel Guité given previously in Table 2 for CATV costs for a 10,000 subscriber system with 50 per-

Table 10. Current and Estimated Market Prices for Selected Terminal Items (in dollars)

Item	1971	Late 1970s
One-way addressed systems		
Address decoder and logic unit	100	
Video tape recorder	1000-1500	
Facsimile machine	1500 (Xerox LDX)	
Subscriber response systems		
Basic control unit	300 (Vicom model)	180-200
Shared voice and video systems		
Microphone/speaker unit	50-100	25-50
Video camera (B/W)	1500	350-500
Subscriber initiated systems		
VTR with framegrabber	1500 (MITRE Corp.)	425
Keyboard	800-1000	
Screen feedback		100-200

Sources: Baer (note 2); Ward (Chapter 9 of this volume); Guité (note 9); N. E. Feldman, The Economies of Scale in Urban Cable Television in the Dayton Area; and Radical Software, # 2-3.

cent market penetration under different systems detail these costs more accurately.

A move to a two-way CATV system, Guité estimates, would add about $25 per subscriber for modification of amplifiers and another $20 at the head end, including the cost of a mini-computer and billing mechanisms. Other estimates, such as those by John Ward for the Sloan Commission, agree that adapting the amplifiers and related equipment for a two-way system would add about 30 percent to the cost.

However, there is far less consensus on terminal costs. Table 10 lists some current market prices and best guesses for selected terminal items with price projections to the late 1970s.

Configurations under current development are described in Table 11. For $85 the user has a four-button response pad. For $200 he can get a considerably different and more sophisticated 12-button pad. Estimates for full text input and output terminals are perhaps three times that. Add to that, or consider as alternatives, VTR or a framegrabber as in the MITRE console, and it seems that $500 in addition to the viewer's present investment in a TV set might provide a modest on-demand type terminal.[29] To do well the kind of job described by these techniques could easily bring the costs to $1000.

However, Noam Lemelshtrich in Chapter 11 of this book has suggested screen-feedback systems that can use either CATV or telephone lines up-

Table 11. Presently Developing Terminals

Terminal Developer	Nonverbal Feedback Number of Buttons	Printer	Microphone	Frame-grabber	Cost of Home Hardware
Hughes Aircraft	12	Strip printer 100 words/ min.	—		$200
A.D. Little	4-6	—	—	—	$85
Vicom	12	—	Allows voice communication through the cable	—	$180-200*
Mitre	Push-button telephone	—	—	Small half-inch video re-corder	$425

*Ward (note 8) estimates $135.
Source: Guité (note 9), p. 24.

stream, can use the face of the TV screen for display, and permit full key-board return at a cost of perhaps $100. With that kind of idea likely to bring costs down, it seems reasonable at present to use $500 as a rough estimate of terminal costs.

Cost to the Consumer

If we assume that with further development, a terminal costing roughly $500 added to the TV set will be able to provide effective two-way marketing, edu-cation, and informational services, then it is clear that American consumers can and would easily pay the hardware costs.

Table 2 (reprinted earlier), from Guité's study, is relevant, though not in-tended to answer this question exactly. He was dealing with CATV systems at 50 percent penetration. The relevant figures are at the two extremes of his scale. Present CATV systems typically charge $5 or $6 a month for the one-way service described on the top row, with none of that money going into pro-ducing the material that gets cablecast. The proposed MITRE frame-grabbing system is priced by Guité at $17.43 a month with a $415 terminal but again without any of that money going into production costs. The terminal is of the same general magnitude as the $500 terminal that we have been considering. Thus these data suggest that subscriber charges of $20 per month or about $250 per family per year would cover the system costs.

That is close to the range of other utility and communication bills American customers pay. The average American household spends only about $5 per month on consumer electronic products, primarily TV sets. The overall national investment, after two decades of the TV era, in 86 million TV sets, a couple of hundred million radios, and $2.2 billion in broadcast equipment depreciated now to $1.2 billion is only about half of what would have to be invested in terminals alone for an on-demand communication system. However, if ordinary CATV comes to be the standard American broadcast system the customers will be paying about $5 in monthly charges in addition to the $5 for equipment, which reduces the increment for on-demand services.

A $20 monthly charge is not out of line with telephone bills per household. The $18 billion per annum in phone bills amounts to about $15 per instrument per month. Recall also that earlier we estimated that "normal" growth in expenditures on consumer electronics at 8.1 percent, the highest rate for any consumer goods sector, would lead to a level of expenditure of $175 per household per annum by 1990. These figures are not totally out of line with the possible prospect of spending about $250 a year for a new type of on-demand service, compared to over $4,000 per household likely to be spent for all durables in 1990, provided the services offered are ones consumers want strongly enough to produce a significant shift in their income allocation. All that is involved is a shift by 1990 of expenditure on consumer electronics from 3.3 percent of expenditure on consumer durables to 4.2 percent, certainly a possible shift if the product is attractive. On the other hand, the figures suggest that 1990 is the earliest that such a new charge would be borne by consumers, and not easily then.[30] It may not be until the year 2000 that consumers have enough income.

However, note that we have so far considered only hardware costs. The software costs may be as much or more. The cost of providing education, news services, checkless banking, logistic controls, or home marketing may sometimes be less than present methods (given a hardware system in place), sometimes much more. Someone has to pay these costs, too. They may be directly chargeable, as with pay TV. They may be publicly supported, as with education. They may be absorbed by the seller, as in checkless banking or charged in the monthly utility bill to the consumer.

The broad picture of what is needed for on-demand communication by way of terminal facilities is now becoming fairly clear. The same is not true for the situation at the head end, neither regarding hardware nor software.

The hardware for CATV and even for two-way CATV involving audience polling or for pay TV is well understood. What is not understood are the implications at the head end for handling truly interactive on-demand services. It is a technology that has a great deal in common with computer time sharing, but serving the mass public is not identical to serving computer users.[31] Among the kinds of devices that need to be considered and subjected to the same kind of preliminary costing and evaluation that we have given to terminals are microfiche selectors, telephone data sets, devices such as the Sheridan equipment for displaying poll results to the human broadcaster,[32] mobile pickup for source materials, etc.

But all these hardware considerations interact with the even more difficult and ill-understood software considerations. How many phone calls of what duration must one assume for effective remote shopping or education? What will be the balance between video and audio material? How will efforts be distributed over the day? How important is picture definition, color, hi-fidelity, speed of machine response, and what do these imply for origination?

In an on-demand system, costs at the head end are variable to some extent with the size of system, but to what extent? What is the balance in the total cost between data sets and other equipment that are linearly related to the number of persons phoning in and transmitters and such equipment that are not so affected at all? It is obvious that until work on hardware pricing is followed up with work on the costs of software and of substantive services in such areas as home marketing and how they will be paid for, the rough hardware cost estimates that have been made can be quite misleading.

Sequencing of Introduction
One way of looking at the whole system, including software as well as hardware, is to consider how different services can rest on the shoulders of other services. For example, what educational uses could be introduced if certain shopping facilities existed? What are the possible relations of checkless banking and in-home marketing? What are the relations of in-school CAI and on-demand in-the-home education? What are the relations of pay TV to citizen participation broadcasting? What are the relations of library retrieval systems with microform storage to the needs of on-demand broadcasting? These are fairly obvious examples, but many hidden relations will appear as one digs deeper.

One also needs to look not only at how various services can use on-demand communication via cable but also at what other alternatives exist for doing the same things. Many things can be done electronically but may end up being done in other ways for convenience or economy.

One pair of alternatives within the general range of on-demand communication is that of cable versus telephone. We have not yet looked at this choice closely. There are political and regulatory issues here as well as economic and technical ones. At the moment the general reluctance of AT&T to diversify its services, and the reluctance of its regulators to permit it to diversify, tend to push consideration toward cable, but it is far from clear under what circumstances modification of the phone system would be part of a superior solution. Clearly the preferred solutions vary with rate-setting policies.

Assaying Consumer Needs
Given the human response to nonexistent systems is extremely hard to predict, one common suggestion is that one run an experiment with a prototype system in a small area. Unfortunately, many of the key questions cannot be tested that way, since they depend upon scale. Thus wiring one town for on-demand communication would prove that it can be done but would not provide the rich library of scripts and experienced artists that make a communication system worth tuning in to.

A few things can be experimented with (those we shall discuss in the next section). Other topics one can study only by looking for indicators operating in analogous situations in a natural setting. The crucial matter of how lifestyles are changing in ways more adapted or less adapted to on-demand communication is an example of the latter.

TWO KEY APPLICATIONS

At several points we have noted that economically the most compelling application of on-demand communication is marketing from the home, but that a
successful marketing configuration would be a fairly advanced one that would
also permit educational and other activities of many kinds.

Before we leave this report it therefore behooves us to take a somewhat
more detailed look at those specific possibilities.

First let us note that there have been various marketing experiments that,
while not electronic, have many features in common with marketing from the
home. Mail-order and telephone shopping also preserve the privacy of the
housewife and save her a trip.

A study should be made of trends in those marketing schemes. We have not
done that. We do not know if the decline of rural reliance on the Sears catalogue
is the dominant trend or if that is offset by the large rise of specialty mail
order promotions. Nor do we know whether telephone and mail-order shoppers
are ones who have first familiarized themselves with the store and its stock in
person and then use remote shopping only for repeat orders. Nor have we assembled the data by field to indicate what types of commodities sell well by
phone or mail. Also, we would like to know where those orders come from:
rural customers, suburban, central city. And we would like to know how telephone orders differ from mail orders in what stimulates them, who places
them, and how much difference the delivery service makes to them.

It is clear that on-demand shopping is differentially adapted to different commodities. Few people will ever buy a house without going there. However, a
remote access real estate service could display the want ads, thus providing
first glimpse data by an interactive retrieval system for pictures, text, and
maps. Nonetheless, people will go to the house before they will buy. So, too,
with automobiles. Clothes are a special case. People often want to try them
on. However, an interactive light pen or other screen feedback for computer
graphics could make the console an important part of the purchase of clothes
or furnishings. That would require that the customer be able to draw modifications as a model, as for example, by redrawing the neckline. Nonetheless,
these items are all ones with substantial difficulties.

The most promising type of marketing for on-demand systems is groceries.
Food and beverage purchases are about one-fifth of the consumers expenditures
(albeit down from one-third in 1946). Shopping for groceries, however, is no
special occasion. The shopper returns repeatedly, buys similar things over
and over from among a familiar range of alternatives.

A rough ranking of the likelihood of using on-demand communication for pur-
chasing of certain commodities is as follows:
1. Food
2. Hardware
   Drugs
3. Books, magazines, records
   Liquor
   Gifts
4. Furniture
   Appliances
   Autos
   Clothing

For most of those categories we have available data on the percent bought in
the central city and in the suburbs. There is an indication (though autos are
an exception) that the things that people seem likely to buy from the home are
the things they do not bother to go downtown for:

	Percentage Bought in Central City
Food	44
Drugs	47
Furniture	56
Appliances	50
Autos	42
Clothing	60

We may well, therefore, concentrate on the grocery business in assessing
the prospects of marketing in the home via on-demand media. However, it
should be noted that if an in-home marketing system develops for that or other
mass markets, one of its likely uses may well be at the other end of the marketing continuum, namely, for selling low-volume items that meet highly specialized demands. The advantage that some types of electronic communication
systems offer for such products is the possibility of pulling together geographically dispersed customers into a single market. For example, collectors
of 19th-century stamps or of Russian icons are not very many in any one area,
yet nationally they represent a significant market. A dealer in such items today is likely to rely heavily on mailed catalogues. Such remote marketing
could be facilitated by low-cost long-distance electronic communication. However, that kind of marketing will require the facilities of a switched system.
Such low-volume uses will not be able to afford one of 40 or 80 cable video
channels even on a time-shared frame grabber basis except occasionally for
scheduled periods. Most of the transmission, whether text, facsimile, or
audio, will presumably be over telephone lines.

Grocery marketing, on the other hand, is ideally adapted to use of a cable
system since it is organized on a neighborhood basis. The local grocery can
display its wares to a geographically partitioned portion of a single CATV system. There is an interaction between market areas and cable areas. The two
systems will each presumably be designed with consideration of the other.

Whether such a home-based marketing system will be cost effective in competition with stores to which customers come depends upon many things, including the alternatives that stores offer to meet the new competition. One can
well imagine supermarkets introducing free coffee, comfortable ladies rooms,
or electrified carts if in-home shopping started to become widespread. That
would be particularly likely if in-home shopping turned out to cost a bit more
for the convenience offered, thus allowing a small growth in the present very
small margin that the competing store owner has to work with.

On the other hand, the electronic marketer has substantial cost advantages
to trade off against the costs of communication, delivery, etc. The areas for
possible cost savings include:

Elimination of parking lots

Reduction of store area by eliminating space for display and shopper movement

Placing stores in low-rental locations

Elimination of separate advertising expenses unrelated to point of sale

Avoidance of handling of perishables

Reduced inventory tied to a flexible delivery system

Against those economies there are other costs that will rise. Delivery cost is a clear case in point. Computer and radio control of mobile vehicles can make delivery much more efficient than now, but it is nonetheless a large cost that must be paid directly instead of being an invisible social cost borne by the consumer driving to the store.

The communication costs will, we assume, be shared between consumers, who will be paying perhaps $20 per month for the communication service, and the marketer, who will be paying production and time charges.

Labor costs are unpredictable without further study. At markets now, each object bought is handled by a checker and a bagger. It is not clear whether more or less labor would be involved in having an employee place the objects on the ordered list into a bag from an automated shelf. The answer probably varies by type of product. Meats, fruits, and vegetables require more labor than cans and would differ in an automated store, too.

Finally, checkless banking would presumably operate to make electronic shopping more economical than otherwise. The same equipment and same message that places the order can without further human intervention serve to order the transfer of funds.

All of these details need to be examined more fully by further research. Present indications certainly point to shopping from the home as a possibility but by no means as a certainty. The calculations are complex.

In any case, grocery shopping by cable will not burst forth full blown. If it comes it will evolve through a series of steps. The earliest marketing efforts using CATV will be (or, more accurately, are now) special promotions. A commercial tied to an entertainment program displays a commodity and also displays the phone number where the customer can order it. The next step on cable is push button or screen-feedback recording of orders, with commercials expanded to a regular shopping service.[33] (One of the Italian TV programs with highest ratings is a half-hour of solid commercials.) How the transition will be made from CATV marketing to an on-demand service is clearly the critical matter. That will take large amounts of capital. That, also, if it occurs, will produce major economic and social changes. That transition is not likely to occur widely before 1990 at the earliest and more likely around the year 2000.

If marketing services do at that time bring soft or hard copy and/or screen feedback into the home, there will be opportunities for use of them in educational applications, too. These applications have great social importance and have been the object of much speculation. A large relevant literature deals with teaching machines, CAI, television teaching, interactive television, and teaching aids. Partly because of the large amount of previous work and partly because educational applications are not likely to be as self-sustaining as commercial ones, we chose not to explore this field as thoroughly in our limited space. Nonetheless, since this is clearly an area of leverage second to none but marketing, we shall sketch some of the considerations that need to be examined further. We shall take up topics in parallel to our discussion of marketing.

There have been many experiments in teaching by interactive computer, some at a distance, for example, the Stanford CAI math experiments and Project Plato. There is more experience to go on here than in marketing.

In-home learning from a communication console is differentially adapted to different subjects. Different equipment is needed for different subjects. Math and languages can be well taught by audio equipment plus two-way text transmission. Art and science need some kind of video. These points have been discussed earlier in this paper. Suffice it here to note that the variety of possible configurations for different needs is favorable to gradual introduction and gradual upgrading of educational facilities.

There is a great opportunity for nationwide transmission, particularly for those courses that will have relatively few students in any one location. Thus there might be one national center for training X-ray technicians to which a student anywhere could connect by long-haul communication. As with marketing, that kind of usage operates more conveniently on a switched system. On the other hand, there are some locally based educational activities, for example, courses on local governmental or social problems. Those and highly popular materials would be better served by CATV.

Education in the home raises some costs and reduces others. There are economies in physical plant and teachers' salaries. There are new costs for software preparation and communication.

All of these are points for extended further study as are the social-psychological issues of student motivation in the absence of other students. It may well be that multiperson hookups will prove to be essential for many purposes.

In this essay we shall go no further, though similar analyses should ultimately be made of every one of the approximately 30 applications listed in the following appendix.

APPENDIX

Among major possible areas for on-demand communication, we may take note of the following:

A. For home consoles:

1. Retail marketing
2. Entertainment
3. Education
4. Information services
    4.1 Notices, want ads, theater and TV schedules, public service information (for example, "no school today"), timetables, traffic advice
    4.2 Weather, tides
    4.3 Stock quotations and other economic information
    4.4 Daily news
5. Community services, for example, social service advice
6. Medical help
7. Protective services, police, fire, surveillance
8. Dial-a-bus
9. Political activity
10. Games

B. For institutional applications:

1. Inventory, logistic, and delivery control
2. Financial, accounting, and cash flow systems
3. Checkless banking (use of home terminals also possible)
4. Credit validation
5. Labor supervision and control
6. Employment services
7. Travel reservations
8. Stock market services
9. Real estate agency services
10. Professional services for
    10.1 Doctors
    10.2 Lawyers
    10.3 Architects
    10.4 Engineers
11. Command and control
12. Governmental community services
13. Law enforcement
14. Traffic control
15. Mobile vehicle control (aircraft, trucking, taxis, dial-a-bus)
16. Licenses, admissions, permissions, authorizations
17. Libraries
18. Information retrieval
19. Education out of the home
20. Market research and testing (uses home terminals also).

GROUPS OF SERVICES

As we noted earlier, there are few noninstitutional on-demand services that would be economic if installed all alone. Most services become possible only if they can use equipment partially paid for by other services too. Therefore, it is important to consider groups of services that could use the same equipment. We are now ready to consider some natural groups.

SERVICES REQUIRING ONLY AUDIO MESSAGES

Group I

The simplest group of on-demand services requires only:

In the home: a conventional telephone

At point of origin: a tape recorder

The services include: weather, tides information, time-of-day, airline arrivals and departures, no-school announcements.

The user dials a number that determines what tape is played to him. Such services exist today. The range of new useful services of this kind is small.

Group II—A Flexible Recorded Message System

This group of services requires:

Either:

In the home: a telephone and a directory of numbers to dial

At the point of origin: multiple tape recorders with direct access extension numbers

Or:

In the home: CATV, button pad signal generator, and a directory or search algorithm

At the point of origin: multiple tape recorders.

The user signals a code that determines what precise message gets played to him.

The services could include:

"No-school" announcements for a particular school

Traffic bottlenecks for particular routes

Opening and closing times of stores and services

Theater programs and times

TV schedules

Sports results for particular events

Daily economic information

Farm information

Such services are only a small addition to what now exists.

Group IIA

This is the same as Group II, except that it includes computer control of taped messages. This adds flexibility and thus a more complex set of messages, for example, trip-specific traffic avoidance advice and mobile vehicle control. Few applications can use the added flexibility without more elaborate terminals.

SERVICES PROVIDING TEXTUAL MESSAGES

Group III — Transmission of Text to TV Screen

In the home: TV set, button pad signal generator or screen feedback, and character generator

At point of origin: computer-controlled digital transmission

This allows for services requiring short text displays and no permanent record, such as the following:

Stock quotations

Want ads

Dial-a-bus

Games

Market research

All those listed under Groups II and IIA.

This group of services is most likely to be psychologically appealing only if it is combined with audio feedback as in II, using text on the screen as a reinforcement or economical supplement to the audio transmission.

Group IV — Two-Way Transmission of Text with Hard Copy Device

In the home: button pad signal generator or screen feedback and facsimile, Xerox LDX, or typewriter

At point of origin: computer, digital transmission

Different systems allowing different bit rates allow for different uses depending on volume of text readily produceable. This group is adaptable to all uses under III plus various forms of information retrieval such as:

News

Electronic mail

Inventory, logistic, and delivery control.

This degree of sophistication quickly leads to a requirement for full keyboard for textual response, that is, Group IV A.

Group IV A—Two-Way Text Transmission

Same as IV except full keyboard by typewriter or screen feedback for upstream communication. This group serves all uses under IV plus:

Remote computing

Some educational applications

Financial, accounting, and cash flow systems

Employment services

Travel reservations

Some law enforcement applications

Community services information

Libraries

Information retrieval

Services for lawyers

Group V—Text Transmission Plus Validator

This is like groups IV and IV A but adds a device at terminal for secure identification of user. Added uses include:

Checkless banking

Credit validation

Licenses, admissions, permissions

Retrieval of private information

Command and control

SERVICES PROVIDING VIDEO MESSAGES

Group VI—TV Plus Aggregated Feedback

At terminal: TV, button pad or screen feedback

At point of origin: cumulator

These devices permit a poll of audience to tell broadcaster how audience responds. Uses include:

Market research

Political response

Feedback on teaching

Group VI A

This is the same as VI, except at the point of origin, a computer is used to record individual responses instead of a cumulator to record aggregate.

Its uses are the same as those of VI, with individual records, as well as receiving orders or pledges in marketing or fund raising.

Group VII — Pay TV

In home: TV, button pad, unscrambler

At point of origin: CATV transmitter, computer, scrambler-unscrambler

Uses: Pay TV

Long-haul implications: networking.

Group VII A — Flexible Pay TV

In home: same as VII plus VTR

Flexible pay TV permits 24-hour use of channels for transmission with playback at user convenience. Uses include:

Pay TV

Electronic delivery of video to classrooms (with or without scrambling).

Group VII B — More Flexible Pay TV

At point of origin: computer-controlled transmission and library of stored programs.

More flexible pay TV permits collecting of requests for simultaneous transmission of identical requests later on; this evens out transmission load and permits optimum transmission rate. Uses cover:

Pay TV

Classrooms.

With this type of service, broadening of market by intersystem linkage becomes profitable. Since it uses cables during slack hours, it evades time-zone problems.

Group VIII — Still Video Plus Text

At terminal: same facilities as for Groups V and VI plus TV, frame grabber or slow scan-receiver.

Uses:

All uses thru Groups V and VI

Retailing

Education

Real estate services

Services for professionals, for example, architects, engineers

We flag this group as of particular importance. None of the simpler configurations are likely to permit effective on-demand retailing or education in the home. This, therefore, is the basic configuration that might have a profound social effect.

Group VIII A

This is the same as Group VIII plus two-way voice. Supplementing a mechanical on-demand system with conversation will improve many marketing and educational services. Another potential use of this service is medical treatment. This facility, which permits the physician to show the patient still graphics and also to converse, is probably the simplest one acceptable for most medical purposes, though most discussions of medical applications have assumed upstream video (Group XI).

Group IX — Light Pen (etc.) Feedback

This is the same as Group VIII plus screen feedback with a terminal with continuous capability. (A light-pen is the most familiar version.) Uses include all uses of Group VIII with special advantages for those benefiting from computer graphics displays, for example, logistics, architecture, education, etc. This facility also provides much better feedback in marketing and education than text keyboard alone.

SERVICES WITH SPECIAL FEEDBACK

Group X — Feedback from Automatic Sensors

At terminals: sensors

At head end: computer

Uses:

Surveillance (fire alarms, burglar alarms, etc.)

Traffic control

Medical treatment

For many uses these devices need to be combined with some of the previous configurations. For instance, traffic information (II-IV) can be combined with traffic sensing or medical treatment (VIII A), with medical sensors on the patient.

Group XI—Upstream Video

At terminal: VTR, video transmission (slow scan or fast scan)

Uses:

Medical observation (of patient)

Labor supervision

Surveillance.

  For many other purposes in retailing, politics, education, traffic control, etc., multiple sources of video transmission are highly desirable, but that does not imply upstream video from all or most terminals. It implies multiple origination, not video feedback.

Notes

1. An example of interactive communication that is not "on-demand" is the situation in which a speaker can poll his audience as to whether they agree with him. In that situation there is feedback from the audience but no control by the audience of what gets broadcast. Another term roughly equivalent to "on-demand" is "subscriber initiated."

2. Walter S. Baer, Interactive Television: Prospects for Two-Way Services on Cable, Rand, R-888-MF, November 1971, p. 18. Paul Baran, Potential Market Demand for Two-Way Information Services to the Home, 1970-1990, Institute for the Future, R-26, December 1971.

3. There are literally millions of books available, yet the great majority of reading is of a handful of current best-sellers, old favorites, and assigned textbooks. It is easier to follow the guidance of book reviews and teachers than to search the whole gamut of titles for one that one would like best. If all the films ever made were available in an on-line library, people would still need guidance as to what one they might enjoy, and that guidance would have to be organized just as book reading is now organized.

4. On the other hand, using conventional TV receivers on hub-type switched cable systems means redundant investment in channel selectors (see Chapter 9 in this book.) Such hub-type systems may ultimately use simplified TV sets.

5. An estimate for one current trillion-bit system is that storage of machine-readable text will cost $3 per million bits, flat fee without reference to duration of storage. The cost for writing and mounting the strip is the dominant part of the total cost. Once written, the small strip of film costs almost nothing to store.

6. Jordan J. Baruch, Interactive Television, Educom, October 1969.

7. Letter from Dean Burch, Chairman of the FCC to Senator John Pastore, Chairman of the Senate Subcommittee on Communications, August 5, 1971, FCC 71-787. "After studying the comments received and our own engineering estimates, we have decided to require that there be built into cable systems the capacity for two-way communication. This is apparently now feasible at a not inordinate additional cost, and its availability is essential for many of cable's Public Services. Such two-way communication, even if rudimentary in nature, can be useful in a host of ways — for surveys, marketing services, burglar alarm devices, educational feedback, to name a few. Of course, viewers should also have a capability enabling them to choose whether or not the feedback is activated."

8. John Ward, Present and Probable CATV/Broadband-Communication Technology, Sloan Commission, 1971 (reprinted as Chapter 9 in this volume).

9. Michel Guité, "New Technology for Citizen Feedback to Government," mimeo, MIT and Stanford, December 1971.

10. The cost figures have been derived from, W. S. Baer, Interactive Television, and W. S. Comanor and B. M. Mitchell, "Cable TV and the Impact of Regulation, The Bell Journal of Economics and Management Science, Spring 1971.

11. The total cost of the Rediffusion system is higher than any of the previous systems, but not because of two-way capability.

12. Norman R. Nielson in an unpublished paper for the Annenberg School of Communications uses almost identical figures for 24 channels with and without keyboard. He adds five dollars, not three, for the frame grabber and five dollars for head-end and development costs, making his advanced system estimate $25 per subscriber per month instead of $17.43. Baran (Potential Market, p. 6) using a quite different Delphi methodology, came up with an estimate of a $20.10 per month median estimate of customer bills.

13. Guité, New Technology, p. 15. Costs shown here are consistent with figures from John Thompson of A. D. Little, Inc. ("The Optimum Telecommunications System," NCTA conference, July 1971), and the Dominon Bureau of Statistics (Community Antenna Television, Catalogue 56-205, 1969). Figures for costs of trunk-type cable systems are somewhat higher than recent estimates by D. A. Dunn of Stanford ("Cable Television Delivery of Educational Services," IEEE Eascon Conference, Washington, October 1971).

14. The Investment Decision, Cambridge, Mass., 1957.

15. E. Mansfield, The Economics of Technological Change, p. 110. For an elaboration of this study and its findings see E. Mansfield, Industrial Research and Technological Innovation (New York, 1968), Chap. V.

16. Lee Rainwater, Interim Report on Explorations of Social Status, Living Standard, and Family Life Styles, to National Institute of Mental Health, August 1, 1971, p. 104.

17. Business Week, December 27, 1969, p. 49.

18. Note that this almost surely understates the increase in real purchases of electronic equipment (compare the cost of a television in 1950 and an equivalent model today) and even more substantially understates the increases in purchases of books, since the spread of cheap paperbacks coincides with the period for the data. Corrections in the normalization of prices would thus tend to reinforce the implications drawn from the figures on the graph.

19. Rainwater, Interim Report, pp. 83-84.

20. Data from the 1961 and 1971 Fortune surveys of the 500 largest U.S. corporations.

21. New York Times, August 18, 1971.

22. Data compiled from Business Census of the United States, 1963 and 1967.

23. Business Week, November 28, 1970, p. 21.

24. Everett M. Rogers, Diffusion of Innovations (Glencoe, Ill.: Free Press, 1966). Elihu Katz and Paul Lazarsfeld, Personal Influence (Glencoe, Ill.: Free Press, 1955). J. C. Mathur and Paul Neurath, An Indian Experiment in Farm Radio Forums (Paris: UNESCO, 1959).

25. The Washington Post business section of April 13, 1971, reports some Bell System discenchantment on this point.

26. Melvin L. DeFleur, Theories of Mass Communications (New York: David McKay, 1966), p. 74.

27. Business Week, April 24, 1971.

28. Baran's Delphi procedure (Potential Market) estimates the new two-way communication business at about $20 billion by 1990.

29. D. A. Dunn estimated a framegrabber terminal configuration at $410 ± $100 in a paper at the IEEE Eascon Conference, October 1971 ("Cable Television Delivery of Educational Services").

30. Baran's Delphi panel (Potential Market) made median estimates that are slightly (perhaps half a decade to a decade) more optimistic than ours.

31. See Nielsen.

32. See Chapter 11 in this volume.

33. Conceivably the services could include the consumer-sponsored ones.
There is a "Consumer Information System Bulletin" published by a Washington
research group that is discussing how to provide a kind of on-line consumers'
union.